In Honor of the 20th Anniversary of the Third World Academy of Sciences

Science Progress in China

Editor-in-Chief

Lu Yongxiang

Science Press

Beijing

Responsible Editor: Yao Pinglu   Qu Yanli

SCIENCE PROGRESS IN CHINA

Published by Science Press
16 Donghuangchenggen North Street
Beijing 100717, China

Printed in Beijing

ISBN 7−03−012244−5

Preface

Research on science and technology has a long record in human history. New breakthroughs have been made continuously, with the development and progress of society, on the way toward the heights of science. Over the past 100 years from the late 19th century to the present, science and technology have successively evolved through the ages of electrification, atomic energy, aviation, and space, and now enter the information age. They are the most important impetus to the development of human civilization. The enormous productivity and abundant fruit for human civilization are mainly attributed to scientific discoveries, technical innovations and continuous progress in engineering technology, as well as to the extensive application of science and technology to human productive practice and social life. Achievements in science and technology made by human beings consist of contributions from multiple nations in the world. The Third World countries, which enjoy long histories and admirable cultural traditions, represent the foremost origins of science, and are recognized for their own contributions to the development of modern science and technology. The founding and development of the Third World Academy of Sciences (TWAS) is a paramount demonstration of the common will and resolution of Third World peoples to promote science and technology.

Modern science and technology have seen rapid development in China in the past century, as a result of both the dissemination and innovation of world science and technology in China and the succession and development of profound scientific achievements of ancient China in every discipline. The dynamic advancement in science and technology in China has provided a strong support for the social and economic development of the country, as well as a significant contribution to the development of science and technology in the world. In the book "Science Progress in China," Chinese scientists have outlined the development and accomplishments across a spectrum of science over the past 50 years. A litany of world-class achievements, such as the first synthesis of crystalline bovine insulin, the publication of the frame diagram of rice genes, the publication of the high temperature superconductor material structure (Y-Ba-Cu-O), and the invention of the finite element method and the mechanical theorem proving technique in the area of multi-variable complex function in mathematics, which preceded by at least ten years the western countries, all represent solid intellectual contributions by the Chinese scientific community. Other examples include the theory of petroleum formation in land phase, which has laid down an important foundation for primary formation and development of the petroleum industry in China, new methods for measuring geological climate change developed during the course of loess study, new evidence of early-state biological evolution provided by the discoveries of Cambrian oryctocoenosis and primitive angiosperm, and so on. These examples, and many more elaborated in the collection of reports, substantiate the improved role of China in the international community, and the great contributions. Chinese scientists and engineers have made to accelerating the modernization of the national economy to improve the livelihood of the people promoting sustainable development in China.

However, as a developing country, China remains behind the developed countries in terms of economic development and will greatly benefit from renewed innovation and the creative spirit of its people. However, the nation once demonstrated, in history, the world's most advanced productivity yielding much creative scientific fruit. The creativity of the Chinese people has proven a powerful spiritual force behind the nation' ssustained survival, development and growth.

Today, China is in a critical period of development facing a series of challenges, such as optimizing the economic structure, rationalizing the use of resources, protecting the ecological environment, raising the quality of life for the population, eradicating poverty and fostering coordinated development of the whole society. These challenges can not be comprehensively addressed without the integrated development of science and technology. The realization of science and technology development in China will fulfill the following objectives:

-To substantially promote the development of science and education,with emphasis piaced on cultivating and nurturing scientific talents;

-To establish an innovation system in China to enhance the nation's scientific capacity, especially creativity and original research capacity;

-To take an active part in international cooperation,for promoting the development of science and technology for all nations and ensuring progress for human civilization.

China will continue to follow the strategies of societal rejuvenation through science and education towards achieving sustainable development, to develop a complete national innovation system, and to promote development and prosperity in China by advancing our science and technology potential and improving the quality of our labor force.

The development of civilization is heavily reliant on the embodiment of human creativity and embracing joint efforts of all nations. The development of scientific undertakings in China can not do without the world, and vice versa. Chinese scientists will continue to take an active part in international scientific exchanges and cooperation, and to make greater contributions to the progressive cause of humankind.

Jiang Zemin
President
The People's Republic of China
January 18,2003

Foreword

Being the first foundation principle for productivity,science and technology is the revolutionary and driving force for the progress of human society,and is the major foundation for human civilization and advanced culture.The Chinese government and its people have arrived at a common understanding to promote the rapid and sustainable development of the economy,to achieve the sustainable development of society,and to ensure the national security with the help of science and technology.

Over the past half century,the people's Republic of China has formed a preliminary and complete system of modern science and technology derived from the first Five-Year Plan.The 12-Year Plan developed in the mid-1950s has laid an initial foundation for the modernization of science and technology(S&T) in China.The S&T structural reorganization,launched since the nation's reform and opening-up at the end of the 1970s,has given strong impetus to the transfer of science and technology into productivity.The Pilot Projects of National Knowledge Innovation Programmes,initiated five years ago,have greatly promoted the development of the nation's scientific undertaking. Science and technology are playing an important role in the economic construction,the sustainable development of society and the national security in China.

At this opportune time,the TWAS 14th General Meeting,TWAS 9th General Conference & 20th Anniversary Celebrations,and TWNSO 8th General Assembly will be held in Beijing this October. Hosting these important events is naturally a great honor for the scientific community of China. Scientists in China feel it great previledge and pride in sharing what we have achieved over the years in scientific research to this Conference,the Third World Academy of Sciences and the world scientists. In this connection,we are pleased to present this book "*Science Progress in China*" to all the readers.This collection represents year's efforts in assessing and documenting our most significant achievements.

The book introduces the Chinese scientific research results of China made over the past 50 years,with special focus on the past 20 years S&T advances. It consists of two main parts.The first part,totaling 7 articles,focuses on macro-aspects of S&T introducing respectively scientific technology,engineering technology, science policy, science education,science funding and scientific organizations in China, *inter alia;* while the second part,totaling 23 articles,gives introductions of development in every sector or discipline of the sciences,including mathematics, physics, chemistry,astronomy,earth science,life science and technological science. The esteemed authors of the book, either Fellows of TWAS,Members of CAS,Members of CAES,experts, or decision-makers in science and education sectors ,have provided their latest research results and personal reflections of their respective fields,

It is a great pleasure and honor to have President Jiang Zemin of the People's Republic of China to preface this special book.It should be recognized by all Chinese scientists and the international S&T community that President JIANG has and continues to give great and persistent support on China's S&T undertaking. Under his auspices,the national strategy of Rejuvenating China Through Science and Education was created. He also maintains close watch on the latest progress in S&T,and attaches great importance to the development of the Third World Academy of Sciences and its impact on world affairs.In March,2001, he readily agreed to share his thoughts for the commemoration for TWAS:"The Third World Academy of Sciences has made a remarkable contribution to the S&T development of the Third World countries,and it is firmly believed to make greater achievements in the new century." Taking this opportunity,we would like to express our heartfelt gratitude to President Jiang for his steadfast leadership and support.

On this very occasion of the 20th Anniversary of TWAS as well as the publication of this book, I would like,on behalf of the Chinese Scientific Community, to convey our compliments and congratulations to the Third World Academy of Sciences. Over the past 20 years, TWAS has enjoyed its brilliant past and is expected to have a more glorious future.Closely together with TWAS, we,the Chinese scientists, are willing to make greater contributions to the S&T development of the Third World and the progress of human society as well.

Prof.Dr.-Ing ***Lu Yongxiang***
President
Chinese Academy of Sciences
February 15,2003

CONTENTS

Science and Technology Progress in China

Lu Yongxiang

Chinese Academy of Sciences

Lu Yongxiang, *was born in Zhejiang Province. He is Vice-Chairman of the Standing Committee of the National People's Congress; President of the Chinese Academy of Sciences; Chairman of the CAS Presidium; Vice-President of the Third World Academy of Sciences; Member of the Chinese Academy of Sciences; Member of the Chinese Academy of Engineering; Vice-Chairman of the Academic Degrees Committee of the State Council; and Professor of Zhejiang University etc. Professor Lu got his Doctor Degree in Engineering (Dr.-Ing) from the Technical University of Aachen, Germany; Doctor of Engineering honoris causa from Hong Kong University of Science and Technology; and Honorary Doctor of Engineering from the City University of Hong Kong. In his academic career, Professor Lu has made important contributions to the development of mechanical engineering, especially in the field of fluid power transmission and control and higher education of engineering. He has achieved about 20 patents in China, Europe and USA, published at home and abroad over 250 papers in scientific research and engineering education and 2 monographs. He received Rudolf-Diesel Medal in Gold in 1997, Alexander von Humboldt Medal in 1998, the Knight Commander's Cross (Badge & Star) of the Order of Merit of the Federal Republic of Germany in 2000. As a professor, he has supervised 30 Ph D students, 25 master degree students and 5 postdoctoral researchers. For his outstanding contributions to education, he was awarded the National Higher Education Prize in 1989.*

Abstract: China's science and technology tradition represents continuity for one of the oldest civilized states since ancient times. And as western culture was connected to the Orient around the end of the 16th century and the beginning of the 17th century, science and technology gradually began to be introduced into China. The "Westernization Movement", promoted by government in the middle of the 19th century, marked the development of the modern period of science and technology. Systematic development, however, began in earnest in the 20th century. With the foundation of the People's Repubic of China, science and technology in China began to more fully develop as part of the nation's modernization, with more rapid development demonstrated in the past 20 years through reform programs and progressive open movement policies.

China's science and technology tradition represents continuity for one of the oldest civilized states since ancient times. And as western culture was connected to the Orient around the end of the 16th century and the beginning of the 17th century, science and technology gradually began to be introduced into China. The "Westernization Movement", promoted by government in the middle of the 19th century, marked the development of the modern period of science and technology. Systematic development, however, began in earnest in the 20th century. With the foundation of the People's Repubic of China, science and technology in China began to more fully develop as part of the nation's modernization, with more rapid development demonstrated in the past 20 years through reform programs and progressive open movement policies.

1. INTRODUCTION

Fifty years ago, political unification provided the foundation for significant independence in China, enabling the nation to plan and commit to economical, scientific and cultural development. In the First Five-Year Plan, the preliminary framework for a modern science and technology system was formed. Subsequently, the following 12-year plan established the foundation for implementing the science and technology modernization in China. Chinese leadership adopted a strategy of sustainable development and promoting prosperity for China through science and education, which resulted in the development of science and technology as a national policy. China's policy of reform and global engagement with the outside world was aligned with the reforms of the science and technology system to accelerate the process of turning science and technology achievements into productivity. Chinese science and technology, as implemented under the guidance of various national programs, made remarkable contributions to the national economic construction, sustainable development objectives, as well as enhanced national security.

Science and technology has significantly changed the prosperity of Chinese society over the past 50 years. A comprehensive deployment and development system has been developed for each basic discipline, from organization to research teams. Industrial technologies have been established, reaching the level of developed countries in the 1980s. Owing to several generations' improvement of farm crop seeds and adoption of proper cultivation techniques, 1.2 billion Chinese people can now have ample food and clothing under per capita land use conditions of only 1/3 that of the world average. Average life span of the people has reached 72 years due to advances in medical science and development of a health care system. A comprehensive system has been established with regard to the exploration and utilization of earth and ocean resources; the forecasting and response of weather and natural disaster; and the protection and renovation of water conservancy facilities and eco-environment. Chinese science and technology is advancing with the times and will make new contributions to the overall construction of a comparatively prosperous society in China.

Chinese scientists and engineers, working in various academic disciplines and technology fields, serve Chinese modernization programs with devotion and diligence. In addition to their significant contributions to increase China's economic position and enhance comprehensive state power, they have also created conditions to alleviate unbalanced ecological systems and deteriorating environmental trends for the sustainable development of economics and society. I would here like to introduce some of the major achievements made by my Chinese colleagues, selecting aspects ranging from physical science to space technology, as well as our international academic exchange, scientific education, and moral construction.

2. PHYSICAL SCIENCE AND RELATED TECHNOLOGY

Physical sciences, primarily physics and chemistry, focus on the study of micro-structures and interactions of the matter forming the basis of all scientific disciplines. A series of new technologies can be derived from these physical sciences to provide new knowledge for developing novel materials and devices. The development of physical sciences in the 20th century tended to rely more and more on huge complicated experimental facilities, while China in the first half of the 20th century could only furnish scientists with relatively simple experimental facilities. Under these more modest conditions, Chinese scientists predicted the transuranic elements (in 1933) and proposed experimental schemes to look for neutrinos (in 1941). But the early experiments on the electron-positron pair generation and annihilation (in 1930), discovery of uranium nucleus triple fission (in 1946) and the discovery of the μ-atom(1949), could only be performed outside China's laboratories by Chinese scientists who had the opportunity to go abroad for research.

In the second half of the 20th century, China began to speed up the construction of experimental facilities. The first experimental nuclear reactor was created in 1958, followed by creation and improvement of large-scale experimental facilities and national labs, especially in the past 20 years. Examples include the Beijing Electron Position Collider (in 1988) which is the most capable of working in the τ-c particle physical energy region, Lanzhou heavy ion accelerator system (in 1988) for low energy nuclear physics research, the 2.16 caliber optical astronomical telescope, the largest in East Asia (in 1989), and the multi-channel Solar Magnetic Field Telescope (in 1994). Dozens of key national laboratories with updated facilities have been established. These labs provide significant improvements in the conditions for further development of Chinese physical sciences.

Modern physics research in China started after the "physics revolution" at the beginning of the 20th century. During the second half of the 20th century, particle physics and quantum field theory, atomic nuclear physics, condensed matter and statistical physics, theory of gravitation and astrophysics became the most important basic research areas for Chinese physicists. Chinese physicists working in various branches of physics have made numerous contributions in theory, experiments and technical applications. In the field of theoretical research, Professor Zhou

Fig. 1 Beijing Electron Positron Collider

Guangzhao (CAS Member) and colleagues proposed "Theoretical Study on Unified Green's Function Approach for both Equilibrium and Non-equilibrium System " (in 1985), Huang Kun (CAS Member) and colleagues presented Huang-Zhu Model of semiconductor superlattice theory of optic-phono model (1988), and Professor Ouyang Zhongcan (CAS Member) accomplished the "Study of Liquid Crystal Model of Biological Membrane " (1990). In experimental and observational studies, there are many significant results such as discovering fifth order symmetry in quasi-crystal (in 1984), synthesizing 220 new nuclides with heavy ion accelerator (1992-2002), completing the accurate measurement for the τ-lepton mass (1992), and research on mass distribution in universe by gravitational lensing effect (during 1994-1999). In the field of technical applications, Chinese scientists and engineers designed and built Qinshan Nuclear Power Plant (in 1991). Independently, Chinese scientists proposed inertial confinement principle in nuclear fusion (in 1961), constructed a driving device for "Shenguang Laser" for experiment (in 1987), constructing magnetic confinement devices, such as superconductor Tokamak (in 1994) and "circulater No.1" device(1986). All of these works paved a way for future energy sources in China.

Chemistry research on structure, property as well as mutual reactions of various substances,and on how to create new materials, not only provides reliable knowledge for mankind's understanding of the physical world, but also provides efficient

Fig. 2 Qinshan Nuclear Power Plant

solutions for synthesizing a variety of usable substances. The start of chemistry study in China was right in phase with the introduction of quantum mechanics which forms its basis. Branch subjects such as biological chemistry, organic chemistry, inorganic chemistry, physical chemistry and analytical chemistry were established in the first half of the 20th century while macromolecular chemistry and radiation chemistry were developed in the second half of the 20th century. Chinese chemists working in various branch disciplines of chemistry have acquired significant achievements on theory, experiments and technical applications. In the field of theory, there are many important achievements, such as string bag model for simulating biological nitrogen fixation enzyme with atomic cluster structural chemistry by Lu Jiaxi (CAS Member) et al. (in 1973), and ligand field theory of quantum chemistry and molecular orbit graphic theory by Tang Aoqing (CAS Member) et al (in 1979). In the innovation of experiment methods there are important achievements such as the new methodology study on new synthesizing methods of carbon-carbon double bonds (in 1989) and new approach in determining element isotopic mass spectrum (in 1991). In the applied research field, the technology for obtaining kalium by separating brine in saline lake and the technology of obtaining boron and lithium resources by direct distilling (in 1980) made great contributions for the development of saline lakes and the construction of northwestern regions. Research on organic fluorine and the fluorine surface activator (in 1960) played an important role in developing highly explosive speed detonators used in plastics bonding. The successful development and production of the lithium ion battery (in 1996) broke the technical barrier for certain countries such as Japan.

Material is the physical basis for creating living entities and producing varieties of facilities etc., in essence, the foundation of mankind' s survival and development. Although material research and development in China started in the late 1950s, China has reached dominance for both resources and technology in the fields of rare earth material, inorganic nonlinear optical crystal, and advanced ceramics. As examples, rare earth-transitional metal magnetic material (in 1983), nonlinear optical crystal barium borate (in 1983) and lithium triborate (LBO)(in 1987), and fiber enhanced ceramic composite materials (in 1970). Additional representative significant achievements have been accomplished. In recent ten years, both high temperature superconductor material and nanometer material have become focal points for Chinese scientists. In the research of superconductor material, after discovery of the high critical temperature superconductor material $YtBaCuO_2$(1986), Chinese scientists have produced many studies regarding the industrialization of high temperature superconductors. With the fabrication of a hundred meters of bismuth series high temperature superconductor tape in 2000, China pushed superconductor research into the production phase. In research of nanometer materials, development of coaxial cable at nanometer level (in 2000) may lay the foundation for producing the next generation of light-guide fiber. Discovery of the 50X super-plastic ductile deformation of nanometer metal copper (in 2000) may forecast the new and significant application prospect brought out by this new property of nanomaterials. Research on completely ordered and fully metal nanometer lattice growth (in 2002) may also have important applications in nanometer electronic technology.

3. LIFE SCIENCE AND TECHNOLOGY

Life sciences in China were initially developed in the first half of the 20th century. By the second half of the 20th century, molecular biology had begun to be applied to the life sciences. Subsequently, biological molecules, enzymatic structure and function, biological films, molecular genetics, and the gene project, among other advances, acquired greater accomplishments. China was the first to synthesize Crystalline Bovine Insulin with relatively high activity (in 1965) and Yeast Alanine Ribonucleic Acid with the same natural chemical structure and complete biology activity (in 1981). This achievement stimulated Chinese life scicntists to develop more profound and broader research programs. These pioneering works with important significance to develop life sciences, agriculture, and bio-medicine were successively published in the scientific literature. Of special note was the sequencing of human genome (in 2000),the research on drosophila selectivity behavior mechanisms (in 2001),and the Theory Topological Perception and Functional Hierarchy of Vision. China's participation in the human genome project was exemplary, the unique member from a developing country. Chinese scientists completed 1% of the working draft of the human genome map, indicating that Chinese scientists are fully capable of joining in significant international cooperation projects. The research on drosophila selectivity behavior led to the discovery that the brain mushroom body plays a determinative role. More important is that the fact that the experimental model has proven to be applicable for any animal, therein blazing a new methodology for studying animal decision making and selection behavior genes, as well as for molecular mechanics.This theory of Topological Perception and Functional Hierarchy of Vision has obviously provided a lively and fresh way of thinking for the development of cognitive science.

Science and technology's contribution to agriculture progress has raised productivity by 35%, primarily as a result of the promotion of high quality seeds and cultivation system reform. The seeds of paddy rice, cotton, soybean, corn and wheat have been improved several times over the past 50 years. Multiple cropping indexes have increased by 25%, and afforesting survival rates have been increased by 20%. Chinese scientists have not only learned comprehensive techniques for transferring plant and animal genes but also have the capability of conducting studies on many plant and animal genomes. Significant progress has been achieved in the study of paddy rice cross breeding and the paddy rice genome. The annual increment of grain yield brought about by the study (in 1970) and promotion of cross breeding paddy rice by Yuan Longping (member of Chinese Academy of Engineering) now provides enough food for tens of million people. The breeding program for super hybrid paddy rice can be expected to increase these benefits. Research activities, such as the completion of Gene-Centric Sequency Map(2002),and the completion of draft and database INDICA rice genome (in 2001), the completion of the precise sequencing of chromosome 4 for the rice genome (in 2002), and progress of subsequent work, will not only disclose paddy rice secrets but also help to study other gramineous crops 'genome, such as wheat and corn, thus bringing along basic and applied research for the whole family of grain crops.

Biotechnology applied to the medical fields is developing rapidly. China's medical biotechnology started in the late 1970s, however, delightful progress has been made with aspects of gene engineering multi-peptide medication, go-ahead medication, and antibody engineering. Research such as synthesizing anti-swamp fever southernwood (in 1974), manual interferon medication (in 1992), ethyl-liver gene engineering vaccine (in 1992), the discovery of the gene of Anti-microbial Peptid and its important biological functions, as well as the discovery that the IHH gene confirmed to be the cause of the genetic disease Brachydactyly Type A-1 (in 2001), discovery of the heritable opalescent dentine gene I disease (in 2001), the study on pathological changes mechanism of nerve degradation disease and new target point of medication effect (in 2001),the Gene Responsible for Children Cataract (2002), a new rind thrombolytic drug of Recombinant Staphylokinase(2003), Dr Chen Zhongwei 's microsurgery, and Dr Hu Mengchao 's achievement in liver and gall clinic science, represent the significant progress and the advancing level of Chinese scientists in the medical fields.

4. INFORMATION SCIENCE AND TECHNOLOGY

The development of information science and technology, along with the generation of the electronic computer, semiconductor electronic devices and lasers, received high attention by Chinese scientists and the government from the very beginning. The studies on computer science and technology, semiconductor science and technology and radio electronic science and technology were carefully planned and deployed beginning in 1956. These technologies were listed as urgent subjects to fill the gaps in the leading science and technology fields. As early as 1957, the first transistor was successfully developed in China, and the first electronic computer was born in 1958. China's development of the laser followed by one year the world's first laser. Chinese information science and technology related to the development of computers has made significant progress in both the hardware and software. Some new directions, such as quantum computing and quantum communications, have received close attention as well.

In the development of the computer, China produced the first generation electronic tube computers (1958-1964), the second generation of transistor computers (1965-1972) and the third generation of middle or small-scale integrated circuit computers (1973-1980), then entered the era of ultra large-scale integrated circuit computer at the beginning of the 1980s. Peak floating point operation speeds for several high performance computers included: the Milky Way Series Parallel Vector Supercomputer that reached 13 GFlops per second (1997); the Shenwei I high performance computer that reached 384 Gflops (1999); the Dawning 3000 Super Server that reached 403.2 Gflops (2000); the Legend Deepcom 1800 Linux Cluster that reached 2 Tflops (2002). Peak performance for the newly released Dawning 4000L Linux Cluster, on Mar 14, 2003, was 4 TFlops(2003). All these computing advances have played significant roles in Chinese weather forecasts, surveys of oilfields and other scientific computing. In software development, Chinese scientists developed a Temporal Logic Language (1983), Intelligent English/Chinese Translation Software (1992), High Performance Distributed Parallel Numerical Linear Algebra Software (1999) and Linux Based Secure Operating System (2002). Chinese Information Processing System, Chinese Text Conversion System and Chinese Intelligent Interface, inter alia, have had special significance for advancing Chinese culture. Chinese laser typesetting system (1985) developed by Wang Xuan (CAS Member) brought the Chinese printing industry into a new era. In communications technology, the introduction of foreign technology was incorporated along with independent Chinese developments; as exemplified by the second generation of mobile communication GSM digital switchers developed by three Chinese companies Datang, Huawei and Zhongxing that was fully put into operation at the end of the 20th century.

A new direction for Chinese information science and technology, quantum computing and quantum communication, began in recent years. In quantum information theory, Chinese scientists have proposed the principle of quantum error avoiding codes (1997), the principle of probabilistic quantum cloning (1998), and a new kind of quantum information processor (2000). In quantum communication, Chinese scientists co-authored published experiment reports on recognition of the quantum concealed state (1997) and achieved 50 kilometer Fiber-optical Quantum key distribution,greatly promoting the development of Quantum Secure Communication Application.Chinese scientists have recognized the importance of basic research of nanometer structures, quantum devices, and integration techniques.

5. MATHEMATICS AND SYSTEMS SCIENCE

Mathematics has its own excellent tradition in China, and Chinese mathematics had joined with the world's mathematicians since the beginning of the 20th century. The progress of collegial mathematics education, mathematics research and international communication within China in the first half of the last century laid the foundation for the further rapid development of Chinese mathematics. From the second half of the 20th century, the studies on mathematics in China went beyond the fields of pure mathematics and began to integrate with scientific computing and engineering design. The Institute of Applied Mathematics and the Institute of Systems Science were established as early as 1979 in the Chinese Academy of Sciences, and the Institute of Computational Mathematics and Science Engineering Computation was built in 1995. These institutes merged into the Academy of Mathematics and Systems Science in 1998. Mathematics, systems science which is a part of applied mathematics, and many other important branch disciplines have developed in China along with a number of major accomplishments.

Representative achievements in mathematics research include the research on Characteristic Classes and Imbedding Classes in Topology (1946), the Theory of Functions of Several Complex Variables in Classical Domain (1954), the initiation and development of Finite Element Method (1965), the research on Goldbach's conjecture (1966), the mechanical proving of mathematical theorems (1978), the Critical Point Theory and its application (1983), and the Symplectic Geometric Algorithm of Hamiltonian System (1984) among many world-class advancements. Among the achievements obtained during the last 20 years, the research conducted by two members of Chinese Academy of Sciences, Feng Kang and Wu Wenjun, drew wide attention and favorable comments from international academia. Feng Kang, having initiated the finite element method independently of western colleagues, created and developed as well the symplectic geometric algorithm of the Hamiltonian system, which solved the long-term prediction in dynamics computational problem that had existed for years. Feng's method was not only applied to astrophysics, molecular dynamics and many other fields successfully, but also stimulated many subsequent research initiatives. Wu Wenjun, after his studies on characteristic class and imbedding class, again made valuable contributions to a completely new field. He used the ideas of algebraization of geometry in Chinese traditional mathematics to the mechanics for proving geometry theorems, and established an elimination method for solving nonlinear simultaneous algebraic equations as well as an ordering method for partial differential equations that can be applied to the mechanical theorem-proving in differential geometry, which ultimately changed the field of automated reasoning research in international academia.

Systems science has a comparatively short history over the past 50 years or so. Its development in China relied on engineering science and started with the studies on operations research and cybernetics. It was Qian Xuesen (Tsien Hsue Shen), a member of Chinese Academy of Sciences, who was the first in the world to apply systems science to the fields of engineering. His monograph "Engineering Cybernetics" (1954) has become a classical work in this field. The studies on systems science in China not only created important academic results, such as the study on Differential Dynamical System Stability (1980), but also produced many applied achievements such as Population Forecast Model (1980), National Grain Output Forecast Model (2001), and so on. These works are of crucial practical significance for the development of Chinese economy and society. In the last two decades, important theoretical advances in systems science research in China have also been made. For example, the Anti-interference Control Theory of Non-linear System (1999) has been successfully verified both in the laboratory and by computation simulation. It has been applied to robots, electric power system, wave filtering and other practical engineering.

6. ADVANCED MANUFACTURING TECHNOLOGY AND AUTOMATION TECHNOLOGY

Manufacturing technology is the backbone of human civilization and forms the base of national competitive edges. Manufacturing technology is the base of all technologies manufacturing various products for the surviving and development of human being. The initial and fundamental meaning of technology is "manufacture". Manufacturing is considered as the mother of industry. Manufacturing technology has experienced the developments of mechanization, precision, automation, intelligence, and environmental life cycling, with the advanced production technology developed in the 1980s as the latest development of manufacturing technology. Advanced production technology has become a model of industry innovation and modernization, which may create extensive and profound influence on the development of the national economy. Significant progress in Chinese advanced manufacturing technology and automation technology are reflected in the fields of modern integrated manufacturing system and robots.

The concept of "computer integrated manufacturing system" (CIMS) was put forward in the US in 1973. In China, it developed and became "contemporary integrated manufacturing system" in the second half of 1980s. The contents of CIMS have extended into comprehensively utilizing information technology and advanced management to realize optimization management and operation of enterprises to improve efficiency and market competitiveness. The development of CIMS in China has experienced three stages: information integration, process integration and inter-enterprise integration. The efforts made by Chinese scientists for the development of CIMS have been recognized by international academia. China has been awarded the top prizes three times from the American Manufacturing Engineer Society CIMS. Tsinghua University won the "leading university award" in 1994, Beijing No.1 Machine Tool Factory won the "leading factory award" in 1995, and Huazhong University of Science and Technology won the "leading university award" in 1999. CIMS has been applied to hundreds of enterprises in China since 1987. These enterprises have been engaged in machinery, electronics, aviation, aerospace, light industry, textile industry, petroleum, chemical industry, metallurgy and other major manufacturing industries. Most of these enterprises achieved distinct economic and social benefits with CIMS, and they have played a crucial influential role on the informatization and modernization of manufacturing for various Chinese industries and regions.

Since the first robot was created 50 years ago, the development of robots has followed two technical lines. One field of development is designed to exhibit human functions for improving human quality of life and working efficiency. These robots are extensively applied in manufacturing and are thus called "industrial robots". The other robotic field is designed to expand the function and living space of humans and are called "advanced robots". With the continuous progress of robotic technology, the two sorts will integrate and develop into intelligent robots. For industrial robots, China has mastered the optimization design technology and the high performance control technology of operating machines. Manufacturing for industrial robots includes serial products such as welding robots, assembling robots, carrying robots, printing and packaging robots,hence furthering the industrialization of the nation's manufacturing and application system of industrial robots. For Chinese advanced robots, China has successfully developed prototypes such as underwater operation robots, robots used in nuclear power stations, wall-climbing robots and pipeline engineering robots. While most advanced robots remain in laboratories, the autonomous deep-sea robot has reached an internationally advanced level. The 6000 meter diving-depth autonomous robot was successfully manufactured in 1995. It was developed from a prototype and completed testing for Pacific Ocean benthos exploration.

Fig. 3 The 6000 meter diving-depth autonomous robot "CR-01"

7. EARTH SCIENCES AND ENVIRONMENTAL SCIENCE

China has a long history in the scientific research of earth sciences. The studies of geology, geography, atmosphere, oceanography, and other fields are closely related to the exploiting of natural resources and the understanding of the environment. Studies on palaeoanthropology and palaeontology are exceptions that exclusively aim at enhancing scientific knowledge. The number of research projects conducted under the guidance of the Tenth Five-Year Plan and the national sustainable development strategy provided a crucial scientific base for the exploitation of soil resources and environment protection. Many important results have been achieved, including the Physical Map Sets of the PRC (1964), the Research on Chinese Layer-control Mineral Deposit Geochemistry (1986), the study of Tectonic Evolution and Formation Process of Continental Shelf and Adjacent Sea Area, the investigation of the Uplift of the Qinghai-Tibet Plateau and Its Influences (1979), The Rotation the study on the Qinling-Dabie orogenic belt and ultra-high pressure metamorphism and so on.

With respect to resource survey and utilization, China has carried out a series of key projects, such as the regional comprehensive investigations in the 1950s, the research on regional planning of agricultural natural resources started in the 1960s, the control program of territorial development initiated in the 1980s and the "21st Century Agenda" put forward in the 1990s. All of these initiatives provided a crucial scientific base for the development and utilization of mineral resources, biological resources, water resources, and soil resources. The Theory of Continental Petroleum Generation (1960-1980) and the Theory of Coal Turning into Gas (Oil) (1979) made great contributions in the exploration and development of oil and gas fields. As a result, China is no longer an "oil shortage" country. The research results on calculation and evaluation of national water resources and the planning and decision-making support system of the Yellow River water resources provided a scientific basis for the planning of the national economic construction.

In the field of environmental cognition and protection, China began to work on investigation, prevention and cures for local diseases, the conservation of soil and water resources in large river basins, the history of climatic change, the forecast of various natural hazards and many other topics in the 1950s. Many studies on the investigation of environmental background, environmental capacity, biological detection, environmental forecasting and so on have been carried out since the 1980s. Continuous research on East Asian atmospheric circulation and Chinese climatic change contributed greatly to the establishment of the Chinese climate forecast mode. The sequential reconstruction of ancient East Asian atmospheric circulation evolution history based on dust accumulation in Loess Plateau has been a crucial guiding factor for the recovery and control of the environment, as well as the prevention and restoration of soil desertification, sand storm, water loss and soil erosion and other geologic hazards in the development of Chinese western regions. It also positions the Chinese loess as one of the three pillars for the study on global environmental change. The two other pillars are deep-sea sediment and polar ice core. These three pillars provide a crucial scientific basis for the cognition of rules and mechanisms governing the Earth's environment change and for the forecast of environmental evolution. Serial research on geology and the environment have also contributed to the Yangtze River Three Gorges water conservancy projects.

In the fields of palaeoanthropology and palaeontology, the fossils found by Chinese scientists have provided important evidence for the study of human and organic evolution. As for palaeoanthropology, the discoveries and researches of ancient human fossils found in Zhoukoudian of Beijing, Lantian of Shaan'xi Province, Liyang and Yuanmou of Yunnan Province provided crucial evidences for the cognition of human origin and evolution. As for palaeontology, the biota discovered in Chengjiang of Yunnan Province, West Rehe of Liaoning Province, Weng'an of Guizhou Province are listed among the top "surprising discoveries" of paleontology in the 20th century. The discovery and research of Chengjiang biota fossil in 1984 revealed the actual world appearance of marine organisms 530 million years ago. This research also provides a precious window to probe the mystery of the "Cambrian Period Firestorm" and to understand the pedigree of organic origin and evolution. Since 1987, many bird and beast fossils have been discovered in Liaoxi. Among them, fossils of 80

Fig. 4 Original shrimp fossil discovered in Chengjiang

million to 150 million years old filled in a more than 100 year gap for the research of early birds (from archaeopteryx to Cretaceous Period birds). China has become the world research center of early birds' origin and evolution as a result of these fossil discoveries. The animal fossils discovered in Weng'an were published in 1998 on phosphate animals that lived 580 million years ago. These fossils are the earliest real body fossils preserved in the world and the research is helpful for unveiling early animal evolutionary processes.

8. SPACE SCIENCE AND TECHNOLOGY

A major objective of space technology is to send spacecraft into outer space and provide experimental platforms for the exploration of space environments, resources, and other studies in multiple disciplines. The development of space science and technology experienced three stages since the first satellite was launched: the earth - space launching and recovering technology of space shuttle, the permanent near-earth orbit technology of space station and the deep space exploration and limited outer space emigrant technology. Shortly after Russia and the US launched their first satellites, Chinese scientists began to develop space research for the development of spacecraft technology and ground tracking and control technology. Satellite research and manufacturing in China began in the mid 1960s. The first Chinese satellite was successfully launched in 1970. Since then, China has launched

Fig. 5 No.1 Bright East, the first satellite launched by China

more than 40 satellites and has mastered the technology of launching and recovering geo-synchronous orbit satellites, sun-synchronous orbit satellites, and so on. China's space program was first meeting the demands of communication and broadcast, general survey of national territory, weather forecast, scientific tests, and so on. China has since begun tests on launching manned spacecraft to prepare for the Chinese astronauts to get on the moon.

The major goal of space science is to study the phenomena of physics, chemistry, life and so on that occur in the outer space under the conditions of micro-gravity, radiation and other surroundings. Research on space science in China has a history of nearly 50 years. Through balloons, rockets, satellites and other space observation methods and facilities on the earth, China has laid a foundation in the fields of space physics, space life, micro-gravity and so on. Studies on space physics in China started from the late 1950s and progress was achieved in both observation and theory. The Theory of Magnetic Field Reconnection Induced by Fluid Swirl Induces (1985) is a most crucial research achievement. The research on space life sciences started at the end of the 1950s. Serialized research was focused on the returned organisms from space aircraft testing for space medicine. Research on micro-gravity in China started from the late 1980s. The recovery satellites have been used as research platforms for micro-gravity. In addition, a "Dropping Tower" was constructed on Barth serving as a micro-gravity laboratory. Collaboration among dozens of Chinese laboratories, concerning many space science subjects, have joined these research initiatives.

While the aim of space program in China is to serve the construction of national economy and promote the progress of science and technology, China has also made great efforts in contributing to the international space program. China is currently implementing two space programs: the manned spacecraft test program and the Double Star Programme. This strategy will propel the Chinese space technology to a new world stage. The manned space flight program started from the beginning of the 1990s. Four unmanned testing spacecrafts of the "Shenzhou" series have been successfully launched since 1999. The "Shenzhou" No.4 testing spacecraft consists of four major parts: the propulsion module, the capsule, the orbital module and the extension part of the orbital module. Included with 52 pieces of scientific experiment equipment in the spacecraft was a"simulation man". Its success laid the foundation for planning the launch of the first manned "Shenzhou" No.5 spacecraft. The Double Star Programme includes the equatorial satellite and the polar area satellite. The two satellites are located in the important active region of the near earth magnetosphere which the present international Solar-Terrestrial physics exploration satellites have not covered yet. That programme, together with other 4 satellites programme of the European Space Agency CLUSTER, will make for the first time in human history, six-point detections in geospace to carry out the research on the global variation rules of the Earth magnetosphere and the mechanisms of storm events.

Fig. 6 Chinese "Shenzhou" No.4 spacecraft

9. INTERNATIONAL SCIENTIFIC EXCHANGE AND COOPERATION

China represents one of the Third World countries. Its scientific research is still at the verge of entering the world science center, with the desire to promote a fine international academic exchange environment. The scientific research and cooperation between Chinese scientists and our international colleagues has continuously expanded from just over thirty countries in the 1950s to all two hundred countries at the end of the 20th century. By the end of the 20th century, there were nearly 20,000 Sino-foreign cooperative projects operating each year, with the participation of approximately 50,000 Chinese scientists and 20,000 foreign scientists. At present, a rapid development program for international scientific exchange has been formed, and multi-level, multi-channel and multi-form international scientific cooperation has been established, which includes various laboratories, joint research projects between China and foreign countries, the establishment of some junior scientists teams and partner groups with the Max-Planck Society. The Sino-Europe Geo-Space Double Star Programme (2001) has already started; the Sino-U.S-Russia Global Advanced Network Plan Cooperative Agreement (2002) has been signed; Chinese scientists have taken the lead in establishing "The International Society for Digital Earth" (2002); and Chinese scientists organized a great number of international conferences, including the 24th International Conference of Mathematicians that was recently successfully held in China (2002).

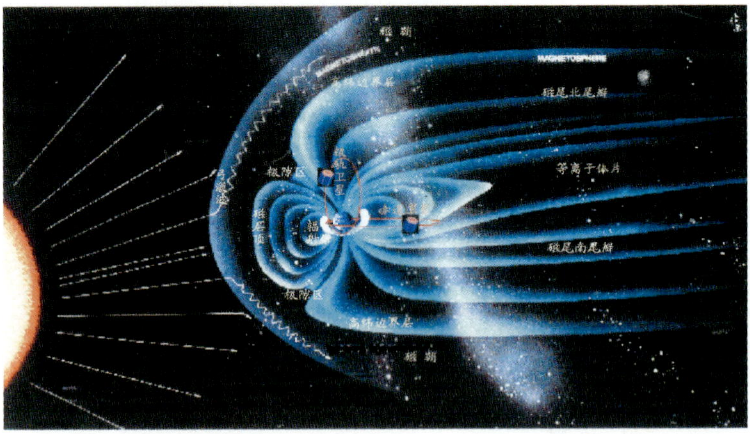

Fig. 7 Important regions in the earth magnetosphere and the track for Double Star Programme

Many cooperative projects have made remarkable achievements. Chinese and Russian scientists discovered anti-sigma-minus hyperons in Dubna United Atomic Nucleus Research Institute (1960). Chinese scientists successfully manufactured germanate bismuth (BGO) crystal, which was used in making energy detectors by

European Nuclei Research Center (CERN) (1983). Chinese scientists participated in international exploration across the Antarctic and obtained abundant new data (1990). Chinese and U.S scientists cooperated in accurate measurements of τ-lepton mass on Beijing Electron-Positron Collider (1992). Chinese scientists provided large-scale permanent magnets for Alpha Magnetic Spectrometer on the "Discovery" space shuttle of U.S. to detect dark matter and antimatter (1997). At an altitude of 7000m on Qinghai-Tibet Plateau, Chinese and American glaciologists drilled and obtain 20,000 ice-core samples that were used to diagnose environment change (1997). Chinese scientists participated in the International Ocean Drilling Program and discovered the abundance of creatures living at a depth of 2722 meters in the southern South China Sea (1999). Chinese scientists completed a working draft of 30 million alkali-base pairs on No. 3 chromosome in the Human Genome Program (2000). Chinese scientists are responsible for accurately mapping the No. 4 chromosome of the international paddy rice genome program.

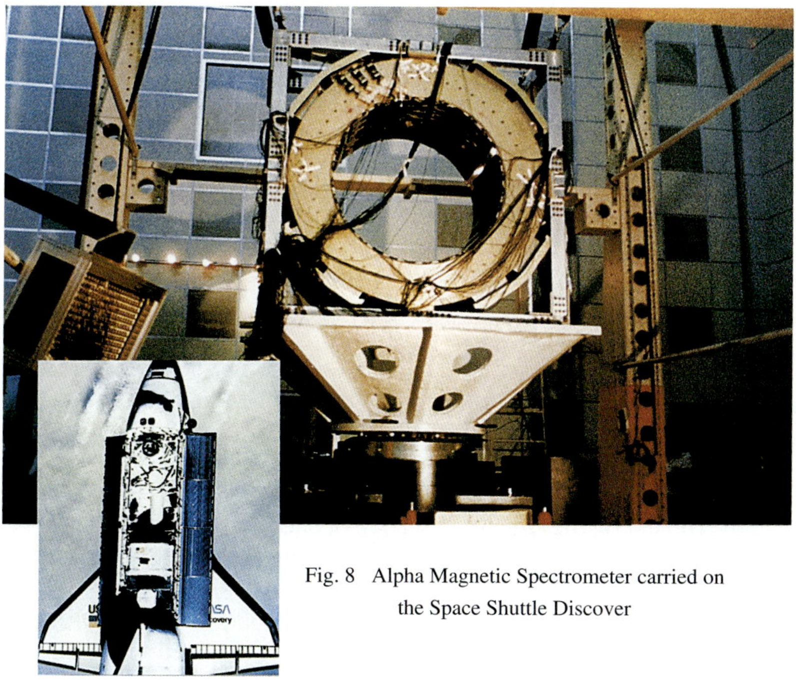

Fig. 8 Alpha Magnetic Spectrometer carried on the Space Shuttle Discover

As a developing country, China shares similar philosophies and ambitions with other Third World countries, and Chinese scientists have always made active efforts in developing South-South scientific cooperation. To improve friendship and seek cooperation, Chincse scientists often visit countries of the Third World. Chinese scientists use South-South Cooperation Funds to educate doctors for the Third World countries; support the scientists of the Third World countries to visit China, participate in training courses and attend international conferences. The China and Brazil Earth Resource Satellite No.1 (CBERS-1) jointly manufactured by China and Brazil was successfully launched on Oct. 24, 1999 and continues to operate normally with advanced remote sensors, including CCD camera with five spectral ranges coverage and resolution of 19.5m, infrared multi-spectrum scanners and wide-angle imager. The satellite is used for resource investigations and environmental monitoring not only for China and Brazil, but also for many countries and regions throughout the world. While increasingly incorporating into the extended international science community, the Chinese scientific community has maintained and fostered close relationships with scientists of the Third World for academic exchange and cooperation.

10. SCIENCE EDUCATION

Modern science took shape in the 17th century, but science education had not found its position in universities until the end of the 18th century. After that the systematical education of scientific and technological knowledge began. Science education in China started during the "Westernization Movement" in the second half of the 19th century. At the beginning of the 20th century, a series of new policies, especially the policy of "abolishing imperial examinations and establishing schools" laid a foundation for the national science education. The establishment of Youmei Tsinghua School (1908), the reform of Beijing University (1912), "democracy and science" flag of the New Culture Movement (1915) and a series of reform measures by the government on the education system opened up the way for the development of science education in China and embraced modern education as its basis. The education reform in the 1950s greatly changed the deployment of science and technology education in China. The strategy of "revitalization the nation through science and education" put forward at the end of the 20th century further enhanced the position of science education. One of the prominent results in the progress of science education in China is that a group of research-oriented universities are growing up in the reform.

As a developing country, China has shortened its gaps with advanced countries from several hundred years to several dozen years in the past century. This is closely related to the continuous science education in China. China established a modern college education system in the 1930s, established a group of scientific institutes, colleges and universities such as the University of Science and Technology of China in the 1950s, set up the Graduate School of Chinese Academy of Sciences and graduate schools of universities in the 1970s and fostered a contingent of scientific and technological experts. At the end of the 20th century, there were several million scientific and technological staff members serving at various positions, who made contributions to the economic and social development of China. Among them there are outstanding scientists who made scientific and technical contributions renowned to the world. Chen Ning Yang and Tsung-Dao Lee who won Nobel Prize of physics in 1957 both received their higher education at the Southwest Associated University temporarily founded during the Anti-Japanese War period. Chen Jingrun, who brought the proof of Goldbach's Conjecture to its frontier position, Wang Xuan, the information scientist who made China say farewell to lead and fire in the printing industry, and Yuan Longping, who is honored as the "father of paddy rice hybridization" are all scientists trained in China.

Although China has made substantial achievements in its science education, it still cannot meet the needs of science development and construction for the country. Science education in modern China is facing severe challenges. China has become acutely aware of the crisis regarding scientific and technological innovation ability, and is making efforts to liberate and energize scientific and technological creativity through a series of reforms on the education system. Primary examples include the "211 Project" for education development, the 21st Century Education Revitalization Plan, Technology Innovation Project, and Knowledge Innovation Program. Research-oriented universities and research institutions engaging in high-level academic research and technological development are incubation environments for fostering innovative talents and the best method to combine teaching and research into one. Science education should lay emphases on fostering students' capability for independent thinking and problem-solving. The education of science history, science philosophy, science sociology, human studies and social science shall be reinforced to make students receive an all-around and balanced development experience so they can be competent for the heavy burdens of science development, country construction and advancing human civilization.

11. SCIENTIFIC ETHICS AND MORALS EDUCATION

Science and technology have profound changes on human life, impact people's traditional concepts and challenge the ethics and morals that were based on the existing industry civilization. The change of social structure and relations caused by network technology, and the challenges on human dignity and belief brought about by biological technology have presented a litany of the world's difficult problems recognized for concern by the international community. Ethics conflicts and the blurring of moral standards aroused in the scientific and technological development have become discussion keystones in the Chinese scientific community. Three aspects have become the focus of our concerns: the first is behavior criterion and the supervision system on scientific research and technology development activities in the engineering fields; the second is ethic conflicts and moral challenges posed by new progress in modern science and technology; and the third is the risk reduction in a knowledge society and the social responsibility of scientists and engineers. We as world scientists should fully consider the difference in social development levels and cultural traditions between countries and understand that disagreement, or even conflicts on some concrete matters, is inevitable. We suggest that the international community take "science and technology serving for the welfare of human beings" as a guiding principle and seek for equality, justice and fairness in human society through "negotiation" and mutual respect. We promise to fulfill our social responsibility and international obligations.

Ethic conflicts brought about by information technology and biology technology have become the focus of people's attention. The Chinese scientific community has been actively participating in international conferences on the aspect, and taking it as a duty-bound responsibility to join the construction for future civilized order and social rules for human beings. In the mid-April 2002, at the 80th meeting of the famous "Xiangshan Conference", the Chinese scientific community discussed the ethic disputes and legal problems incurred by the life sciences and called on the government to constitute relevant laws and regulations to standardize life science research, to establish a national life ethic committee as soon as possible, and guide the public to launch extensive discussions on life science ethics. The deep concern of the Chinese scientific community on ethic issues in science and technology shows our humanistic concern for science and technology itself, our responsibility to others and ourselves, our yearning for peace and safety, and our concern with the destiny of the entire human race. Confronted with the challenge of science and technology on human values and ethics, the scientific community in China is thinking about a question of significance for future human survival: to what extent should the present culture support the belief of the human race's future?

The science community is a unique social group that masters special knowledge. The moral character of scientists and engineers is the premise for them to carry out social responsibility and take on public moral obligations. Because of enhancements in social prestige and enticement of various kinds of rewards and honors, the immoral behavior that happens occasionally in the scientific community not only ruins scientific reputation but also corrodes the scientific organism. In the early 1980s and 1990s, promoted by scientists such as Prof. Zou Chenglu (CAS member), the scientific sector in China launched two free discussions on the moral issues in science. In the year 1997, the CAS Academic Divisions established the "Moral Construction Committee", and it stipulated the "Self-discipline Rules for Academicians" in 2001. The Chinese Academy of Engineering and China Association for Science and Technology has developed similar organizations and carried out corresponding activities. The Chinese scientific community has expressed in a special announcement to its countrymen and international colleagues its resolution on maintaining the dignity of science and adhering to scientific morals. The Chinese scientific community has been enjoying high prestige with their elevated status among the public, which is related to its consistently adhering to stringent moral and academic standard. We believe that although the competition in the scientific and technological fields become fiercer and fiercer, the scientific community shall always stick to the moral principles of truth, faith, openness, neatness, tolerance and rational inquiry. These are not only the criteria of value on which the existence and development of science depends on, but also the behavioral standards for mankind. Scientists should not only rectify the style of study inside the scientific community, but also popularize scientific knowledge, advocate scientific methods, propagate the scientific thought and carry forward the scientific spirit to the public.

12. CONCLUSION

Humankind has passed the 20th century that was extremely extraordinary in scientific and technological developments. We are now confronting the rising tide of scientific and technological inventions in the new century. During the one hundred years of the 20th century, people of the Third World deeply suffered from backwardness of science and technology. Differences in the development levels of science and technology are one of the major factors that has caused the contrast of "poor south and rich north" in the global village. The prosperity and power of the Third World countries depends on science and technology. In the globalization era of a knowledge economy, science and technology becomes the key factor of overall national strength and international competitive edges. The developing countries should rely on science and technology to develop their economy and improve people's living standard, and increase scientific and technological levels through international scientific exchange and cooperation. The scientific exchange and cooperation between the Third World countries has a special significance, because many of the problems these countries confront are common. Science is an international cause. The problems such as the protection of the ozone layer and outer space environment, the exploitation of underground and deep-sea resources, the construction of information superhighway and network safety, the acknowledgement and protection of scientific and technological patents, the human genome sequencing cannot be solved by one country or one government alone. The exploration of universe, Earth, substance and life, the concern with eco- environment, the arrival of the information era, and the influence of globalization on science and technology make the scientific cooperation between countries necessary. The international exchange of science and technology is especially important for the Third World countries. The Third World Academy of Sciences acts as a bridge to link scientists of the developing countries. Let's make joint efforts to build TWAS into an irreplaceable bridge to promote South -North cooperation as well as South-South scientific and technological exchanges and cooperation, and jointly create the bright future for the mankind.

China's S&T Development – Strategies and Policies

Xu Guanhua

Ministery of Science and Technology

***Xu Guanhua**, male, Han nationality and a native of Shanghai, was born on December 16, 1941. He is currently a researcher, a member of the Chinese Academy of Sciences (CAS), a member of the Third World Academy of Sciences (TWAS), and a foreign member of the Sweden Royal Academy of Engineering. He is Minister of China's Ministry of Science and Technology (MOST). Xu is mainly engaged in research on resources remote sensing and geographic information systems. He studied in Beijing Forestry University from 1959 to 1963, and further studied remote sensing in the Department of Natural Geography in the University of Stockholm, Sweden from 1979 to 1981. He was successively the Assistant Researcher, Associate Researcher, Researcher and Director-General of the Research Institute of Forestry and the Institute of Resources Information, Chinese Academy of Forestry Sciences; Director-General, Institute of Remote Sensing, Chinese Academy of Sciences; Vice President of the Chinese Academy of Sciences; Vice Chairman of the State Science and Technology Commission (SSTC). He is a member of the Presidium of CAS, Director-General of the Department of Earth Sciences of CAS, Chairman of the Appraisal Committee for the National Natural Science Award, Chairman of the China Society of Image and Graphics, Director of the Academic Committee of the National Key Laboratory on Resource and Environmental Information System, Director of the Academic Committee of the National Key Laboratory on Remote Sensing and Aviation Mapping, and Editor-in-Chief for the Journal of Remote Sensing.*

Abstract: This article discusses emphasis the thinking way of the adjustment of Chinese S&T development strategies and the suppleness of great policies. Firstly readjusting Chinese S&T development strategies and putting greater emphasis on original innovation to achieving leapfrogging progress in S&T progress; secondly readjusting the concept and management system of S&T innovation, and cultivate a sense that human factor comes first; thirdly readjusting the principle of S&T innovation, concentrating resources on the most important and shelving others for the time being; fourthly readjusting the mode of S&T innovation, emphasizing the shift from single-item innovation to integrated technologies and the need to develop competitive products and industries thereon; fifthly readjusting the policy target in S&T innovation, shifting from establishing S&T institutions to wider S&T resources in entire society. Coupling with technology the way of thinking of the strategy adjustment, the nation set out a series of policies, including launching the special S&T items and improving the core competitiveness of Chinese key industries; Progressively replying the challenge of entering WTO, launching the strategy of talented person, patent and technique standard; deeply reforming the system of S&T research, and putting forward the system of national innovation; Constructing the good environment, and accelerating the progress of the indoctrination of high-technology; Extending to open outward, and Enhancing the international technology exchanges and cooperation.

As the new century unfolds, governments of all countries are pondering on and mapping out new strategies for economic and social development. As the world's largest developing country, China has made faster S&T progress and innovation a priority in its economic and social development, with a view to maintaining a sustained, rapid and sound development of the economy in response to the opportunities and challenges after its WTO accession. This major decision conforms to the laws of economic growth and S&T development in the present-day world.

The Sixteenth National Congress of the Communist Party of China called for building a well-off society in an all-round way, a new approach to industrialization and higher requirements for S&T work, all of which have greatly inspired our people working in S&T fields. Since the founding of the People's Republic more than half a century ago, China's S&T cause has grown from scratch and made gigantic strides. Particularly since the early 1990s when our government attached greater importance to S&T development and laid down the strategy of "invigorating the country through science, technology and education", China's science and education have been brought onto a track of rapid growth. Reform to S&T structure has made important headway. Scientific research and technological innovation have yielded remarkable achievements. S&T innovation has become a principal driving force behind economic expansion. The R&D/GDP ratio nationwide increased from 0.68% in 1990 to 1.09% in 2001, with the total R&D spending reaching an all-time high of 100 billion yuan. China's S&T development level has already ranked among the leaders of the developing world. A brisk S&T development driven by a vibrant progress of the entire nation – this is the common experience of all our scientists and technology workers.

1. READJUSTMENT OF CHINA'S S&T DEVELOPMENT STRATEGIES

We are soberly aware that despite our rapid progress there are still many contradictions and problems in China's S&T development calling for prompt solution. For example, China still lags considerably in making the shift from imitating others to making original innovation, from emphasizing single technological breakthrough to strengthening integrated technological upgrading, and from stressing technology orientation to promoting product and industry orientation. What is more, although our spending on S&T has gone up quite a lot in recent years, it still falls short of what the developed countries, the newly industrialized countries and even some developing countries have achieved. Our S&T progress is yet to have a significant impact on and give greater impetus to our economic restructuring, the overhaul of traditional industries and development of new industries. The coordinated management of S&T resources nationwide and their integrated utilization need further improvement. Such a state requires that we make a scientific response in the areas of strategies and policies. A necessary readjustment to the way we approach S&T development is highly called for.

First, we should readjust the guiding concept for S&T innovation so that it will put greater emphasis on original innovation with a view to achieving leapfrogging progress in S&T progress. Chairman Jiang Zemin once pointed out, *"original innovation, as it gives rise to qualitative S&T change and development,* *is an important embodiment of a nation's contribution to human civilization and also the commanding height of S&T competition in today's world.*" The pattern of international S&T and economic development has been undergoing profound changes, with battlefields of industrial competition moving from assembly lines into research labs. Original innovation, as the major source of new technologies, will not only bring about technological breakthroughs but give rise to new industries and new economic structures, offering unlimited opportunities for late starters to overtake the frontrunners. It has thus become a basic determinant for one's position in international division of labor. Strengthening original innovation and endeavoring to shift from simple imitation to original innovation is the important principle for China's S&T development strategy in the new century. On the one hand, we should support the unfettered probe by our scientists and encourage their innovative work in basic research where the need of the country and frontier sciences are closely connected, thus opening greater possibilities for both technologies and production and making due contributions to humanity. On the other hand, we must enhance our capabilities for independent innovation in hi-tech R&D, acquiring more proprietary technologies of our own, upgrading the technological level of our industries, and gradually freeing ourselves from dependence on foreign technologies in industrial development and restructuring. In terms of institutions and policies, we must establish a new evaluation system consistent with original innovation, introduce incentive mechanisms that accord with varying laws of S&T development, and encourage greater participation of our scientists in international academic exchanges and joint research programs. It is our abiding conviction that in S&T development, China must not only give greater scope to its institutional advantages, but more importantly, it should foster a strong sense of national confidence, carrying forward its pioneering spirit demonstrated by the successful development of the atomic and hydrogen bombs and man-made satellites, firmly grasping the good opportunities presented by the new S&T revolution, securing original innovation in the key areas of economic lifeline, national security and indigenous advantages, and achieving leapfrogging progress in technology and productivity by capitalizing on the advantage of being a late starter. We are convinced that a more open and more relaxed scientific environment will give birth to more S&T achievements and bring forth more scientific giants.

Second, we should readjust the concept and management system of S&T

innovation, and cultivate a sense that the human factor comes first. It is safe to say that talents, especially tiptop talents, are the precondition for innovation. In this time of fierce contention for brains, the government should focus its main attention on creating an environment for innovation, improving required services and fostering an innovation-friendly culture. We must change the old mentality that things are more important than people, and let the human factor come first in all our endeavors. To do so, we must deepen reforms to our S&T management system, encourage S&T workers to work more innovatively under the market conditions and bring about an open, vigorous and competitive situation in our research institutions, thus giving greater scope to the initiative and creativity of our professional talents. "Human factor coming first" also requires that we create an environment favorable for potential scientists to come forward, including such measures as free access to public S&T infrastructure and sharing of scientific data, so that even the "small potatoes" could still be treated fairly in their innovation efforts. Finally, the management system must be reformed if human factor does come first. S&T workers should be steered to care about the economic and social value of their research achievements, which should become technology that can be applied in production or merchandize that can be sold for a profit, rather than something that can only stay on paper as essays or reports. In a word, knowledge should become patents or standards, with its value eventually realized in the marketplace. Commercialization research must be oriented towards turning out marketable products and creating competitive industries. S&T workers should be encouraged to use the market yardstick rather than pure academic and technological ones to measure their success. Technological innovation should be rewarded more in the form of market incentives rather than academic prizes.

Third, we should readjust the work principle of S&T innovation, concentrating resources on the most important and shelving others for the time being. In this time of rapid S&T development, no country is able to hold to the leading position in all fields of S&T endeavors. To a developing country like ours, how to prioritize our S&T work in basic research or hi-tech research – what to do and what not to do – always comes first in our strategic decision-making. As early as in 1990, President Jiang Zemin laid down the guiding principle for S&T work as "leaving some things undone in order to do other things". However, owing to lack of understanding and ineffective institutional discipline, that principle has not been properly implemented. At present, impatience for success remains a problem in our S&T development. Trying to do more for the country is certainly admirable, but it will easily lead to scattering of valuable resources. With too little food but too many mouths to feed, the country could barely keep all the projects alive, let alone accomplishing something more substantial. In addition, as a developing country without much S&T resources to begin with, we should be all the more careful not to scatter our R&D capability thin but make good use of our limited resources. We should use new products and new industries to bring together the strengths of the central government, the localities and enterprises, and accomplish a few large and important projects. At the same time, we should properly handle the relationship between centralism and decentralism. At the beginning of a particular innovation effort, we should give scope to the initiatives from all quarters mindful of their different approaches and technological styles, so that we may use relatively sufficient fund to workout the key S&T problems and identify the points of state support accordingly.

Fourth, we should readjust the mode of S&T innovation, emphasizing the shift from single-item innovation to integrated technologies and the need to develop competitive products and industries thereon. For a long time, our S&T planning tended to go after single-item technologies, which was unavoidable in the early stages of technological development. However, given the inherent need to combine S&T with the economy, R&D of single-item technology could hardly result in competitive products or industries due to its lack of connection with other technologies. In fact, core competitiveness is derived not only from innovation but also from reorganization and integration. If we could integrate the relevant single-item but scattered technologies to produce competitive products and industries, it would be a far greater success in terms of our own innovativeness, the competitive edge of the enterprises using them and their implication to the country's S&T innovative power. To this end, we must do away with the outdated rules imposed during the time of economic planning while strengthening government macro-regulation, and achieve optimized reorganization and rational allocation of S&T resources. All key national S&T programs must highlight the integration of scientific disciplines and the coordination of research institutions, stress the shift from developing single-item technologies to integrated innovation involving single-item and other technologies, and underscore the need to achieve such technological integration centered on competitive products and industries.

Fifth, we should readjust the policy targets in S&T innovation, shifting from established S&T institutions to wider S&T resources in entire society. Thanks to popular support for the strategy of invigorating the country through science, technology and education, the whole society is now paying greater attention to S&T development. Today's S&T is no longer the individual endeavor of the scientists, but conscious and popular acts of mass participation. S&T work at localities is figuring more prominently in the national S&T system. Since the beginning of reform and opening up, the localities have grown stronger economically and felt the urgent needs for S&T advancement. They have been able to correlate such needs with their economic development as their governments paid increasing attention to S&T innovation in the areas. This requires that we, from a macro perspective, enhance the guidance on local S&T development, and promote an optimized allocation of S&T resources. Institutes of higher learning are the principal players of scientific knowledge dissemination. With their distinct humanistic environment that stimulates creative innovation, their great many academic disciplines that intertwine and blend, and the huge number of innovative talents that they produce non-stop, universities and colleges tend to have a unique advantage in original innovation. Closely combining its S&T with education will be of great significance for China to have a sustained S&T development. Enterprises are the main practitioners of technological innovation. In recent yeas, a large number of small and median-sized technology enterprises have grown rapidly, bringing profound changes in the overall pattern of China's S&T development. The national innovation system requires that enterprises play the principal role in R&D and technological progress. In short, we should mobilize all localities, all institutions of higher learning, all enterprises and, indeed, all sectors of society to go for S&T development, so as to create a heartening situation characterized by a mighty army of S&T workers and vibrant development of China's S&T cause. The matter bears on the success of our strategy of invigorating the country through science, technology and education rather than just the S&T management itself. We should break out from the traditional work pattern where government departments, localities, schools and enterprise tended to compartmentalize their functions and roles, and bring about a new S&T management system characterized by greater interaction and accommodation among them.

2. LAUNCHING THE SCHEMES OF TALENTS, PATENTS AND TECHNICAL STANDARDS IN RESPONSE TO THE CHALLENGES AFTER CHINA'S WTO ACCESSION

"Creativity is the soul of a nation and an inexhaustible source of a country's prosperity." In the new century, we will press ahead with the strategy of invigorating the country through science, technology and education so as to provide China's economic and social development with a long-term and reliable engine. To meet the opportunities and challenges presented by China's WTO accession, we will, in the near future, launch three schemes of talents, patents and technical standards.

First, the scheme of talents aimed at winning the worldwide battle for higher-end human talents. International competition as we know it is in essence one over human talents with tiptop talents at the core. The Global Competitiveness Report of the Swiss takes a developing country's ability to retain its S&T talents as a main index for evaluating its S&T competitiveness. Governments can protect their home industries and control the movement of factors in their countries with tariff or non-tariff barriers, yet they cannot control the flow of human talents. With this scheme of talents put in place, we can better train, attract and give greater scope to talented people as a major national cause in the new historical period. People of top professional aptitude have always been a target of the developed countries and their multinationals, which look to China as a source of supply. First, it is due to the overall situation of availability as top talents have always been in short supply. Second, it is due to the quality of the Chinese talents. Data have it that a third of department deans in American colleges and 35% of CTOs and laboratory directors at the Silicon Valley software firms are Chinese. Third, it is due to the rapid and steady growth of the China market in recent years. Multinationals look at China with favorable business expectations. They have made sizable investment in China and followed a localized personnel policy in line with international practices. Human talents feature prominently in China's post-WTO competition. It is an obvious fact that behind the products and market competition there is a competition over talents. Market can be gained and lost, but loss of human talents poses the greatest challenge to China. From the viewpoint of the state, years of investment in bringing up the talents cannot be recouped. From the viewpoint of the enterprises, the cost of finding a qualified replacement can be very high. The big gap between China and the developed countries in S&T innovation and commercialization is, in the final analysis, due to the inadequate quantity and quality of its human resources. Yet continued brain drain will make this gap even wider. Therefore, to bring up more qualified people, keep them and attract more from other sources is a matter that bears on the destiny of the nation. Instead of overlooking it, we must view and go about this matter from a higher plane of strategic thinking and regard our human talents a resource of paramount importance.

The first and foremost concern in the talents scheme is to ensure an ability to attract sufficient human talents and maintain their healthy vitality. At present, we must reform the S&T evaluation system, making identifying, training and stabilizing talents key criteria in national S&T programs and projects. We must reform the S&T management system, introducing open, vigorous and competitive mechanisms, increasing the proportion of personnel allowances in S&T expenditure and overcoming neglect of the human factor in R&D management. To meet the needs of the special S&T items, we must step up efforts to bring in top talents from the overseas, including groups of higher-end talents, provide them with all necessary conditions and promote an inbound surge of overseas talents. We must encourage hi-tech firms to reform their employment and remuneration systems. While paying greater attention to the human factor and tapping human resources, we should study the various incentive possibilities, including ownership of stocks and futures for S&T and managerial personnel, so as to help more talents to come forward. Efforts should also be made to create a tranquil environment where talents can go about their innovative work with an ease of mind.

Second, the scheme of patents aimed at bringing out more innovation results with China's own proprietary technologies. With the application of the TRIPS of the WTO, China is faced with serious challenges in the area of IPR protection. What is happening right now is that the developed countries are quickly turning their technological advantage into market monopoly. In China, many foreign firms, large multinationals and corporate groups in particular, have used huge volumes of patent applications to secure a beachhead for their ambitious penetration of the Chinese market. In IT and telecommunications, aviation and space, pharmaceutical manufacturing and other hi-tech areas, foreign patents have chalked up approximately 60 to 90 percent of the total applications. This will put

China's future industrial development and economic restructuring increasingly at the mercy of foreign patents, primarily those from the developed countries. What we should do, generally speaking, is not so much raising our awareness of IPR protection but more importantly, improving our capability to create, protect and well utilize IPR. To this end, we must encourage our S&T workers, enterprises and colleges to register their patents, invention patents in particular, while giving the benefits of invention to the research institutes and rewarding the S&T workers accordingly. The US "Bayh-Dole Act" of the 1980s awarded the invention patent of government-financed projects to the research institute and the inventor it hires. The Act went a long way to galvanizing the patent work in the US. China has formulated similar policies lately, and I believe they will exert a positive and far-reaching impact. As for projects under the major national S&T programs, we have asked them to keep an eye on the application of relevant proprietary technologies at home and abroad both before and during the implementation and come up with suggestions to bypass foreign patent barriers. What is more, we have made acquisition of invention patents as criteria for project selection and approval under major national S&T programs.

Third, the scheme of technical standards aimed at setting up and perfecting China's technical standards system. Under an open international environment, the developed countries often use their safety standards as the justification and their proprietary technologies as the shield to neutralize the developing countries' competitiveness in productive cost, thus completing the transition from erecting simple tariff barriers to creating a more sophisticated technology barrier. The post-WTO China would find more such technology barrier in the way of its exports. Since many of our existing technical standards fall short of those in the developed countries, China's exports have encountered more restrictions and impediments. With respect to imports, given its inadequate and poorly coordinated technical standards and backward methods to enforce them, China has found it hard to provide rational and effective protection to national industries. For example, some developed countries have worked out 114 standards for rice alone, and failure to meet any one of them would result in a rebuff. According to a rough calculation, technology barriers in other countries cost China over 45 billion US dollars a year in direct and potential loss of exports, more than a quarter of its total annual export. When we talk about setting up China's own technical standards system, we do not mean to get even with the other countries. What we want is to build, in a legitimate and reasonable way, a system that can reflect China's special characteristics and advantages. To this end, we must pay close attention to the global trends with respect to technology barriers and watch out for policies and potential shifts in major developed and developing countries. We must set up, through reform, a national standards research body to be charged with organizing, planning and coordinating standards research activities by enterprises, colleges and research institutions from across society. MOST has made standards research one of the 12 special S&T items. The purpose is two-fold. One is to address the standards formulation in hi-tech areas. The other is to tackle without delay issues relating to technical standards and technical certification in traditional Chinese medicine, Chinese language data processing, and other areas where China enjoys an inherent advantage. In any rate, standards have become a part and parcel of economic competition, technological competition included, between countries. We should press ahead with the development of China's national innovation system so as to give greater play to China's own standards.

3. LAUNCHING THE SPECIAL S&T ITEMS AND IMPROVING THE CORE COMPETITIVENESS OF CHINA'S KEY INDUSTRIES

Driven by technological progress and economic globalization in recent years, world economic structure has undergone an accelerated reorganization with important changes taking place in the allocation of economic and social resources. On the one hand, we see rapid development of high and new technologies and related industries, which has become the most dynamic area of global economic growth and the principal determinant for international division of labor. On the other hand, we see continued infiltration of traditional industries by new and high technologies, which has facilitated technological upgrading of industries and readjustments of the economic structure. These new trends are of equal practical importance to the developed and developing countries alike.

The process of S&T advancement in the past two decades shows, in a sense, that controlling one's market is more important than occupying its territory. Right now, China is faced with severe challenges in its economic and social development. In order to quickly secure a number of commanding heights in the 21st century hi-tech industries and achieve breakthrough in certain major technologies and their commercialization during the post-WTO grace period, the Chinese Government has decided to launch a number of special S&T items and look forward to their substantive success in five years. These include super large IC and software, information security, e-government and e-finance, electricity-powered automobile, functional genomics and biochips, dairy products, new drugs and modernization of traditional Chinese medicine.

The 12 special S&T items give expression to the selective targeting policy of the Chinese Government during the 10th Five-Year Plan period and its resolve to make leapfrogging progress in certain critical areas of technology bearing on China's economic development and national security. In IT area, efforts will be made to fundamentally change the basic monopoly of core foreign technologies by developing our new concept CPU, NC and network software platforms. With breakthroughs in SOC technology and key IC manufacturing equipment, we expect to make China's IC designing and manufacturing a world leader within five to ten years. By setting up a unified and secure platform of e-government and e-finance, we expect to build a cyber-security infrastructure system in synchrony with national informatization while stimulating some related technology industries. In areas of bioscience and pharmaceutics, we will keep pace with international biotech competition and focus on achieving some commercially important patents in R&D of functional genomics and biochips so as to ensure a Chinese presence in biotechnology industries. At the same time, we must reverse the awkward situation of not having drug products with China's own proprietary technologies by stepping up R&D of new drugs and modernization of traditional Chinese medicine. In the area of transportation, we must take full advantages of being a "late starter" and aim at technological breakthroughs and commercialization of electricity-powered automobile, thus gaining an upper hand in the new round of competition. In agriculture, we must focus on intensive farm products processing, increase farmers' income and achieve a sustained and healthy development of animal husbandry, dairy farming in particular. Moreover, we should promote readjustments in the agricultural structure by developing some proprietary technologies with respect to water-saving agriculture and farm products processing. The successful implementation of these special S&T items, we hope, will have a far-reaching impact on China's industrial restructuring, improved market competitiveness and higher income for the farmers.

China's economic development has reached a critical juncture where, unless we carry out the structural readjustment in good time, we may not be able to ensure a sustained economic growth and participate on an equal footing in international competition and division of labor. In keeping with the WTO agreement on government subsidies, the Chinese Government will cut back on its direct subsidies to industries, gradually withdraw from market competition and concentrate itself on supporting public undertakings while ensuring a sound business environment. This means greater government role in support of S&T innovation and institutional reform. On the one hand, the government can enhance market competitiveness of the country's key industries through S&T programs of support to basic research, frontier hi-tech research and other competitive research activities. On the other hand, it can take advantage of the "green box" policies allowed under WTO rules to support R&D and services of agriculture, new industries and other disadvantaged industries. We must rally greater human and financial resources without delay to step up R&D and commercialization of the major and key technologies.

Given the current circumstances, we will have to make some readjustments to the way S&T plans are organized and managed. During the 10th Five-Year Plan period, MOST will see to it that resources be concentrated on the strategically important areas, with the view to expeditiously enhancing the country's global competitiveness in key areas and improving the government's capacity to organize strategic S&T innovation efforts under market conditions.

We are of the view that the launching of these special S&T items accords with the needs of progressing with the times. It is China's response to the challenges after its WTO accession, a measure to act on the "three strategies" and the country's economic revamping, a crucial way to improve the core competitiveness of China's key industries and a step to reform its S&T planning and management system. Science and technology are the primary productive force. They are also the crystallization and manifestation of the advanced productive force. The successful implementation of the special S&T items will make this concept even better understood by the people.

The 21st century is a century of science and reason, a century of innovation and development. Half of the past century saw the Chinese nation suffering in hardship and tribulation; another half saw it rising in bold pursuit of success. History tells us that humiliation imposed from the outside was not to be feared; lacking innovative spirit can cost a nation its last hope. Though China is still lagging behind the developed countries because of its relatively recent encounter with modern sciences and technologies, we have every reason to believe that it can come from behind with the hard and productive work of its scientists and achieve the great national rejuvenation.

Centennial Eulogy of Achievements of Engineering

Song Jian

Chinese Academy of Engineering

Song Jian, *born on Dec. 29, 1931 in Shandong Province, is a distinguished Scientist in the fields of control theory, system engineering and aerospace technology. Dr. Song has received a PhD degree from the Moscow National Technical University. Over the last four decades, he has gained creative achievements in the fields of control theory, guided missile and aerospace technology, and population control theory, and made significant contributions to the development of S&T and environmental protection in China. His academic titles include Academician of both the Chinese Academy of Sciences and the Chinese Academy of Engineering; Honorary Professor of the Academy of Mathematics and System Sciences of CAS; Foreign Member of the US National Academy of Engineering, the Russian Academy of Sciences and the Royal Swedish Academy of Engineering Sciences; Corresponding Member of the National Academy of Engineering of Mexico, the National Academy of Engineering of Argentina; Member of the Euro-Asian Academy of Sciences and the International Astronautics Academy.*

Abstract: This paper reviews the progress of engineering in the world over the 20th century, and describes the important role of engineering in promoting global economic and social developments. Connecting with the 25 achievements selected from "Outstanding Engineering Achievements in China in the 20th Century" sponsored by the Chinese Academy of Engineering, the paper points out the great contributions made by the Chinese technical personnel to the invigoration of

Chinese nation and the national economic development. It speaks highly of engineers' creative work that is recognized and respected by the whole society. The paper reiterates the fact that a country cannot rank among the strong counties in the world without enough scientific and technical strength. The paper points out that there will be a new upsurge in the respect of China's industrialization and modernization drive in the 21st century, especially in the coming 50 years.

Science and Technology made great strides in the 20th century. It is estimated that eighty percent of global scientific discoveries, technological innovations, and engineering constructions were attributed to scientists and engineers of the 20th century. A great amount of new knowledge and engineering conveniences rushed into the world and people's lives, leaving their marks on every corner of the society. Engineers have created equipment and machinery that never existed in history, brought benefit to the whole world, improved the efficiency and the ability to create wealth, improved the quality of life, and elongated the life expectancy of people.

Engineering achievements in the 20th century surpassed the wildest imagination of people in the 19th century.

Ever since Homo sapiens appeared on the Earth 2 million years ago, survival has been the top necessity. Generation after generation, people were fighting against hunger, coldness, and epidemics for food, clothes and shelter.

Before the 20th century, agriculture relied largely on climate and natural conditions. The Irish people, who lived on potatoes, suffered from hunger in 1845 because of crop disease. One million people died of hunger and another one million left their homeland because of starvation. In China, severe shortages of food occurred several times in the 19th century and early 20th century.

Since the 20th century, modern science and technology have comprehensively penetrated agriculture and animal husbandry.

Water conservation engineering constructions increased the agricultural ability to fight against drought and flood. The type of nutrition needed for the growth of plants was discovered in the 19th century. However, large scale production of fertilizers (1909-1019), pesticides (1938-1942), and herbicides (1944) were all engineering innovations of the 20th century. Crop selection engineering produced better species. Research on hybridization began for wheat in 1900, for corn in 1917, for broomcorn in 1950, and for rice in 1960. As a result, output has increased and crops have a better ability to withstand diseases. Engineers designed and made hundreds of types of machines, with the tractor in 1907, combine in 1915, thresher in 1943, binding mechanism and milking machine in 1940, portable sprinkler in 1948, cotton stripper in 1949, and various kinds of machinery for animal husbandry together with processing equipment. Productivity was raised tens of times for agriculture and animal husbandry. The world population increased from 1.6 billion to 6 billion. Agriculture guaranteed the basic food production for the world. Engineering science made the lion's share of the contribution.

The growth of the petrochemical industry was the great contribution made by the chemical engineering technology in the 20th century. Because of the progress made in physics and organic chemistry, engineers developed synthetic fiber in 1903 and artificial cotton in 1912. Engineers mastered the technology of distillation and pryolysis of oil from 1913 to 1936, made plastics from 1909 to 1918, chemical fiber in 1912, and artificial rubber in 1930. At present, plastics and artificial fiber are materials that people cannot do without in clothing, housing construction, transportation, industrial production, and agriculture and service sectors.

The design and construction of electric power distribution networks was a great engineering undertaking in the 20th century. Electromagnetic dynamics started in 1865, the generator and electric motor were innovations made in the 19th century. However, large-scale power plants were not constructed until the 20th century. Electricity has lit all cities and villages, providing means for social industrialization and electrification with flexible, convenient, and potentially inexhaustible power. Communication networks and mobile phones have connected all continents and countries. Modern transportation can send you to any corner of the planet you need.

The radio telegram in 1901, broadcasting station in 1920, television in 1926, radar in 1935, semiconductor in 1915, transistor in 1948, computer in 1946, video cassette recorder in 1956, integrated circuit in 1959, laser in 1960, computer assisted image foundation in 1972, personal computer in 1975 and internet in 1969 were all innovations achieved by engineering technology directed by electrodynamics and quantum mechanics. New technology was initiated and the human society was brought into a brand-new intelligent information era.

In the early 20th century, the train was not used greatly. Most people moved in a scope of hundreds of kilometers. They were confined to where they were born. The industrialized production of automobiles in 1908, construction of freeways in 1940, railway networks, and the high-speed train in 1960 have made a broader space for life and production and allowed the circulation of commodities.

Engineers invented the airplane in 1903, the helicopter in 1939, and the supersonic plane in 1947. Inter-continental airlines were opened in 1939. A great number of large planes were made between 1950 and 1960 and airlines covered the whole globe. You can now reach any city in the world in one day. By the end of the 20th century, the total number of civil airplanes in the world reached 20,000, with an annual number of 1.64 billion passengers boarding airplanes by the end of the 20th century.

Important breakthroughs in medical science occurred in the 20th century. The discovery of anti-biotics in 1928, sulfanilamide in 1932, insulin in 1921, vitamins in 1928, poliomyelitis bacteria between 1952 and 1957, arteannuin in 1970, and their production on a large scale saved the lives of tens of millions of people. Small pox was eradicated in the 1970s. This is the contribution of chemists and pharmaceutical engineers. The design, production and use of the cardiogram in 1903, electroencephalogram in 1929, artificial lung in 1927 and 1953, together with the artificial kidney in 1945, artificial heart valve in 1950, kidney transplant in 1945, heart pace-maker in 1957, tomography in 1973, artificial heart in 1982, and the contact lens in 1985 have provided patients with more and more reliable engineering technology for diagnosis, treatment, and for raising the quality of life of patients. The average life expectancy has been increased from about 30 or 40 years at the beginning of the 20th century to over 70 years now.

The atomic bomb, the guided missile and the satellite, the application of nuclear energy and space technology are the most brilliant achievements made by humankind in the 20th century.

Atomic energy, the rocket, the basic theory of aerospace, quantum mechanics (1889-1929), the theory of relativity (1905-1915), the fission of the uranium atom (1935-1938), and rocket dynamics are all discoveries, which were established in early 20th century. The atom bomb in 1945, hydrogen bomb in 1952, nuclear reactor in 1942, submarines driven by nuclear power in 1954, atomic energy electricity generation(1951-1957), the rocket from 1915 to 1942, the satellite in 1957, man entering space in 1961, astronauts landing on the Moon in 1969, the space shuttle entering space in 1981, space craft landing on Venus in 1982, and a robot landing on Mars in 1997, all of them are great achievements of the highest significance in science and engineering technology.

Generations and generations of humankind have looked into space, without knowing what stars are and what they are made of. In the Zhanguo period in Chinese history, Zhuang Zi (369 BC-286 BC) once wondered: Why is the sky so far away? What is supporting the land? And why do the Sun and the Moon go up and down? Qu Yuan, a poet from 340 BC to 278 BC, asked the heaven: Where do the Sun and the Moon belong? And why are stars so arranged? And why does the night light perish and come out again? Another poet Su Dongpo from 1037 AD to 1101 AD asked the Moon whilst drinking liquor: Which year is it in heaven? And it might be very cold high up there on the Moon with red jade buildings and magnificent housing.

Xin Jiaxuan, apoet from 1140 to 1207, once asked: What a pity that I have no idea where this evening'smoon will go. Is there another world? Do people there just see the light in the east at this moment? There is no root for the flying mirror (the Moon) and I wonder who is holding it? If Fairy Chang E does not marry, who will keep her?

In the middle of the 20th century, from July 1969 to December 1972, the spacecraft Apollo sent a total of 18 people to the Moon with an accumulated time for investigation of more than 600 hours that ended thoroughly people's questions and dreams about the Moon.

Before the 20th century, astronomers did not know definitely, where the sun light comes from and why the stars shine. The advantage of atomic physics and nuclear reactors and H-bomb explosions have irrevocably proved that light and heat of stars, including the Sun, come from the fusion of light atoms.

With the construction of high-energy particle accelerators and detector engineering, physicists are able to do research and have discovered a series of subatomic particles, also proving that the proton and neutron are not basic particles. They are made of six kinds of quark, and more minute structures.

Radar engineering and advanced telescopes provided the key instrument for discovering the homogeneous space microwave background radiation, providing evidence for the plausible theory that the universe originated from the Big Bang.

The great wave of science and technology in the 20th century indicates that science has guided engineering. Engineering technology has been the major engine of the progress of material civilization. Engineering technology has also provided new tools and opened new ways to advance science.

The steam engine initiated the first industrial revolution in the 18th century. At the end of the 19th century and during the first half of the 20th century, electric machines and chemical industry started the second industrial revolution. Mankind then entered a period of electrification, nuclear power, and space industry. Social productivity and the quality of people's life were a great deal improved. The space and time gap between countries, regions, and people was narrowed. The globe has become a village. In the latter half of the 20th century, information technology, like a raging fire, has launched the third industrial revolution. Production and consumption turned from industrialization towards intelligent automation. Yet again, productivity and production efficiency increased greatly. Among the three industrial revolutions in the history of mankind, two occurred in the 20th century.

The records of China's history are completely different in comparison with that of the world. For the Chinese, the 19th century was an era of humiliation and suffering. European powers and Japan invaded the country four times, and killed millions of people. China had awakened in the 20th century, and waged and won the protracted war against imperialism and colonialism. When People's Republic of China was founded in 1949, the economy was very exhausted. People started all from scratch, 200 years later than Europe, to begin the industrialization of the country.

Recently, the Chinese science and engineering communities made a retrospect of the past. We are gratified to see that Chinese scientists and engineers made historical contributions for their motherland even with twist and turns. From the Jingzhang railway built by the Chinese in the early 20th century to Yumen oilfield, from Qiantang bridge in the 1930s to Daqing oilfield in the 1960s, from the Yangtze River bridge in Nanjing, to the nuclear bomb, guided missile and satellite, and together with hybrid rice in the 1980s, all these great achievements opened the way for China's industrialization and modernization. China has succeeded in establishing an independent industrial system within 50 years.

The opening up of the country initiated by Mr. Deng is like a spring breeze that raised the spirits of people, especially engineers and scientists. Engineers made critical contributions to the high-speed economic growth of 8%-9% annually and marked social progress was recorded in the past 20 years.

At the beginning of the 21st century, people see the fast-paced development of science and technology, industrialization, agriculture, commerce, and transportation. The overall productivity of the country is increasing day by day. The quality of life is improving continuously. China is under going great changes. Scientific and engineering technology has entered a prosperous new era.

In retrospect, it is very important for people to understand the historical position and function of engineering technology in the past century. Science is to declare truth to people. She answers the question "why". The task of engineers is to modify nature according to scientific principles and construct a new society, and to tell people how to do it. We must tell people of the country that the construction of railways and bridges, dams and irrigation systems, planes and spacecraft, satellite communication, vehicles and ships, tens of thousands of kinds of machinery, automatic production line, television, telephone, and washing machines, microwave stove, and air conditioning, are all masterpieces of engineering technology. The material civilization that people enjoy in modern society is mainly created by engineering technology. Therefore, engineers are irreplaceable creators of new industry. The creative work of engineers should be recognized and respected by the whole society.

At the beginning of the new century,the Chinese Academy of Engineering organized an activity to select "important engineering technology achievements in China in the 20th century". The activity received an enthusiastic support from CAE members, associations, and societies under the China Association for Science and Technology, together with agencies affiliated to the State Council. A peer ballot has selected 25 items as most significant for the country. The achievements will be compiled into books so that the stories will last together with the Earth. The achievements are also a solace to those deceased contributors of the country. The books will also provide the society with a comparatively accurate document of the importance of engineering technology.

A decision was made between 1955 and 1956 to develop atomic energy and space technology. It was made by Chairman Mao Zedong as the state leader. It was a decision of great significance. When the decision was made, it was a period when the People's Republic of China had been established for only several years. There was a lack of qualified people. Science and technology were backward, and the industrial system was not established. However, it took only 15 years for scientists and engineers in China to complete the design, construction and put into operation of atomic facilities, including nuclear power plants, ships powered by nuclear reactors, rocketry, space vehicle, and satellites. These achievements swept away the cowardice of some Chinese in front of industrialized power countries. It also put an end to the dispute about whether China could develop its own industry, science, and technology by itself.

On the information technology, word processing of Chinese characters and printing industry was entitled the second achievement while telecommunication engineering was the 12th, broadcasting and television was the 15th, and computer was the 16th. Laser photo-typesetting system using Chinese characters enabled Chinese language and culture to enter a new and brilliant period. China now is a big producer of optical cables, digital trunk switchers of communication, mobile telephone, and modern communication equipment. The first black and white television was produced in 1957. In 1999, 39 million color televisions were produced. China has become the largest producer in the world for production and export of television and videodisk players.

Design, production, and application of computers are close to world level. China produced 8.6 million PCs in 2000. China has produced the teraflops high performance computer and the teraflops PC cluster in the domestic market.

On chemical industries, petroleum was of the third place of the achievements and inorganic chemical industry was the 20th, and rare metal was the 22nd. Petroleum is the major energy and raw material for clothing, food, housing and transportation industries. China, started from almost zero, established its own petroleum industry. 160 million tons of crude oil was extracted last year from mainland oil deposits. Petroleum is regarded as the blood and black gold of modern industry. After refining, pryolysis and processing of oil, various kinds of products are produced such as agricultural plastic sheet, chemical fertilizer, plastics, synthetic fiber, and medicine, various kinds of chemical products, tire, and gum material. Increasingly abundant and valuable materials were supplied to industrial and agricultural production and to the need of the consumers.

A factory with an annual output of 80,000 tons of synthetic rubber produced the same amount of products as a natural rubber farm with the acreage of 100,000 hectares. A factory, which produces an annual output of 10,000 tons of synthetic fiber, equals to 20,000 hectares of cotton field or 2.5 million sheep.

The country's petro-chemical industry, started from zero, has got a strong momentum of development. In the year 2000, output of synthetic fibers reached 7 million tons, synthetic plastics 4.7 million tons, synthetic rubber 732,800 tons. The inorganic chemical industry turned from weak to strong. In the year 2,000, output of chemical fertilizer was 31.86 million tons, alkali 8.34 million tons, sulfuric acid 23.65 million tons, and cements 600 million tons. All of them ranked the first in the world.

On the mining industry, abundant rare-earth elements and rare metals could be used as functional materials to produce light, electricity, magnet, and catalyzer, and could be applied widely to the frontier areas such as electronics, nuclear energy, aircraft, and space industry. China is a country of the richest resources of rare-earth elements in the world, with 80 per cent of the total amount of deposits in the world. Chinese scientists and engineers developed new technologies on metallurgy, separation, purification, and processing, and established their own industrial engineering, with output and technology at the leading position in the world.

In the agricultural sector, technology to increase output of crops is in the 4th place. The 14th is the technology for raising of poultry, animal husbandry, and cultivation of aquatic products. The 24th is the light industry and textile. Scientists and experts have selected, modified, and put to extension 5,600 new breeds in 41 lines of farm crops and more than 1,000 new traits of fruits and vegetables. Major crops changed traits for four to six times. There has been enough supply of new chemical fertilizers and pesticides. Farming practices have been improving year by year. Output per hectare has increased 3 to 5 times in most areas. Average per capita supply of cereals increased from 280 kilograms in 1952 to 406 kilograms in 1999. In spite of the population size increased from 400 million to 1.3 billion today, in 1999, average per capita consumption of meat was 50 kilograms, egg 18 kilograms, aquatic products 33 kilograms, fruit 50 kilograms, an increase of 5 to 15 times compared with 1978.

The light industry and textile industry also made rapid progress. Output of textile products ranks the first in the world. Annual output of synthetic fiber constituted 24 per cent of the total of the world. China has become number one in the world in food and garment production. The abundant storage and production capability of grain and clothes are more than enough to satisfy our own people but also to accommodate the United States, Russia, Japan, and Europe. The situation, which we now feel gratified for, never existed in the past 1,000 years.

The fast progress of medical care and production of pharmaceuticals heightened life expectancy of people from 34 years before 1950 to 70 years in 1997. China's population in the 20th century increased from 426 million in 1901 to 1.2 billion in 2000, an increase of three times. The population growth of the country once roused concern and worry.

In the past 20 years, family planning programs have been carried out and a remarkable achievement has been made. On the average, the number of childbearings per woman decreased from 2.86 in 1982 to 1.8 in 1999, much below the critical fertility rate of 2.1. China's population is expected to be stabilized at 1.6 billion in 2040 and stop growing afterwards. This would be a great contribution to the future sustainability for development of the nation and the welfare of the coming generations. It has been an overdue problem, though the dispute and worry about the population explosion problem can be put to an end now.

By the year 1949, the nation's electricity generation was only 8 kwh per capita. In the year 2000, the capacity reached 320 million kilowatts, a per capita consumption of 1,094 kilowatt-hours. More than 98 percent of the families in rural areas can enjoy electricity. Eighty-one power plants with capacity of more than a thousand megawatts each are being constructed this year, including hydraulic, firepower and nuclear power plants.

The Three Gorges water conservancy project, the largest in the world, started generating electricity in 2003. Once completed, it will have a newly increased capacity of 18.2 million kilowatt, with an annual output of electricity of 84.7 billion kilowatt-hours. An era of electrification throughout the country is coming.

The management of big rivers is number seven of the achievements. Geology exploration and development of resources is the 13th, urbanization is number 23rd, and coal mining is number 25th.

In the last century, China trained a strong team of geologists and engineers who turned China into one of the richest countries in identified mineral resources, second only to the United States. The total deposit of coal is estimated 4,500 billion tons while 600 billion tons are identified. Output of coal in 1999 was 1.2 billion tons. Mineral exploration and mining have guaranteed the economic growth and social progress.

The number eight is railway construction. Road construction is the 17th. From the end of the 19th century to the beginning of the 20th century, transportation relied on ships in the south and horses in the north. It took 27 days from Beijing to Wuhan,

56 days to Guangzhou, 59 days to Yunnan, and more than three months to Xinjiang.

In the latter half of the 20th century, 1.4 million kilometers of roads were constructed, among which, more than 20,000 were super-highways, second only to the United States. In the year 2000, the railroad mileage was 68,000 kilometers. Speed for trains was raised. It became the major way of transportation for the country. All kinds of equipment can be made in China such as locomotives, vehicles, rail, and communication equipment. The export of the equipment has also been started.

Design and building of ships is the 9th item. In the 50 years after 1949, engineers designed and made passenger and cargo ships in 1954, oil tankers under 150,000 dwt between 1992 and 1996, frigates in 1957, submarines in 1965, chasers' in 1971 and nuclear submarines in 1974.

In the year 1890, Zhang Zhidong established China's first Hanyang Iron Factory. By 1949, the highest annual output was 1.78 million tons for iron and 920,000 tons for steel. In the 50 years after the founding of the People's Republic of China, 36 iron and steel plants of million tons each had been built. China produced 150 million tons of steel in 2001, thus became the biggest producer in the world.

The 18th item is manufacture of machinery. In the latter half of the 20th century, a comparatively complete machine-building sector was established. And China has the ability to design and produce pressure machines above 10,000 tons (1962-1971), the complete set of equipment for steel factories (1974), turbines (1981-1999), electricity generators from 300,000 kw to 600,000 kw (1981-1988), nuclear reactors (1991), and mining and petrol-chemical installations.

Aviation is the 19th item. In the first half of the 20th century, China's aviation industry was still at her infancy. The new China established its own design and manufacturing ability and produced a batch of fighter planes (1956), cargo planes (1974), helicopters (1985), greasers (1998) and large passenger planes (1980).

Looking back, we have justifiable reasons to be gratified by the achievements that the Chinese engineering communities made in the 20th century. Looking forward, that is only a prelude of a big industrialization drive. The climax of industrialization and modernization construction is still to come.

China is still a developing country with per capita GDP of only \$900; a low-income country. China has a long way to go to become a middle-developed country. As the late Mr. Deng warned, every Chinese must be keenly aware, whatever accomplishments we made, and China is still a developing country. We should behave modestly, work hard, avoid confrontation with all nations, keep peace and friendship with neighbors and distant countries as well, and bide our another 50 years and beyond to realize the dream of modernization.

In the 21st century, we need to build an even stronger engineering and technology force, bring about more and better engineers and technicians. This is the strategic task of the country and of the basic interests of the future people.

China's industrialization lagged behind Europe by 200 years. After 100 years' uffering and struggle, we have come to the consensus that without strong scientific, technological, and engineering ability, it is impossible to realize the dream of industrialization of the nation.

As we stepped into the 21st century, the infrastructure construction and all industries have entered a new period with a new strategic target. To build a prosperous country, science and engineering communities shoulder the most important and historical task. Only if generations of more and more qualified young people join the grand work and devote themselves, a modern China can be built in the 21st century.

China: Brilliant Engineering Achievements over 50 Years

Xu Kuangdi

Chinese Academy of Engineering

Xu Kuangdi, *male, born in Chongde, Zhejiang Province in December, 1937. Ethnic: Han. Professor, doctoral student supervisor, member of Chinese Academy of Engineering. He currently is President of Chinese Academy of Engineering.*

He graduated from Beijing University of Steel and Iron Technology in 1959. Working Experience: Assistant of the Department of Metallurgy in Beijing University of Steel and Iron Technology; Assistant and Deputy Director of Teaching & Researching Office in Shanghai Institute of Technology; Assistant and Lecturer of Shanghai Institute of Mechanics; Deputy Dean of the Department of Metallurgy in Shanghai University of Technology, later Dean; Deputy Chief Engineer & Technical Manager of Scandinavian Lance Corporation, Sweden; Managing Vice President of Shanghai University of Technology, Deputy Director of Shanghai Education & Health Office and Director of the Bureau of Advanced Education; Director of Shanghai Municipal Planning Commission; Administrative Vice Mayor of Shanghai; act as Mayor of Shanghai since Feb 24, 1995. Reelected as Mayor of Shanghai in the first meeting of the 11th People's Congress on Feb. 17, 1998; elected as President of the Chinese Academy of Engineering Sciences, 2002.

Xu Kuangdi went to Imperial College, England, for further study and research in the mid-80s, and was a visiting professor of Royal Institute of Technology, Sweden; gave lectures in the universities of the countries, such as Finland, Sweden, Norway, Australia, German, Japan, etc. by invitation. Major-

ing in research of special steel ,secondary refining and injection technology; owning two patents in Britain and Sweden. Seven major publications: Refining of Stainless Steel, Smelting Technology for Special Steel Abroad, Water-cooling Lines and Slag-hanging for Electric Furnace, Phenomenon of Fluid Flow in Metallurgical Process, etc. Published about 170 academic papers.

Abstract: This article begins with a brief review of the 5,000-year history of the Chinese civilization that shows the prosperity of the nation has been closely linked with its engineering strength. The article then describes the major engineering achievements that helped China achieve gigantic economic and social development in its 50 years history. It is concluded that developing countries should put great emphasis on developing engineering science and technology in order to achieve their modernization.

In 1949, the new China was born after over a century of hard and heroic struggle against imperialists and domestic reactionaries for independence and liberation. The Chinese people stood up again in the world on their own will and rights.

China has a long history of civilization over 5,000 years. Up to the 15th century, China held the lead in the world in many engineering areas, including agriculture, animal husbandry, food processing, textile, papermaking, printing, pottery and porcelain, metallurgy, medicine, water conservation, vehicle, shipbuilding, oceanic navigation, weaponry, among others. The Four Great Inventions of papermaking, movable type printing, compass, and gunpowder gained worldwide fame. All these great engineering achievements had made great contributions to advancing the civilization of both China and the world.

After the 15th century, however, China's feudal monarchs adopted the policy of closing its door to the outside world and suppressed political and social changes. As a result, engineering development in China moved slowly Since then. In the meantime, giant changes happened in Europe, where the Renaissance, the Great Exploration and the birth of capitalism boosted up the liberation of thoughts and brought about the birth of modern science and technology. The first technological revolution in the late 18th century was marked by the steam engine and the second technological revolution in the late 19th century was marked by electric power that brought capitalist countries into strong economic and military powers. The old China was thrown far behind. In 1840, China was defeated in the Opium War by the then biggest world power, Great Britain. Since then, old China miserably experienced long-lasting invasions, bullies, losses of land and humiliations by imperialist powers armed with advanced powerful weapons, culminating in the Japanese invasion that killed over 30 million Chinese people. Although China started to learn and introduce modern science and technology from capitalist countries, and made efforts on establishing modern industry and education system, the process of modernization was simply a failure due to corruptive political systems and continuous wars. By 1949, China's engineering science and technology was terribly weak and economy and society were terribly under developed.

After the birth of new China in 1949, Chinese people immediately started their ambitious struggle for modernization. Overcoming enormous difficulties and hardship, China laid its foundation of industrialization in the first 30 years. In late 70s, China adopted an open and reform policy and entered a new development era. Since then, China has made successful efforts on reforming its economic system and building socialist market economy, developing science and technology as well as education to support economic and social development. As a result, China has achieved giant progress with continued high-speed economic growth and social progress. Today, China is one of the few countries in the world with an industrial system that contains all major industries and the overall national strength is among the top of the world. According to the statistics of the World Bank, China's GDP in 2000 reached $1,080 billion, ranking sixth in the world. If converted with purchasing power parity (PPP), China ranked second, only next to the United States (see Table 1).

During the past 50 years, China has made brilliant achievements in all key engineering fields and these achievements provide strong support to the country's economic and social development. In 2001, the Chinese Academy of Engineering, with support and participation of the Chinese Academy of Sciences and other organizations, carried out a large project to select 25 greatest engineering achievements of China in the 20th century. All these achievements, as shown in Table 2, are the historic contributions made by over 10 million Chinese engineers and lay a solid foundation for China's industrialization.

From the history of blood and tears of the old China, Chinese people learned that

Table 1 GDP ($Billion) and GDP Per Capita ($) in 2000

	GDP	Rank	GDP(PPP)	Rank	GDP/cap	(PPP)
USA	9,883	1	9,883	1	35,045	35,045
Japan	4,677	2	3,617	3	36,828	28,478
Germany	1,870	3	1,867	5	22,806	22,766
Britain	1,413	4	1,359	6	23,557	22,647
France	1,286	5	1,296	7	21,802	21,963
China	1,080	6	5,038	2	856	3,995
Italy	1,069	7	1,248	8	18,422	21,514
Canada	690	8	895	12	22,242	28,872
Brazil	588	9	1,206	10	3,456	7,092
Mexico	575	10	997	11	5,862	10,171
Spain	555	11	712	14	14,231	18,256
India	479	12	2,474	4	472	2,435
Korea	457	13	783	13	9,728	16,653
Russia	251	19	1,216	9	1,720	8,332

we must have a strong national defense with advanced military technology in order to live a peaceful life. In the 1950s, China started its independent research and development on A-bomb, H-bomb, missile and satellite, and overcame a huge number of unimaginable difficulties. In 1960, the first short-ranged missile was launched; in 1964, the first A-bomb was successfully tested; in 1967, H-bomb was exploded; and in 1970, the first satellite was launched. Since early 1990's, China has devoted itself in the manned space technology. By now, China has launched four experimental space vehicles without man and will soon send Chinese astronauts

Table 2 Greatest Engineering Achievements of China the in 20th Century

1. Atomic bomb, hydrogen bomb, missile and satellite	13. Geological prospecting and resource exploration
2. Chinese character processing and printing revolution	14. Animal husbandry, aquatic farming and poultry raising
3. Petroleum prospecting, drilling and production	15. Broadcasting and television
4. Agricultural output increasing technology	16. Computer
5. Prevention and treatment of infectious diseases	17. Road
6. Electrification	18. Mechanization — major complete equipment
7. Harnessing large rivers	19. Aviation engineering
8. Railway	20. Inorganic chemical engineering
9. Shipbuilding	21. Surgery diagnosis and treatment
10. Iron and steel	22. Rare metals and advanced materials
11. Family planning	23. Urbanization
12. Telecommunication engineering	24. Home appliance and textile industry
	25. Coal mining

into space. China's development of military technology, including A/H-Bomb, missile and satellite, is solely for the purpose of defense. All these achievements not only greatly inspire the spirit of Chinese people, but also make great a contribution to the world peace.

China has the largest population in the world and feeding its people has always been the top priority of the country. Since 1949, tremendous efforts have been made in developing advanced agricultural technologies and a large number of major achievements have been made. With genetic seed breeding technology, over 5,600 new high-grade, high yield crop species in more than 40 crop categories have been developed and extensively planted. From 1949 to 1996, a verage grain yield was increased from 1,035kg/ha to 4,890kg/ha. By large-scale water conservation efforts, the ability of reducing the damage of disastrous flood and drought to agriculture has also been greatly strengthened. In 2002, China produced 38 million tons of chemical fertilizer, ranking first in the world. Growing and effective use of fertilizer contributes 8%-15% increase in crop production. The wide use of biological and chemical herbicide, pesticide, bactericide and raticide has effectively prevented crop from damage and decreased crop loss from 30%-45% in the 1960s to 10% now. China has promoted various advanced cultivation technologies. Today, over 2/3 of the total cultivation area is intercropped and multiple cropped, either two crops a year or three crops a year, boosting crop output with increased land utilization ratio. Today, China has a big agricultural machinery industry, producing over 83% tractors of the world and many other machines. Plastic tunnels and greenhouses are very popular in China today, covering over 10 million ha of cultivable land. In 2002, grain output reached 457 Mt, meat 66 Mt, aquatic products 45 Mt, together with egg and fruit, all ranking first in the world. With remarkable engineering achievements, China today feeds over 1/5 of the world population with less than 1/10 of the world's arable land.

In the past 50 years, public health conditions in China have been greatly improved. The life expectancy of Chinese people doubled from 35 years in 1950 to 71.4 years in 2000. Since the very beginning of new China, it has taken the prevention and treatment of infectious diseases as its vital task in public health. China has developed and inoculated people with various vaccines, including smallpox, tuberculosis, cholera, plague, osteomyelitis, diphtheria, pertussis, tetanus, measles and hepatitis. Such diseases like smallpox, osteomyelitis and plague have been either totally wiped out or strictly controlled in China.

The mineral prospecting in China was almost nil 50 years ago. Since then, huge efforts have been made and a cross-disciplinary, comprehensive, world-level technological system of geological prospecting has been well established, helping China made significant achievements in geological prospecting. By now, China has 157 minerals with proven reserves and the number ranks top of the world. There are more than 20,000 mining areas throughout the country and many minerals, including coal, titanium, niobium, tantalum, tungsten, tin, molybdenum, antimony, vanadium and lithium, take leading positions in the world. China is now the second largest ore producer in the world with over 200 thousand mines and an annual output of 4 billion tons of ores, providing over 90% mineral energy and 80% industrial minerals for the country. Iron and steel are primary materials for industrialization. Since 1949, China's production has amazingly increased from a tiny 0.16 million tons to 181.5 million tons in 2002, being the largest producer in the world for seven successive years. A large number of advanced technologies have been introduced and reinnovated, including oxygen converter, continuous casting, coal spurting in blast furnace, and floss protection in converter. China today can produce thousands of kinds of steels, and energy consumption is only 700 kgce per ton of steel being close to the world's best level. In 1949, China produced no more than 10 kinds of non-ferrous metals, and the number expanded in 1963 to all 70 non-ferrous metals. Particularly, China has made great achievements in making good use of its rich rare earth resources and developed first-rate technologies in smelting, segregating, purifying, and fabrication. The development of nonferrous technology has enabled large amounts of high quality materials to be produced for China's economic and social development.

One of the most glorious engineering achievements for China in the past 50 years is to establish a large petroleum industry. In 1949, China produced only 0.12 million tons of crude oil, less than 10% of the domestic consumption. With the theory of oil reserve formation in land sediment by Chinese geologists, China found and developed large oil fields, like Daqing and Shengli in the 1960s and 1970s, lifting China as one of big crude oil producers in the world and making great contributions to China's economic and social development. In 2002, crude oil output was 167 million tons. China has rich coal resources. Coal production in 2002 reached 1.38 billion tons, constituting 2/3 of primary energy production. Since 1970, China has been developing comprehensive mechanized coal mining and other advanced mining technologies. As a result, output, working efficiency, coal recovery and underground working safety have all been significantly improved. The widespread use of electric power is one of the greatest engineering achievements by human beings in the 20th century. In 1949, China's installed generating capacity was only 1.85 GW with an output around 4.31 TWh. In the last 50 years, China has devoted giant efforts to developing advanced power technology, helping China manufacture heavy power equipment, build large power station and design, construct and operate the world's second largest power system with world level thermal, hydropower, nuclear power plants as well as transmission and distribution systems. At present, China is building the Three Gorge Hydropower Station, which is the world largest hydropower plant with a generating capacity of 18.5 GW. China had 338 GW installed generating capacity and electricity generation reached 1,654 TWh in 2002. The huge advancement in electric power industry gives a giant power to push forward China's industrialization and modernization and improves everyday lives of all Chinese people.

Petrochemical engineering in China has grown with the development of petroleum industry. In 1960s, Chinese engineers mastered a number of major refining technologies and manufactured many petrochemical equipment. In the past 20 years, China has further developed many new refining technologies. Together with advanced technologies introduced from foreign countries, petrochemical technologies in China reaches or is close to the world's best level. Today, China is capable of designing 10 million ton refinery plant and manufacturing most major equipments, and annually produces over 90 Mt and more than 800 kinds of petrochemical products, like chemical fibers, plastics, rubbers, chemical fertilizers, medicine, among others. The inorganic chemical industry in China has also achieved significant progress in the last 50 years. Today, China has the full ability to design and manufacture all kinds of large complete equipment for the industry. In 2002, cement production reached 725 Mt, sulfuric acid 30.5 Mt, soda10.3 Mt, all ranking first in the world. China is also a major producer of many new inorganic non-metal materials, including glass reinforced plastic, glass fiber, quartz glass, special crystal and special ceramics.

With the technological breakthrough and market expansion of microelectronics after early 1970s, personal computers after the early 1980s and the internet after the early 1990s, information technology made an astonishing advancement in late 20th century, bringing the world into the information era. China has closely followed the world trend by introducing advanced information technology and making great efforts on R&D of its own in all areas, including microelectronics, computer, software, internet and telecommunication. China's IT industry has developed rapidly, especially after early 1990s, and today it has grown into the third largest manufacturer of IT products, next to US and Japan. In 2002, China produced 14.64 million personal computers, occupying over half of the domestic market. China has also developed high performance computing system and the three series, Shenwei, Yinhe and Shuguang, all have an operating speed of over 400 billion times per second, being one of the few countries in the world with batch production capability to produce supercomputers and top-end servers. Microelectronics technology and industry is also developing fast in recent years. Enterprises being engaged in VLSI design and manufacture have been established one after another and the CPU and DSP chips with China's own intellectual property rights have been developed, showing that the gap between China and the top world level has been narrowed. In telecommunication technology, China has self-developed a large number of high quality and low price equipment for telephone exchange, optical fiber telecommunication, mobile communication, digital communication, satellite communication and Internet access, providing technological and hardware support for the automation and digitalization of telecommunication network. The scale and extent of China's telecommunication network have been expanding continuously. Today, Internet users exceed 60 million, ranking second in world, and telephone users reach over 400 million, being the first in world. China today is the world No. 1 manufacturer of the telephone set, telephone exchanger and mobile telephone, and telecommunication has become China's fast growing pillar industry. China's broadcast and television technology has grown rapidly in the last 50 years. The radio broadcast started before 1949 and monochromic television program was first broadcast in 1958. Colorcast began in 1973 and soon replaced the monochromic broadcast. In 1980s, technological development accelerated. In 1985, China started using satellites for broadcast. In the beginning, China rented or purchased repeaters and then shifted to use self-developed and self-launched satellites. By now, all central and provincial radio and TV stations are equipped with world-level digital transcription, playback and broadcast systems. Since the early 1990s, cable television has developed rapidly and programs are now transmitted through high capacity digital optical fiber networks. Radio and TV broadcast covered over 90% of the population by the end of 2000 and cable TV subscribers reached 96 million households in 2002. China now has a giant radio and TV broadcast system and is the world's largest manufacturer of the color TV (51.55 M sets in 2002), radio receiver, radio-recorder, and VCD player.

China has actively promoted wider application of information technologies to revolutionarize its industrial technologies as well as the management of government and firms. One remarkable example is the Chinese character processing and printing. In the past, all printing works in China used movable lead type printing technology that was inefficient, labor intensive and harmful to worker's health. In the early 1980s, China successfully developed the laser typesetting printing technology that can electronically process and print Chinese characters and then applied it to the printing industry. The technology has been continuously upgraded to include more applications, such as color printing and long-distance printing. The widespread use of the new technology has revolutionarized the printing industry in China.

Before 1949, manufacturing of complete equipment was almost nil. In the past 50 years, China has developed a large number of advanced technologies and established an equipment manufacturing industry. Now, China is capable of manufacturing complete equipment for a 10 million ton oil refining plant, 600 MW thermal, hydro and nuclear power plants, 500 kV AC power transmission system, 300,000 ton ammonia plant, 520,000 ton carbamide plant, large oil field, 20 Mt strip coal mine, 5 Mt shaft coal mine, and large port, large iron and steel plant, large non-ferrous smelting and processing plant, large machine tool manufacturing plant, large water treatment plants, large construction project, and many others.

Fast and convenient telecommunication is another great achievement of human beings in the 20th century, and it significantly reduced time and space distances for people, bringing about faster advancement of human society. During the past 50 years, land, water, and air transportation systems in China have all undergone remarkable development. In 1949, China had only 21, 800 km of commercial railways with outdated facilities and technologies, while the total length reached 71,500 km in 2002 and 17,000 kilometers of it were electrified. Creative Chinese engineers solved numerous major technological problems, enabling China to construct a great many important railways in mountains, plateaus, deserts and canyons in the wild West China. Today, Qinghai-Tibet railway, the highest railway in the world, is under construction in the frozen earth area of Qinghai-Tibet Plateau with altitude of over 5,000 meters. Since 1997, China has completed four speed-lifting projects with over 13,000 km major railway routes and the speed reaches 140-160 km/h. Right before the New Year of 2003, the world's first commercial magnetic levitation railway demonstration line, with introduced German technology, was put into operation, running at a speed of as high as 450 km/h. Early this year, China successfully carried out a test run of a high-speed train with a speed as high as over 300 km/h. All these progresses shows a bright future for the high-speed railway transportation in China. Road transport is very important to a nation's economic and social development. In 1949, China had only 50,000 vehicles and 80,000 km roads that were in very poor conditions. Furthermore, roads were mostly concentrated on east and southeast coastal areas and hardly seen in mountainous and inland areas. Today, the road network reaches all parts of the country with a total length over 1.7 million km, among which 19,000 km are express ways, ranking second in the world. Chinese engineers have been carrying out tremendous R&D in road and bridge construction technology and a large number of world-class highways and bridges have been designed and constructed by Chinese engineers with China-made equipment and materials. In the mid 1950s, China started its automobile industry with truck manufacturing. After the mid 1980s, the industry had entered a new development stage through making cars with introduced technology in joint-venture companies. The variety and output have been steadily increasing. In 2002, China produced 3.25 million automobiles and became one of the largest automobile maker in the world.

Over the long history of China, serious disasters of flood and draught were always big threats to people's lives and properties. In 1949, there were only 23 reservoirs in China and total hydropower generating capacity was 360 MW. Most rivers were in no control or little control. As soon as new China was founded, it put huge efforts on harnessing the Yangtze River, Yellow River and many other threatening large rivers. A large number of flood control projects, large reservoirs, dikes and dams, water course renovation projects, hydropower plants, and soil and water conservations projects were completed, forming a nationwide flood prevention and control system that not only effectively control flood, but also help make good use of water resources, improve environment and promote economic and social development. By the end of 2000, China had built 85,000 reservoirs, 260,000 km dikes, 98 flood detention areas and many hydropower plants with 77 GW installed generating capacity. In the 1990s, China made use of its rich experiences in the design and construction of various complicated hydraulic projects and conquered a large number of difficult technological problems in constructing several giant hydraulic projects. For example, the Three Gorge project, the largest hydro project in the world today, is very important to managing and harnessing the Yangtze River. Its dam, hydropower plant and 5-level shiplock all rank the top of the world in size and technological complexity. The 50-year glorious achievements in harnessing large rivers have significantly reduced the loss caused by disastrous flood and draught and protected over 500 million people, 33 M ha cultivation land, 600 cities and numerous facilities, and achieved huge economic and social benefits.

The urbanization in China has been proceeding rapidly since the New China was born. The number of cities increased from 136 in 1949 to 663 in 2000, while the urban population expanded from 58 million to 502 million in 2002. China has been engaged in large-scale construction of urban infrastructure, including roads, public transportation, sewage disposal system, water supply, fuel and power supply, housing, service and office buildings. By 2000, the area of urban residential buildings and houses reached over 5 billion m^2, water supply increased from 1 billion tons in 1949 to 46.9 billion tons, roads increased from 10,000 km in 1949 to 160,000 km, 84% of the urban population used gas as fuel, and sewage treatment reached 11.4 billion m^3 per day. Urbanization process has accelerated since 1980s, and some well-developed cities have entered the rank of world metropolis.

Home appliance and textile are two of the most competitive industries in China. The blooming era of the home appliance industry started in 1978 when China introduced advanced foreign technology and equipment. In the intense market competition, the industry has invested heavily on technological innovation and achieved rapid development. Today, the industry has remarkable competitiveness in both domestic and the world markets. In 2002, China produced 16 million refrigerators and 31.4 million air conditioners, ranking first in the world. When new China was founded in 1949, the textile industry was its largest industry sector. Over years, it has made remarkable contributions to China's economic and social development. The industry is capable of manufacturing all production equipment for spinning, weaving, printing & dyeing, and finishing, and its products cover 13 different sectors, such as cotton, silk, linen, woolen, knitting, chemical fiber, among others. Today, the outputs of cotton fabric, silk fabric, woolens and clothes rank first in the world.

As a large developing country, China has made great progress in the past 50 years. However, we have even greater tasks to accomplish in the years ahead. China has set its development goals to realize industrialization in 2020 and modernization in 2050.

In the way of modernization, China is facing gigantic challenges. China today is still in the intermediate stage of industrialization and GDP per capita ranks only around 120th in the world. China's engineering science and technology, a pillar of modernization, is still far behind such world engineering powers, like the US, Japan, Germany, France and UK, either in general criteria like personnel, funding, and technical support, or in the number and quality of achievements, which is clearly reflected from the fact that the productivity and resources utilization rate of Chinese industries are far below the world best levels.

In manufacturing sectors, China has a giant manufacturing industry, but the overall technology level is 5-15 years behind the world best level, and productivity is about 1/30 of the US and Japan. China is incapable of manufacturing many important technology-intensive equipment and devices and relies heavily on import from other countries. For example, China has a huge market for civil airplanes, but we still do not have the ability to manufacture, so China has to import from western countries; advanced VLSI chip and precise medical appliances like heart pacemaker are also terribly needed, but China has to spend billions of dollar to buy them from other countries, because China either cannot manufacture them or the quality is poor.

In agricultural sectors, China has successfully provide its huge population with food, but agricultural technology is still several decades behind that of the developed world. Today, huge surpluses of rural labors crowd over limited lands, constraining the upgrade of farming technology, utilizing farming machinery and adopting modern information technology and biotechnology. In 1999, each Chinese agricultural labor generated an added value of about $2,000 (PPP) and feed 3.6 persons, which were only 1/15 and 1/27 that of each US farmers, respectively.

In service sectors, the application of advanced information technology in China is far behind that in developed countries and industries with high added value and significant importance to national economy like finance, insurance, consulting and law Service are far from developed. The development of science and technology as well as education cannot meet the urgent needs of modernization. In 2000, the US had 50 million people with diploma, about 35% of the total labor force, while the number for China was 38 million, which only represents 5% of the labor force.

China today is also facing tremendous opportunities. In the coming 20 years, China aims at building an all-around well-off society. To this end, China will continue to implement the strategy of rejuvenating the country through science and education and that of sustainable development, persist in using IT to propel industrialization, which will in turn stimulate IT application, blazing a new trail to industrialization featuring high science and technology content, good economic benefit, low resources consumption, little environment pollution and a full display of advantages in human resources. In the path to industrialization, engineering science and technology will play a most important and most direct role and it will determine the level and speed of industrialization. The endeavor of China towards industrialization and modernization provides a historic opportunity for the development of China's engineering science and technology.

China today has an advantageous environment for sustained development. First, as a country in the midway of industrialization, China is in a positive process of economic growth, that is, starting from economic growth, followed by larger investment capability, industrial expansion, larger capacity to absorb surplus rural labor, urbanization with higher demand for industrial products, further industrial expansion, and then further economic growth. Secondly, China today is in a world with swift science and technology development, continued economic globalization, industrial adjustment and redeployment by developed countries, all of which provide China with a favorable environment for its engineering and economic development. Thirdly, China has a stable society. Chinese people focus their major efforts on economic and social development and support the open and reform policy of the government. With sustained economic development, the purchasing power of Chinese people has been steadily increased, forming a giant market for goods and technologies. Fourthly, science and technology as well as education have made rapid development, the mechanism of converting technological achievements into industrial applications has been enhanced. China today has over 10 million engineers and the number will continue to grow with the fast development of education that will generate excellent engineering talents to meet the needs of modernization. It can be anticipated that China will experience a sustained economic expansion and social development for decades, and a massive engineering construction will occur in the first half of the 21st century. We are fully confident in achieving our ambitious goal of industrialization and modernization.

Learning from the experiences of the past 50 years, we realized that in order for developing countries to shake off poverty and backwardness, they must put great efforts on developing engineering science and technology. The task of science is to understand the world. Science is a powerful weapon to conquer ignorance and the basis for technological development, while the task of engineering is to transform scientific knowledge into technology, industry and wealth. Engineering, to a great extent, is a decisive force determining the economic and social development as well as the overall strength of a country. In the next 20 years, China will put most of its S&T resources into developing engineering science and technology and catch up with the world advanced level in those technological fields that are essential to economy growth and social development . Under the current situation of technological advancement and economic globalization, China will further enhance its exchanges and cooperation with the international engineering community. Being a developing country as well as a member of the Third World, China knows that a large number of Third World countries are facing tremendous problems of poverty, hunger, disease, information gap, environment pollution, ecological degradation and slow economic and social development. To achieve the goal of survival and development is still the giant challenge to all Third World countries. China's experience in the past 50 years shows that developing engineering science and technology is essential to Third World countries in their way to modernization. China will continue its long-term friendship with all Third World countries and the Chinese engineering community will expand its exchange and cooperation in engineering science and technology with all Third World countries in order to achieve mutual benefits and common development in the years ahead.

China's Science Education

Chen Zhili

Chen Zhili, *studied in the Physics Department of Fudan University, majoring in solid-state physics from 1959 to 1964; postgraduate at Shanghai Institute of Silicate of Chinese Academy of Sciences (CAS) from 1964 to 1968. Associate research fellow at the Shanghai Institute of Silicate of CAS from 1970 to 1980; visiting scholar at the Material Research Institute of the University of Pennsylvania, United States from 1980 to 1982. Vice Minister of State Education Commission of P.R. China, from 1997 to 1998; Minister of Education of P.R.China from 1998 to 2002; State Councilor of P.R.China from March, 2003. Specializing in ferroelectric and piezoelectric materials and devices,she achieved systematic results and published research papers on diffuse phase transition of relaxor ferroelectric material.*

Abstract: The paper introduces the system and its development of science education in China, ever since it's founding in 1949. It focuses on the science education in basic education and higher education as well as the popularization of science and technology in China in relation with the development goals, policies and strategies of the country.

I. SURVEY OF THE ESTABLISHMENT AND DEVELOPMENT OF SCIENCE EDUCATION IN NEW CHINA

Upon the founding of the People's Republic of China, the Chinese government made vigorous programmes to reform, create, or develop the system of culture and science and technology education, focusing on the development of culture and science education, enhancing the people's overall cultural quality, and cultivating scientific and technological personnel. In this way various types of educational systems were either re-created or improved in a brief period of time, thus revitalising China's educational, cultural and scientific and technological undertakings.

Since the founding of the Republic more than fifty years ago, especially since the adoption of the policies of reforms and opening up, with science and technology playing a more and more important part in social life, the Chinese government has been paying great attention to science education to the Chinese citizens, particularly teenagers, by formulating a series of policies and regulations and taking various measures to ensure the smooth development of science education and sci-tech popularisation. This has led to the setting up of a network, devoted to school education, and dedicated to the enhancement of the people's scientific and cultural capacity and quality, and to the extensive participation of the whole society in science education and popularisation.

Science education aims mainly at teenage students, with the purpose of helping them to develop scientific social values and world outlook, so that they learn to advocate science, seek truth, and be creative and innovative in their work and study. As major venues where science education is conducted, schools play a key role in the development of students' overall quality, with the emphasis laid on science education. In the past five decades, educational networks have been established, and improved, at all levels, thus forming a complete and balanced educational system which consists of pre-school, primary and secondary, polytechnic, and higher education and adult and professional education of various types. All sorts of approaches and means have been mobilised to strengthen the education to teenagers of scientific knowledge, skills, ways of research and values.

In order to promote and guarantee the development of China's education and the steady enhancement of the Chinese people's overall quality, the Chinese government has, through a strengthened legal system, issued a series of educational policies and regulations. Among these are the *Decision Concerning the Reforms of Educational System, the Programme for the Reform and Development of Education in China, the Law of Education of the People's Republic of China, the Law of Compulsory Education of the People's Republic of China, the Law of Higher Education of the People's Republic of China, the Decision by the CPC Central Committee and the State Council on the Deepening of Educational Reforms and the All-Round Promotion of Quality Education*, etc.

The campaign of All-Round Promotion of Quality Education is one of the major events in the development of China's education. The replacement of teaching knowledge to students by focusing on the enhancement of their overall quality symbolises a major turning point in educational conception. To enhance the scientific quality of the people is the important context of quality education and has been a centre of attention in every stage and chain of education. In this new situation science education has made still greater progress.

It is China's basic policy to pursue the sustainable development strategy. The implementation of the strategy requires that education serve as a basis on which the quality of the people shall be enhanced, the resources shall be reasonably utilised, and the environment shall be protected. China's education of sustainable development has been centred around the enhancement of the people's environmental awareness. After nearly 20 years of exploration and practice, the country's primary and secondary schools have seen a complete system where classroom teaching, complemented by extracurricular activities, has combined with social and family education in providing suitable, varied and comprehensive environmental education. With reference to higher education, it has made major contributions to the country's environmental protection by taking measures to strengthen environmental education, to vigorously promote environmental scientific research, and to actively develop environment-friendly, practical technology.

While taking school education seriously, the Chinese government has attached great importance to the mobilisation and organisation of social strength for the popularisation of science and technology. This has taken colourful and varied forms and has aimed at the acquirement of scientific and technological knowledge, promotion of scientific approaches, dissemination of scientific ideology, and enhancement of the scientific spirit.

In China's science education and popularisation, extensive efforts have been made to draw experience from advanced approaches to science education in foreign countries. In recent years the Chinese government has laid great emphasis on foreign exchange in the fields of education, science and technology and culture. Such events as the Forum for University Presidents, and cooperation and exchange for scientific, cultural and educational projects at various levels, have been conducted in order to find out about scientific and educational developments in other countries, and learn and study their successful experience in carrying out science education. This has been put into use for the improvement of China's science education and has thus strongly pushed forward the scientific and educational undertakings in China.

In retrospect of the past 50 years, we can clearly see the achievements made in China's science education and popularisation. At present, the popularisation rate for the nine-year compulsory education in China has reached 90%, with the illiteracy rate for young and middle-aged people lowering to below 5%. In higher education, the enrollment rate has reached 14%. Meanwhile, the scientific and cultural quality of the nation has been greatly enhanced. All this has resulted from the fact that the central government has paid great attention to and put enormous energy in the organisation and leading of science education; it has benefited from the extensive participation of the society, with schools at the centre, in the undertaking; and it has also benefited from the formation and improvement of a science education and popularisation network involving on- and off-campus activities.

2. SCIENCE EDUCATION IN THE STAGE OF PRIMARY AND SECONDARY EDUCATION

The Chinese government has always laid emphasis on the development of primary and secondary education. According to the *Decision Concerning the Reform and Development of Elementary Education* issued by the State Council, primary and secondary education is "the foundational engineering for the rejuvenation of the nation by science and education"; it plays an overall, fundamental and leading role in "enhancing the Chinese nation's quality" and providing fundamental scientific and technological knowledge to the Chinese people.

In the stage of primary and secondary education, science education is realised through in-school teaching at various levels and extra-curricular activities and is guaranteed by the training of teachers and construction of facilities.

2.1 Science Education in Elementary Education

Recent years have seen the addition of, besides such basic courses as natural science and languages, comprehensive scientific knowledge and the course of Science which focuses on hands-on activities.

The Science course for the primary school stage requires that the students know simple and plain scientific knowledge concerning their immediate surroundings, be able to apply the knowledge in their daily life, and develop behaviours and habits compatible with science. Students are also required to understand the process and methodology of scientific research and to apply them in scientific explorations, learn to look at and ponder on things in accordance with science, and develop curiosity and interest in science.

The course for the secondary stage requires the students, through further study, to retain their curiosity and interest in natural phenomena, develop a sense of compatibility with the nature, acquire basic scientific knowledge and learn some basic skills in order to be able to explain commonly seen natural phenomena, solve actual problems and develop the habit of scientific exploration and the capability of practice and innovation.

In junior middle schools in rural areas the "Green Certificate" education programme should be implemented, incorporating agricultural technology, operation and management and formulating the new educational models, familiarising the students with one or two practical agricultural techniques. In the stage of senior middle school a comprehensive course of practical activities will be offered so that more efforts are made on research-oriented study; and technical courses will be offered so that general technical education is strengthened. In this way, students will, through their own personal practice in social life and in nature, learn, accumulate and enrich their personal experience, so as to develop an innovative spirit and a capability of practising and learning throughout their life.

2.2 Extensive Extracurricular Activities Featuring Science and Technology Education

The New China has always attached great importance to scientific and technological events intended for teenagers and has planned and organised miscellaneous and colourful events for them. In 1981 the State Leading Group for National Scientific and Technological Activities for Teenagers was set up, taking charge of the coordination, research and organisation of national scientific and technological activities for teenagers. The Group maintained that "in organising scientific and technological activities for teenagers, more efforts shall be made to enhance teenagers' scientific and technological quality". After that the central government issued a series of rules and regulations concerning the organisation and contents of, and the construction of and funding for the venues for, teenage scientific and technological activities nationwide. The *Programme for Science and Technology Popularisation and Related Activities Among Teenagers from 2001 to 2005* was a step further in this direction in that it targeting the tendency of scientific and technological and educational development in the early 21th century, sets out the aims for science and technology popularisation among China's teenagers and related activities. These are: various scientific and technological activities are to be conducted in accordance with the young people's physiological and psychological characteristics and levels of education, so that they gradually learn about scientific and technological developments and have a mastery of necessary knowledge and skills; measures are to be taken to help them develop interest in and keenness on science and technology and consolidate their values and outlook; efforts are to be made to raise their scientific and technological quality so as to help them become qualified workers for the country's construction in the new century; and programmes are to be made for the discovery and cultivation of science-and-technology-savvy young people.

2.2.1 Colourful extracurricular scientific and technological events

The Ministry of Education and the China Science and Technology Association have so far jointly organised many large-scale scientific and technological activities, among which are the compilation and publication of the Science Education for Young People book series, the National Computer-Aided Innovations by Primary and Secondary School Students, the National "Science and Technology Stars" TV Contest, National Creativity and Innovation Contest and Symposium for Young People, the National Biological and Environmental Practice for Young People, Little Scientists for Tomorrow Awards, Olympic Contests for Senior Middle School Students, Young People's Science and Technology Dissemination

Behaviour, etc. These activities have encouraged young people to draw scientific knowledge from the immediate surroundings and to experience the process of scientific exploration, so as to get a deeper understanding of the relationship between science and technology and the society; the activities have aroused their interest in science and helped cultivate their sense of social responsibility; and they have also taught them to understand science and technology, master scientific methodology, develop a sense of creativity, and acquire the scientific spirit through their own practice and exploration.

The central government has attached great importance to the construction of venues for scientific and technological activities for teenage students. In many places of the country the overall planning for local development has incorporated the construction of science and technology stations/museums and science and technology centres for teenage students. In other places bases for science and technology education have been set up for teenage students. At present there are around 10,000 venues nationwide which are dedicated to extracurricular scientific and technological activities for teenage students. In rural areas scientific and technological demonstration parks have been set up and many schools have scientific and technological activity rooms.

2.2.2 Emphasis laid on the development and utilisation of social science and technology resources

Many approches have been adopted in coordinating various enterprises and institutions, scientific research institutes, parks and scenic areas, hi-tech parks in order to expand and enrich the scientific and technological activities for teenage students. Since colleges and universities have rich scientific and technological and human resources, the Ministry of Education has demanded that, starting from the year 2000, the key laboratories at colleges and universities be open to secondary and primary school students. In recent years, many colleges and universities have become bases for science popularisation education to young students and to the public.

2.2.3 Information technology education at secondary and primary schools

Recent years have seen fast development of computer networks and information technology, which have in turn become an important means to push forward various educational activities, especially science educational activities. Due to China's geological expanse, economic and educational development in different parts of the country has been entirely disparate and there is a great gap in the levels of education in these areas. For this reason, it has been imperative for modern distance educational networks to be a means to provide opportunities of life-long learning for young people and other members of the society, and to realise a balanced development of elementary education. In order to fasten the development of the information technology-based modern distance educational engineering, the Ministry of Education has stipulated that in the 5 to 10 years since 2001, the "Computer for All Schools" project should be carried out at primary and secondary schools nationwide, and information technology education should be popularised. Up to the end of 2002, the primary and secondary schools nationwide owned 5,840,000 computers, with an average of one for every 35 people; and over 26,000 campus computer networks were built.

2.2.4 Strengthening the cultivation of researchers and teachers for science and technology education

A full- or part-time science and technology education team consisting of teachers, scientists and other science workers means an important guarantee for science and technology educational activities for teenage students. The educational administration departments and schools at all levels, in order to adapt to the development of science education, have allocated plenty of human and financial resources to organise planned training of teachers. In the meantime, the Government has organised educational reform and pedagogical research so that the content and forms of science education have seen a remarkable improvement. From1998 up to 2000, the Ministry of Education, together with some other related departments, conducted twice the "National Social Investigation and Countermeasure Research Series on the Development of Creativity Among the Young People". A symposium with the theme of Development of Creativity Among Young People was held; discussed at the symposium were such topics as an international comparison of the development of creativity of young people, what reforms to carry out on science and technology in order to adapt to the development of young people's creativity, the effect of the popularisation of scientific methodology in extracurricular activities on the development of creativity of teenagers, the ways for families, communities and other related social structures to create an atmosphere beneficial to the development of teenagers' creativity.

3. SCIENCE EDUCATION IN THE STAGE OF HIGHER EDUCATION

Higher education, an important base in which to provide professional personnel and achievements in scientific research and to carry out knowledge innovation, propagation and application, serves as the source for the national scientific research and economic and social development. As stipulated in the *Decision by the CPC Central Committee and the State Council on the Deepening of Educational Reform and Overall Promotion of Quality Education*, in higher education attention must be paid on the development of college students' creativity, practical ability and pioneering spirit and on the overall enhancement of their humanistic and scientific quality. A series of reforms have been carried out as far as institution construction, academic disciplinary construction, and personnel cultivation are concerned and remarkable progress has been made.

3.1 Important Projects and Institution Construction Implemented in Colleges and Universities

During the Ninth Five-Year Plan period, in order to push forward the overall quality education, the state implemented the *"Programme for the Reform of Content of Courses and of Curricular Systems in the 21st Century"* and in 2000 carried out the nationwide *"Pedagogical Reforms in Higher Educational Institutions in the New Century"*. In order that the higher educational institutions become bases for knowledge innovation and high-level, creative personnel cultivation, the state organised and implemented the "High-Level, Creative Personnel Project" and the 211 Project aimed at the construction of key higher educational institutions and branches of learning; established the system for the exchange of domestic and foreign visiting scholars; set up the "Scientific Research Fund for Doctoral Programmes at Universities", "Award for Excellence of Performance of University Teachers", "Special Ph.D. Scholarship", "Ph.D. Dissertations Honours Award", and "Programme for Financial Aid to Outstanding Backbone Teachers at Colleges and Universities" and gradually built up over 12,000 key branches of learning, 200 personnel cultivation bases for liberal and science disciplines, 100 teaching bases for various elementary subjects, and 20 bases for university students' cultural quality development, all of which are to become the state-of-the-art teaching demonstration bases in the country. Those institutions and measures have remarkably enhanced the colleges and universities' level of scientific research, scientific research conditions, and the way of cultivation of high level scientific and technological personnel.

3.2 Pedagogical Reform of Science Education at Colleges and Universities

3.2.1 Carry out the structural readjustment of undergraduate branch of learning and strengthen disciplinary construction

Priority shall be given to hi-tech disciplines that are in urgent need in society; more efforts shall be made to develop applied technologies closely related to local economy; encouragement shall be given to the establishment of trans-disciplinary and cross-disciplinary subjects and to the expansion of discipline areas; and the number of undergraduate branches of learning shall be reduced to 100 from the now 200. Teaching plans and a credit system of free optional courses under the guidance of supervisors shall be called for as an institutional guarantee for the development of students and of various branches of learning.

3.2.2 Deepen the reform of curricular systems and structures and realise the synthesisation, personalisation, and serialisation

Trial projects for cultural quality education have been conducted which integrates science education with humanist education. Curricula at all the colleges and universities are directed towards those consisting of elementary courses and specialties plus synthetic courses; students are encouraged to opt for courses across university or school or departmental boundaries; and the concept of "taking into account both science and humanist subjects and imbibing extensively from multiple courses" is advocated. Key courses are selected and optional courses added to synthesise and improve the curricula, so as to realise the synthesisation, minimisation and personalisation of the curricular systems. Serial lectures, such as "Science and Technology Forum", "At the Cutting-Edge of the Discipline", etc. have been conducted at many colleges and universities to popularise the latest results of modern science and technology.

3.2.3 Consolidate the teaching of various practice courses, develop modern science and technology education, and make great efforts to cultivate students' creativity and practical ability

The state has systematically organised the renewal of the content, methodology and experimental, technical means in practice courses, adding experiments intended for the students' ability to design, to synthesise and to create. Steps are to be taken to accelerate the construction of on- and off-campus practice bases, bases integrating production, learning and research, and bases that facilitates students' creation and practice. Measures are to be taken to expand and push forward the construction of university-based national key laboratories, engineering research centres and university scientific and technological parks, to encourage the combination of teaching and scientific research in order to realise the sharing of resources and to lead students to participate in scientific work as early as possible.

Since 1992, the state has implemented the production-learning-research project, encouraging colleges and universities, scientific research institutions and enterprises to cooperate in miscellaneous forms in technological innovation and the development of hi-tech industry, and to strengthen pioneering education to teachers and students. Encouragement and support were given to postgraduate and undergraduates students in participating in scientific research activities, and independently setting up enterprises of high and new technologies, so that they, through production-learning-research practice, develop a sense of pioneering, risk prevention, and the ability to use the knowledge acquired.

3.2.4 Reform the curricula and pedagogy of the institutions of higher learning, and help students consolidate their sense of being the main body and their ability to study independently

Curricula shall be renewed, textbooks shall be revised, scientific spirit shall be given emphasis, and scientific ideology and methodology shall be developed. "Bilingual teaching" shall be piloted. Pedagogy shall be reformed, adopting elicitation, discussion, research, case analysis as methods of teaching. Modern educational technology, such as multimedia and network, shall be made full use of in order to develop and enhance students' ability to study independently.

3.3 Organise Multiple Extracurricular Scientific and Technological Activities

Since 1989 the Ministry of Education, in cooperation with the Ministry of Science and Technology, the Central Committee of the Communist Youth League of China and other related departments, has organised the biennial "Challenge Cup" National

Competition for Extracurricular Academic Works by College and University Students, the National Contest for Electronic Designing by College and University Students, Mathematics Teaching Aid Contest, "Creation and Innovation Competition", etc. Scientific and technological festivals, scientific and technological activities weeks, scientific and technological saloons, academic saloons, scientific research papers competitions, scientific and technological contests, scientific and technological achievements shows, etc., all served as a stage for college students to show their creative talents.

Social practice activities, such as those dedicated to enterprises tackling difficult problems, construction of urban communities, and lifting of poverty by intellectual means in rural areas, have enhanced college students' sense of social responsibility and their scientific attainment and ability to combine theory with practice.

3.4 Make Full Use of Networks of Educational Information to Strengthen the Development and Application of Multimedia Software for College Anduniversity Curricula

Great efforts are made to strengthen the construction of network infrastructure and informatised campuses (connected to broadband networks). At present nearly one thousand colleges and universities have access to the Internet, many of them have online institutes that vigorously push forward new teaching models and construct new, open, online educational systems. Many of the college and university libraries are actively making for network-based, digital development, sharing international, domestic and campus library resources. Electronic reading rooms and teaching materials are open to students so as to enhance the utilisation ratio of libraries' information resources, cultivate students' awareness of information utilisation, and enhance their competence for literature retrieval. Software institutes, which are aimed at the promotion of large numbers of high-level information development, application and management personnel, have been established on trial at dozens of colleges and universities in the country.

4. SCIENCE AND TECHNOLOGY POPULARISATION IN CHINA

The Chinese government has attached great importance to sci-tech popularisation and has regarded it as one of the key measures for science and technology education to the whole nation, dissemination of scientific knowledge and the enhancement of the people's scientific quality. In the past fifty years the government has increased the input of resources and launched vigorous campaigns for sci-tech popularisation in the whole society.

4.1 Construction of Science and Technology Museums

Science and technology museums serve as an important component of the country's scientific and technological undertakings, as the infrastructure for the "revitalise China through science and education" strategy and as the public cultural facilities for science popularisation, promotion and education. They have seen many activities and events held, such as science popularisation exhibitions, reports, film and TV shows, personnel training, etc., for the popularisation of scientific knowledge. Nowadays China has 29 provincial-capital-level museums where over 3 million audience is received.

4.2 Construction of National Bases for Science Popularisation Education

In order to bring into full play the role of science and technology museums and other museums in science popularisation, and actively push forward the socialisation, accessibility and regularisation of science popularisation, the China Science and Technology Association has, since 1999, established"National Science Education Bases" throughout the country. The now 201 Bases in the country have played a very positive role in science education to the people.

In order to implement the policy of ruling the country according to law, the Chinese government has attached great importance to the legalisation of sci-tech popularisation. For the purpose of pushing forward and regulating sci-tech popularisation by legal means, the 28th Session of the Standing Committee of the Ninth National People's Congress adopted the *Law on Sci-tech Popularisation of the People's Republic of China* on 29th June, 2002. The promulgation of the law indicates a major step taken to consolidate sci-tech popularisation and to enhance the citizens' scientific and technological quality. The law, on the basis of the summary of the experience gathered in sci-tech popularisation since the founding of the Republic, clearly defines such significant issues, in the form of law, as the fundamental principles, organisation and management, social responsibility, the relationship between the main bodies, measures of guarantee, legal responsibilities, etc., in sci-tech popularisation. Various governmental departments, especially science and education administrative ones, have moved in the direction of world scientific and technological development and social progress by attaching great importance to and pushing forward science education, and has thus provided legal insurance to China's sci-tech popularisation and science education.

Scientific and Technological Organizations in China

Zhang Yutai

China Association for Science and Technology

***Zhang Yutai,** was born in Tancheng, Shandong Province on September 22 of 1945. He is member of the 9th National People's Congress(NPC), and the Law Committee of 9th NPC. He also shoulders responsibility of the Vice President and Chief Executive Secretary of China Association for Science and Technology(CAST), as well as the First Secretary of the leading Party group within CAST. Prof. Zhang graduated from Beijing University of Aerospace in 1968 and received the Bachelor's Degree of Computer Science. His past and present appointments include: Division Chief and Deputy Director of the General Office of the State Science and Technology Commission, Deputy Director of the General Office of Scientific Leading Group under the State Council, Deputy Secretary General of Chinese Academy of Sciences(CAS) and Director of the Joint Office of Academic Branches, CAS, Proprieter and Chief Editor of "China Science", Vice Chairman of the Steering Committee of the State Scientific Popularization Joint Conference, member of the Steering Committee on Socialist Culture and Ethics Development, member of the National Joint Conference for Scholars, Senior Advisor of the International Union for Scientific Productivity, Advisor of the Academic Branch Presidium, CAS, etc. Prof. Zhang has engaged in scientific managerial work and scientific policy-making over a long period of time. He has also established very broad relationship with scientists and technicians as well as other parts of the profession.*

Abstract: The earliest modern scientific and technological organizations in China dates back to the late Qing Dynasty. Entering the 20th century, a number of societies of natural sciences, either for single subject or multi-disciplines, have sprang up. After the foundation of People's Republic of China, China Association for Science and Technology (CAST) was founded as an umbrella organization of scientific and technological workers. Since its formation in 1958, CAST has been carrying out academic exchanges extensively, strengthening science popularization, furthering its role of " Home of Scientific and Technological Workers". At the same time, it also has been promoting scientific exchanges and cooperation with foreign countries as well as China's Taiwan, Hong Kong, and Macao. CAST has made great contributions to China's scientific and technological advancement and economic and social development.

1. HISTORY OF SCIENTIFIC AND TECHNOLOGICAL ORGANIZATIONS IN CHINA

1.1 Scientific and Technological Organizations before the Founding of New China

Early in 1895 in the late Qing Dynasty, Tan Sitong, the representative of the Chinese reformers, established Liuyang Arithmetic Society in Hunan province. To the reformers, the first important thing in China at that time was to create scientific societies. It was just due to such ideas and with the publicity and organization by the reformers that a number of societies of natural sciences sprang up like mushrooms after rain. The Agricultural Society was one of these. Dr. Sun Yat-sen, the forerunner of the Chinese democratic revolution, drafted for the Agricultural Society a circular on soliciting for members, in which he described various societies existing in the West world and the tasks of the Agricultural Society, and placed high hopes on the society.

Entering the 20th century, the constitutional and the revolutionary movements rose up energetically in China. Various kinds of meetings and lectures became popular after the 1911 Revolution. The May 4th Movement gave further impetus to the dissemination of the ideologies of science and democracy in China. The ideology of "save the nation with science" was all the rage for a time. A group of personages in the circle of science, represented by those who had studied in Europe and the United States and returned later, in the belief that science should play a key role in bringing prosperity and strength to the nation, advocated creation of societies for all scientific disciplines. Against this background, a lot of scientific and technological organizations were set up for comprehensive or individual disciplines.

The Chinese Science Society was established in 1915 by some returned students from the United States. It organized various activities in China. In addition to publishing journals such as *Science Monthly, Science Pictorial, Academic Proceedings,* as well as books such as *Science Series and Science Abroad*, the Society held 26 annual meetings from 1916 through 1948. It cooperated with some other organizations, 20 organization at maximum, to jointly convene the annual meetings since the 20th annual conference. For publicizing science and advocating study and research, the society hosted lectures or exhibitions every year. Many foreign scientists, such as physicist Paul Langevin (1872-1946) , were invited to China and gave lectures to members of the Society. In 1922, the society established the first scientific and biological research institute in China – the Biological Research Institute, which contained a zoological division and a botanical division. In 1929, they set up the Mingfu Library in Shanghai, which covered an area of $600m^2$, subscribed to more than 140 journals from UK, USA, Germany, Japan, etc. and obtained through exchange more than 40 journals from overseas institutions. The Chinese Science Society also set up a number of awards for young scientists. Its effort was also involved in activities such as standardization of scientific terms, scientific education and scientific consultancy. It even set up a company, running the business in scientific books and instruments. Before the establishment of Academia Sinica, the national academy, the society sent representatives to attend quite a few international scientific conferences. During the period of its existence from 1914 through 1949, the membership of the Society increased from 35 at the very beginning to 3,776. The Chinese Natural Science Society founded in 1927 also experienced rapid development, with its membership, starting from only 4, expanding to more than 2,600 in later years. Besides lot of society activities, the most outstanding contributions of the Chinese Natural Science Society were: 1. it promoted the popularization of science; 2. it advocated the spirit of democracy; and

3. it criticized the selfish ideas of some scientific workers and promoted the spirit of selfless and industrious service. Other influential comprehensive natural scientific organizations included the Chinese Institute of Engineers founded in 1912 and the Chinese Technology Society founded in 1916.

More organizations concentrated their interest on individual disciplines or scientific fields. There were about 30 such national organizations, including the Chinese Pharmaceutical Association, the Chinese Nursing Association, the Chinese Medical Association, the Chinese Society of Forestry, the China Association of Agricultural Sciences Societies, the Chemical Industry and Engineering Society of China, the Geological Society of China, the Chinese Astronomical Society, the Chinese Chemical Society, the Chinese Society of Physics and the Chinese Mathematical Society, all of which were founded before 1949 and are still active today. These organizations have the following characteristics in common: they were all created by scientists themselves and have been growing steadily; they have stringent organizational regulations, including sound statutes, clear purposes and formalized organizational structures. The Geological Society of China even has its own logo and anthem. Their activities range widely, covering the annual meetings, the publication of academic periodicals, the examination and definition of professional terms, unifying industrial standards (e.g. the standards for meteorological observation unified by the Chinese Meteorological Society), organizing rewarding activities, translating foreign academic works, organizing study tours and observations (such as observation of solar eclipses and the variable stars organized by the Chinese Astronomical Society), discussing about improving education in universities and high schools (e.g., the Chinese Chemical Society and the Chinese Society of Physics set up special committees in this regard), participating in international activities, joining international organizations, popularization of science, publication of popular science journals.

In the 1930s and 1940s, some scientific and technological organizations were established under the support of the Communist Party of China. For example, the Chinese Association of Scientific Workers was founded in 1945 based on the Science Forum initiated in 1939 by Zhou Enlai and Pan Zinian, the chief of the Chongqing-based newspaper *Xinhua Ribao*, an effort to expand the patriotic united front. Within the short period of one year after its founding, its membership expanded to 700-800. And then, the Association and its counterparts in UK, USA,

France and Canada jointly created the World Federation of Scientific Workers, of which the China Association for Science and Technology (CAST) is still a member to this day. A few scientific organizations, either comprehensive or special, were born in the Communist-led anti-Japanese democratic base areas in the Shaanxi-Gansu-Ningxia, Shanxi-Chahar-Hebei border regions and in northwest Shanxi, Shandong, and Northeast China. The Border Region Defense Society and the Shaanxi-Gansu-Ningxia Border Region Natural Science Society in Yan'a n, the Natural Science Association in the Shanxi-Chahar-Hebei Border Region and Shangdong Natural Science Club in Shangdong province, as well as some other single-discipline societies were a few examples. At the founding ceremony of the Shaanxi-Gansu-Ningxia Border Region Natural Science Society, Mao Zedong said:"I am entirely in favor of the establishment of the Society because science is such a good thing that solves a lot of problems such as clothing, food, housing and traveling. So, everybody should welcome it and study science." The Northeast China Natural Science Society was the last scientific organization founded in the liberated areas before 1949, whose purpose was to unite all scientific workers to work for the creation of a new China.

1.2 Post-Liberation Scientific and Technological Organizations and the Establishment of the China Association for Science and Technology (CAST)

On the eve of the founding of new China, like the people all over the country, all the scientific workers felt overjoyed and came to realize that it was their sacred duty to contribute to the construction of new China. The Chinese Science Society, the Chinese Natural Science Association, the Chinese Association of Scientific Workers and the Northeast China Natural Science Society jointly sponsored the national congress of China's natural scientific workers. In an open letter to all its branches and members, the Chinese Association of Scientific Workers expressed that "the new era we have eagerly been expecting for years has come up today. The significance of the new era to us is the new awareness of science and the new attitude towards work." "An urgent and practical task in front of all the Chinese scientific workers is to respond to the call from the central government and Chairman Mao Zedong to restore and enhance production. That is our duty and we must rise up and make our best endeavors to answer the call." In July 1949, a meeting was held to take preparations for the National Congress of Scientific Workers. The 15 full delegates and 2 alternate delegates elected by the meeting attended the first Chinese People's Political Consultative Conference as the representatives of the circle of natural science and thereby participated in the creation of new China. Hence the predecessors of CAST became a constituent body of the Chinese People's Political Consultative Conference, the top political advisory body of the country. The National Congress of Scientific Workers was formally held in Beijing in August 1950. Chairman Mao Zedong met with the participants and other Party and government leaders like Zhu De and Zhou Enlai spoke at the Congress to extend congratulations. At the Congress, the National Federation of Scientific Societies (NFSS) and the National Association for Popularization of Science and Technology (NAPST) were inaugurated, with Dr. Li Siguang, a geologist, and Dr. Liang Xi, a forest scientist, as their presidents respectively. Mr. Wu Yuzhang was elected the honorary president for both organizations. With the birth of these two brand new national scientific and technological organizations, a new chapter was opened for the development of scientific and technological organizations in China. After their establishment, the two organizations enjoyed great care and support from the Party and the government. During the first Five-Year Plan period (1953 - 1957), the two organizations made great efforts to promote the advancement of science and its popularization. NFSS organized the scientific workers to take an active part in the campaign to support the war of resistance against US aggression in Korea and in the movement of land reform and the economic construction. It also helped the founding of new academic and professional societies. During that period it published 94 academic journals to support research and developed cooperative relations with foreign countries in areas of publications and personnel exchange, thus played an important role in upgrading the country's scientific standards. As an organization oriented towards the public, NAPST focused its efforts on disseminating scientific knowledge to both urban and rural people by means of organizing lectures, exhibitions, movie and slide shows, etc., contributing to the enhancement of the people's scientific and cultural level. In September 1958, the two organizations decided to merge and form the China Association for Science and Technology (CAST). Prof, Li Siguang was elected President of the first National Committee of CAST.

For almost half a century, the development of CAST, and its predecessors alike, has always been closely linked to the destiny of the nation. The National Congress of Scientific Workers and the two organizations it founded – NFSS and NAPST were born at about the same time as the People's Republic. Since its formation in 1958, CAST has conducted academic exchanges and science popularization activities. In 1964 and 1966, it successfully organized the Beijing Science Symposium and Summer Physics Symposium, respectively. During the 10 years of turmoil of "the Cultural Revolution", CAST and all its affiliated societies were forced to suspend their operation. It was not until on the eve of the Third Plenary Session of the 11th Central Committee of the Communist Party of China that things returned to normal under the intervention of Mr. Deng Xiaoping. During the last 20 years, CAST held five National Congresses and Zhou Peiyuan, Qian Xuesen, Zhu Guangya and Zhou Guangzhao were elected presidents of the 2nd, 3rd, 4th and 5th National Committees. Zhou Guangzhao was re-elected president of the 6th National Committee at the 6th National Congress in June 2001.

2. CONTRIBUTIONS TO CHINA'S SCIENTIFIC AND TECHNOLOGICAL ADVANCEMENT AND ECONOMIC AND SOCIAL DEVELOPMENT

2.1 Taking Economic Construction as the Core Task, Promoting Integration of Science, Technology and Economy, and Continuously Exploring New Areas in the Service of Economic Development

Bearing in mind the mission of supporting the national economic construction, CAST and its affiliated organizations have paid great attention to uniting all scientific workers to participate in the country's modernization drive. To implement the National Poverty Alleviation Program, CAST elaborated an action plan to help the country's underdeveloped areas by means of science and technology. The plan has been implemented through over 3,000 projects in 1,317 villages in 214 townships in 39 counties of China, as a result, 240,000 households benefited from the plan. In the minority ethnic areas, demonstration projects have been implemented to publicize scientific knowledge. Moreover, the Community Non-Regular Education Project has been executed in 120 poor counties in cooperation with United Nations Children's Fund (UNICEF) and more than 20 million people have benefited from this project.

To promote the coordinated regional development, 34 national societies and 10 provincial branch associations jointly organized a symposium in 1996 and came up with the *Proposal on Accelerating Development in Western China*. In 1998, CAST, in cooperation with the Chinese Academy of Engineering, initiated a forum on science, technology and socio-economic development in Western China, with the output of a series of proposals on the development and optimum utilization of energy and resources, protection of ecology and environment in Western China that were submitted to the Communist Party Central Committee and the State Council. In order to implement the strategies for exploitation in Western China, the 2000 annual meeting of CAST held in Xi'an focused on the theme "Exploitation in Western China: Science, Education and Sustainable Development". At this meeting, over 3,500 scientific workers exchanged their ideas on the exploitation in Western China and put forward a series of important policy-related suggestions. The work of science popularization is also carried out with more stress on the western area through "Science Popularization Project in Western China".

To push for the better implementation of the national strategy of sustainable development, CAST and its affiliated organizations have carried out various activities that have attracted close attention of the general public and helped widespread understanding of the sustainable development concept.

2.2 Carrying out Academic Exchanges Extensively and Promoting the Development of Various Disciplines and Upbringing of Scientific Talents

To bring into full play its role as the main channel for academic exchanges, CAST has developed a system of encouraging and managing scientific meetings so as to further enhance the quality of academic exchanges. Since 1996, CAST and its affiliates have held more than 77,000 domestic scientific meetings with 6.5 million participants in total and more than 3,400 international meetings with a total number of over 600,000 people participating. In these activities, the principle of "Let a hundred flowers blossom and a hundred schools of thought contend" has always been advocated, so that a favorable academic atmosphere has been created and innovative thinking encouraged, and thus, benefiting the development of various disciplines and scientific and technological innovations.

Starting from 1999, CAST has held its annual meeting on a regular basis, which is characterized of integrated and interdisciplinary topics and widespread participation. Each of the annual meetings that have been convened so far was attended by thousands of scientific workers. These annual meetings have promoted the interaction, permeation and fusion of different disciplines and impelled knowledge innovations, generated positive influence on the development of economy, society, science and technology in the host places. These meetings have gradually become academic pageants with extensive influence in the scientific circles. The local science and technology associations also organized in conjunction of the event activities with their own features. These activities have become the outstanding examples of academic events with unique local characteristics and great influence.

CAST lays stress on the integration of academic exchanges, policy advice and scientific verification so as to render services to the decision making. It organizes relevant member societies to hold serial seminars, drawing up "White Paper on Natural Disaster Reduction" and "Green Paper on Pest Control" on annual basis so as to provide forecast and proposals for disaster reduction and prevention as well as pest control. To respond to the need of the central government, CAST organized experts in studying the issue of IT application in the national economy and other key issues and submitted *Proposals on Issues Concerning the IT Application in the National Economy*, which were considered seriously by the Government. Many of the proposals and suggestions set forward by CAST and its affiliated organizations were adopted, making the decision-making process more scientific and democratic.

To promote the development of scientific disciplines, CAST organized 68 member societies in compiling and publishing *The 21st Century Book Series on the Development of Scientific Disciplines.* Some member societies conducted serial academic activities on long-term basis for new emerging disciplines and thus effectively promoting the development and construction of these disciplines.

To bring up young talents, CAST created a young scientist forum, which has so far held 72 sessions. Approximately 2000 young academic bellwethers participated in these activities, and thus, actively promoting the cultivation of young scientific talents. CAST also held four National Symposia of Young Scientists in 1992, 1995, 1998 and 2001, with an attendance of more than 2700 young scientists, including over 600 Chinese scholars from overseas.

2.3 Strengthening Science Popularization, Striving to Raise Scientific and Cultural Quality of the People, and Serving the Construction of Material and Cultural Civilization

In April 2000, President Jiang Zemin wrote the important inscription "Uphold the spirit of science, publicize scientific knowledge, and disseminate scientific thoughts and scientific methods"for the inauguration of the new wing of the China Science and Technology Museum, pointing out the direction of science popularization. In recent years, CAST and its affiliated organizations have been actively implementing the Science For Every Citizen Project so that a new situation has come into being in which science popularization work is carried out in a more extensive, socialized and constant way.

To disseminate scientific knowledge among rural grass-root cadres and farmers, CAST carried out various kinds of practical skills training in rural areas, with an attendance of 69.53 millions people, and nearly 50 millions rural cadres have mastered one or two practical skills. Together with other organizations, CAST co-sponsored the campaign for "delivery of culture, science and technology, as well as health care to the rural areas". Moreover, other science popularization events include "Science Popularization Winter", "Spring of Science and Technology" and "Ten-thousand-li Tour of Science Popularization" and so on. CAST has been conducting the science popularization demonstration throughout the country and denominated 103 science popularization demonstration counties, 98 science popularization demonstration bases in rural areas, 17,000 science popularization demonstration bases at all levels and 1 million science popularization households throughout China. These activities have accelerated the spreading of science in rural areas. To popularize science in remote areas, specially renovated trucks or vans, "Caravan for Science Popularization – Mobile Science and Technology Hall" as so called, are used, which is popularly received by the broad mass of the people.

To popularize science among administrative officials, CAST has organized more than 100 members of the Chinese Academy of Sciences and the Chinese Academy of Engineering to give over 240 lectures to some 100,000 officials at various levels throughout the country.

CAST carried out the science communication activities among the children all over the country, with a general name of "The Elder Helping the Younger". There are several millions of children participating in activities such as the summer camps, winter camps, field investigations, small inventions. In addition, CAST organized the National Juvenile Scientific Innovation Contest, Young Student Biological and Environment Science Experiments and the National Challenge Cup for college students. These activities gave prominence to cultivating innovating spirit and ability to practice. Take the three sessions of the scientific innovation contest for instance, they attracted a 46.8 million kids and proved to be an effectively push to the development of quality-oriented education. CAST has sent 109 middle school students to participate in the International Science Olympics and they obtained outstanding results in successive years. In September 2000, the 12th International Olympic of Informatics was held in Beijing, and President Jiang Zemin sent a congratulation message to it.

CAST and its affiliated organizations organized a lot of large-scale science popularization activities, such as "Science and Technology Popularization Week", "Science Festival", "Science Popularization Walking with You" at railroad stations and on board the trains, etc. In May 2001 and May 2002, CAST, in cooperation with the Ministry of Science and Technology, and other departments, sponsored two sessions of nationwide Science and Technology Week. Colorful activities were carried out during the event, which aroused strong reverberation in the society and attracted the active participation of the masses. While commemorating the 40th anniversary of its founding, CAST organized simultaneous large-scale science popularization activities in 96 large- and medium-sized cities all over China, which have brought extensive social effects. CAST also united social forces and established a batch of science popularization bases and demonstration bases throughout the country. It has set up demonstration communities in cities for science popularization to strengthen the science popularization work in the urban communities. CAST initiated a Science Popularization Gallery Project, aimed to build in 100 cities science popularization galleries with a total length of 10,000 meters and actually the total length of galleries already built which is longer than ten meters has reached 15,000 meters. Since 1996, CAST and its affiliated bodies have held 240,000 science popularization exhibitions in total with about 390 millions of visitors. These exhibitions aroused the enthusiasm of the broad masses, and the children in particular, to learn and practice sciences.

CAST has been further strengthening the infrastructure construction of science popularization. The Science and Technology Museum is a facility for implementing the strategy for revitalizing the nation through science and education. The construction of science and technology museums was accelerated thanks to the attention and support of the Party and government leaders at various levels. Since the new wing of the China Science and Technology Museum opened to the public in April 2000, the number of visitors has exceeded 4 million. CAST cooperates with the press and publication departments as well as various kinds of mass media and has published more than 8100 varieties of books and sponsored the publication of over 2000 scientific magazines. The Chinese Public Science Website has been set up and put into operation. A series of survey has been conducted on the science quality of the Chinese public, which has been included in the state statistical index system and provided the basic data for making the policies on science popularization and education. CAST established the first research organization of science communication and education jointly with the Beijing Normal University, which promoted the infrastructure construction for science popularization.

The promulgation and implementation of the *Law for Popularization of Science and Technology of the People's Republic of China* marked the beginning of the new era in which science popularization has been brought into the legal system, and will have great significance for the healthy development of China's science popularization.

2.4 Strengthening Scientific Exchanges and Cooperation with Foreign Countries as well as China's Hong Kong, Macao and Taiwan

CAST fully capitalizes on its unique advantages as a scientific and technological organization and works hard to strengthen scientific exchanges and cooperation with foreign countries as well as China's Hong Kong, Macao and Taiwan regions. It has thus further enhanced the influence of China in the international science and technology communities and formed an all-facet, multi-level and multiform opening pattern. As the representative of China, CAST and its affiliated organizations are members of more than 240 international scientific and technological organizations. Our connections with the International Council for Science (ICSU) and the World Federation of Engineering Organizations (WFEO), the two most influential international organizations in science, technology and engineering, have become even closer. While bringing into full play the roles of senior scientists of the older generation, CAST is actively intensifying its efforts to push young scientists and engineers to the forefront of international cooperation.

CAST has furthered international bilateral science and technology exchange activities. Since 1996, CAST has received approximately 100,000 visitors from various countries. It sent more than 6,800 groups of scientists and engineers abroad to attend international conferences and participate in other exchange activities. CAST has also sent, through various channels, management staff of enterprises to take professional training courses or advanced technical study programs abroad. The relevant national societies affiliated to CAST have in the past few years hosted the 30th International Geological Congress (IGC), the 62nd General Conference of the International Federation of Library Associations and Institutions, the 47th Congress of the International Astronautical Federation, the 20th World Congress of Architects, the 16th World Computer Congress, the 14th Congress of the International Federation of Automatic Control, the 24th International Congress of Mathematicians, etc. All these were extensively praised by the world's science and technology community as great success.

As a non-governmental organization, CAST has broad links with science and technology community in Hong Kong, Macao and Taiwan. Its affiliated organizations have organized various forms of cross-strait exchange activities and strengthened their relations with their colleagues in Taiwan. On the occasion when Hong Kong returned the Motherland on July 1, 1997 CAST held an exhibition entitled "Science Creates A Beautiful Future" in Hong Kong, which displayed in an all-round way China's resplendent achievements in science and technology since the nation took on the road of reform and opening-up and the exhibition became a popular event among the all walks of life in Hong Kong. After China resumed its execution of sovereignty over Hong Kong and Macao, the exchanges and cooperation between CAST and scientific and engineering organizations in Hong Kong and Macao have continued to develop with momentum.

2.5 Actively Participating in and Deliberation of State Affairs, and Furthering Its Role of " Home of Scientific and Technological Workers"

As a distinct circle attending the Chinese People's Political Consultative Conference (CPPCC), CAST and its local branches have often organized scientific and technological workers to actively participate in the democratic management and supervision of the state and social affairs. Over the past few years, they have put forward over 33,000 proposals, of which about 20,000 have been adopted.

CAST and its affiliated organizations have conducted serious investigations on the conditions and opinions of the country's scientific and technological workers,

The National Natural Science Fundation of China

Chen Jia'er

National Natural Science Fundation of China

Chen Jia'er, *an Academician of the Chinese Academy of Sciences and member of The Third World Academy of Sciences (TWAS), Fellow Institute of Physics and Chartered Physicist of U.K. and Member of New York Academy of Science, was born in October 1934 in Shanghai. He worked in Peking University after graduated from Jilin University in 1955. As a visiting scholar, he did research work at the Rutherford Laboratory and Oxford University, U.K. during 1963-1966, and at the State University of New York at Stony Brook and the Institute of Laurence-Berkeley, U.S.A. during 1982-1984. He has been a professor of Peking University since 1984 and was the President of the university from 1996 to 1999. He was honoured as Doctor of Sciences, honoris causa, of Loughborough University in UK in 2002, the Chinese University of Hong Kong in 2000, Waseda University in Japan in 2000 and Menlo College in U.S.A. in 1999. Now he is the President of the National Natural Science Foundation of China (NSFC), the President of Chinese Physical Society and the Chairman of Beijing Association of Science and Technology.Chen Jia'er is an internationally known physicist on particle accelerators with pioneering achievements in the research and development of superconducting linear accelerators in China. His research areas include cyclotron, Van de Graff, accelerator mass spectroscopy, beam bunching system, radio frequency quadrupoles, and RF superconducting acceleratory cavity. With his team he established China's first AMS users' facility at Peking University that played a key role in the accurate radiocarbon*

dating of the Xia-Shang-Zhou dynasty chronology. He has published about 140 publications and got a dozen of awards.

Abstract: The establishment of the National Natural Science Fund ation of China (NSFC) marked the beginning of the system for science funding transforming from a central planning system to a competitive system compliant with the socialist market economy in China. Its main responsibility is in accordance with the guiding principles, policies and plans for the development of science and technology in China, by appropriate management of the National Natural Science Fund from the central government, to support basic research and some of applied research, identify and foster talented researchers in the realm of science and technology, accelerate the progress of science and technology, and promote the socioeconomic development in China by giving full play the guidance and coordinating role of the national natural science fund.

The creation, the management and operating system of the national natural science fund, and the role of NSFC in promoting basic research in China are expounded in the paper.

The NSFC is one of the main sources of supporting basic research by the Chinese government. The budget of NSFC, starting from 80 million yuans RMB when it was first established, has been increasing at an annual rate of 26%, and has reached 2.0 billion yuans in 2002. In 1980 to 2002, NSFC has supported a total of more than 60,000 projects of basic research. The overall level of Chinese basic research including mathematics, physics, chemistry, astronomy, geo-sciences, biology and material sciences has been raised significantly. NSFC maintains a wide range of support to encourage the development of interdisciplinary research and new disciplines and boundary disciplines by means of priority setting, special fund, major research plans, etc. Using a competitive mechanism, NSFC continuously support the team of basic research consisting of about 60 thousand people, and maintain a reasonable scale of basic research in China.

I. THE ESTABLISHMENT OF THE NATIONAL NATURAL SCIENCE FUND OF CHINA

The natural science funding system is a scientific research supporting system different in nature from the administrative appropriation. Its prominent features are the practice of democratic management of the funding based on the scientific community itself, and the introduction of competition into basic research, thereby optimizing key elements of knowledge generation. The natural science funding system was implemented in the early eighties of the last century as a result of the reform on the scientific and technological structures in China. It was started in 1981 when 89 Members of the Chinese Academy of Sciences proposed to set up a national fund for natural sciences, which was then endorsed by the Chinese government. In the following year, the national natural science fund was formally inaugurated. In March 1985, it was pointed out in the *Decision on the Reform of the Science and Technology System of China* that the science fund system would be gradually put in practice for supporting basic research and part of applied research. The organizations mainly carrying out basic research will have to obtain their scientific revenue through applying to the science fund.

Thanks to the special care of Mr. Deng Xiaoping, the State Council approved in February 1986 the establishment of the National Natural Science Foundation of China (NSFC) to manage the national natural science fund. The main responsibility of NSFC is in accordance with the guiding principles, policies and plans for the development of science and technology in China and by appropriate management of the national natural science fund from the central government, to support basic research and some of applied research, identify and foster talented researchers in the realm of science and technology, accelerate the progress of science and technology, and promote the socioeconomic development in China by giving full play the guidance and coordinating role of the national natural science fund.

The creation of the national natural science fund and the establishment of its administrative organization marked the beginning of the science funding system transforming from a central planning system to a competitive one compliant with the socialist market economy in China.

2. THE MANAGEMENT AND OPERATING SYSTEM OF THE NATIONAL NATURAL SCIENCE FUND OF CHINA

Since its founding, NSFC has been following the rules of the development of basic research, exploring ways of supporting basic research compatible with the socialist market economy, and doing its best to create a good environment for fountainhead innovation and the development of innovative culture with distinctive features of the science funding system. A science funding system having Chinese characteristics is now evolving to its matured stage.

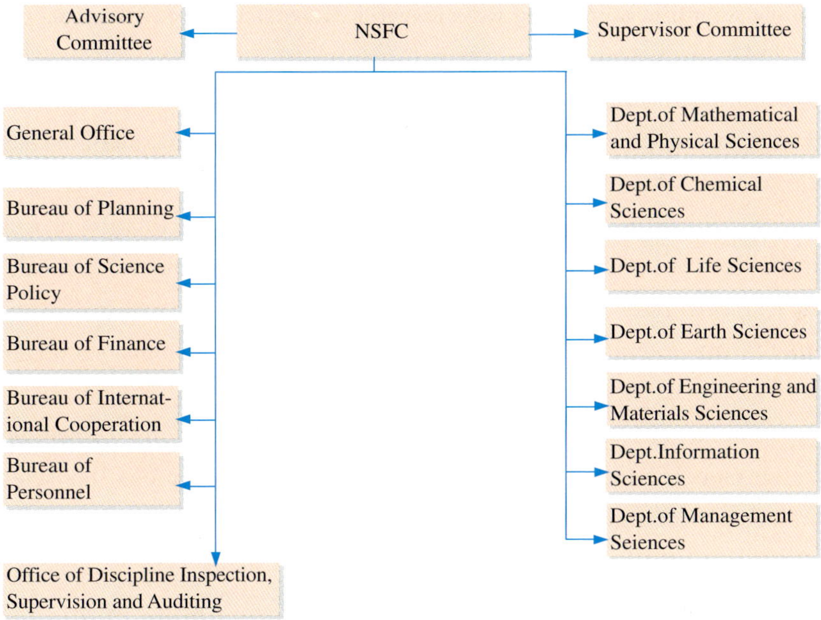

Fig. 1 NSFC Management Structure

Through continuous exploration and experiments, a sound management system has been established which integrates the decision, implementation and supervision functions (see Fig. 1: Management Structure of NSFC), a good mechanism of stimulating innovation, scientific democracy and fair competition has been maintained, two funding blocks of projects and talents have taken shape which are interrelated and coordinated (see Fig. 2: NSFC Funding Structure), and the evaluation principle of relying on experts and developing democracy to select best proposals for support in a fair and reasonable way has been formed. NSFC tries its best to advance with the times and make great efforts to strengthen management innovations in its work. In order to adapt to the constant integration and division of research disciplines, steady improvement on the funding blocks has been made. NSFC works hard to increase the funding intensity for General Program projects and the approval ratio to create a flexible environment for free exploration by Chinese scientists. To deal carefully with non-consensus projects to protect high risk and innovative projects, effective measures have been taken and short-term exploratory research projects have been introduced. To encourage the intercrossing of disciplines and the impact of different academic ideas, the Major Research Plan has been initiated. The mechanism of continuous support has been set up to encourage scientists doing long-term research. To foster a large group of research talents and teams for innovative research in frontier areas, NSFC has developed a whole set of talent-funding programs and initiated on the trial basis the Fund for Innovative Research Groups. To defend the scientific and fair nature of the national natural science fund, NSFC has made consistent studies and investigation on the evaluation criteria for scientific exploration, improved the evaluation procedures and methods, standardized the evaluation practice and strengthened the evaluation teams. To further improve the contact and communication with external experts, an expert advisory committee for each scientific department has been set up and the advisory system has been enhanced. A series of regulations and policies for project management have been formulated so that the management of the science funding can better meet the requirements of creating a good environment for scientific research and encouraging fountainhead innovation.

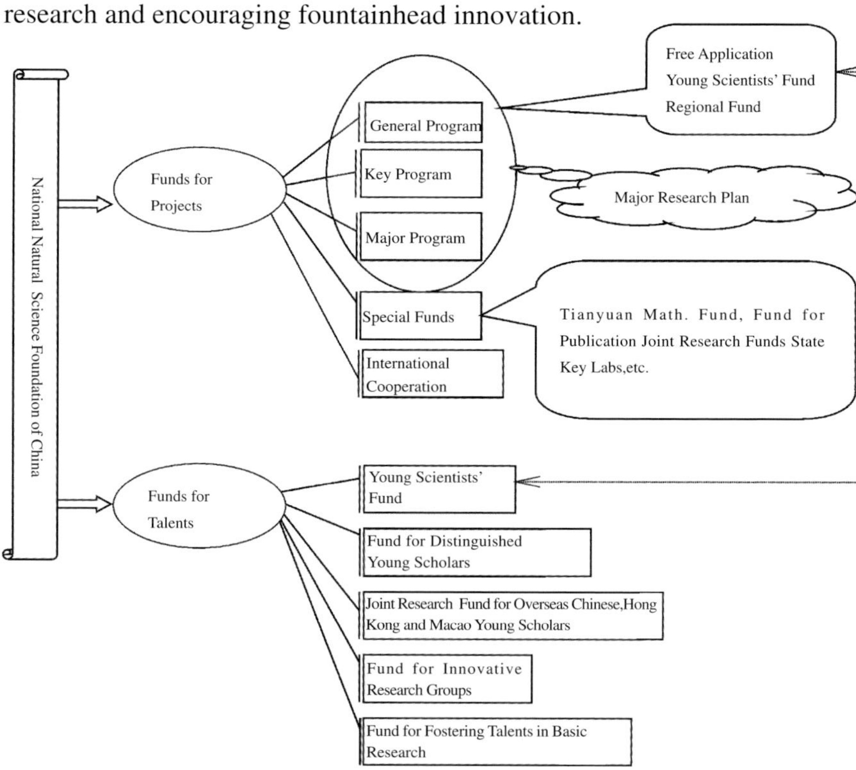

Fig. 2 NSFC Funding Structure

Good environment for innovation is a kind of culture. The work described above has indeed reflected the contents of that culture. Therefore, great efforts have been made to promote the establishment of a culture for innovation that has the distinctive characteristics of the science fund. Through its evaluation and funding work, NSFC has advocated the spirit of seeking truth from the fact and"daring to be the first in the world". In project evaluation, NSFC insists on the principle of being practical and realistic, discards non-scientific factors, and defends the fairness and justice of the national natural science fund. In the assessment and management of research achievements, it tries hard to avoid using simply the number of papers published in journals listed by SCI and the impact factor for the judgement of the academic value. Through a variety of measures, such as academic seminars and workshops and scientific reviews and criticism, it has implemented the policy of letting a hundred schools contend in order to facilitate extensive academic exchanges and the impact of novel ideas for sparks of innovation. Through the strengthening of the supervision and auditing, it has sternly dealt with misconducts in scientific research and encouraged the scientific spirit. In the internal management, it has stressed on developing the consciousness of providing good service and keeping close contact with the broad masses of scientists.

We have gained some experiences in the management of the national natural science fund. Among them, the most fundamental and important is maintaining a good mechanism of "encouraging innovation, being scientific and democratic, and funding through fair competition". Originality in scientific discovery and exploration is not only the highest state sought after by scientists, but also the number one criterion for judging the effectiveness of the national natural science fund, and the soul of the work of NSFC. The mechanism of encouraging innovation means doing best to create a favorite environment for innovation, and encouraging scientists to conduct research by integrating personal interests with the national needs, so as to give full play to the initiative of the vast number of scientists. The mechanism of being scientific and democratic means to maintain the closest relationship with the scientific community, to have a continuous and systematic participation of scientists in the project selection, evaluation and supervision and the strategic study of priority areas, so that the funding decision can be made on a scientific and democratic basis. The mechanism of fair competition means to insist on everyone being equal in the competition, and the selection of project by using creativity as the criterion instead of personal" identity" and " fame", so as to provide an equal opportunity for every scientist doing basic research to display his or her talents.

3. THE ROLE OF THE NATIONAL NATURAL SCIENCE FUND IN PROMOTING BASIC RESEARCH IN CHINA

During the past 17 years after its founding, NSFC adheres to the mission of supporting basic research and fountainhead innovation and has made substantial achievements in promoting the development of basic research and high-tech breakthroughs, and in discovering, fostering and gathering outstanding research talents.

3.1 By Giving Full Play to the "Initial" Driving Force, the Natural Science Fund Has Become the Headstream for Scientific Innovation

The national natural science fund is one of the main sources by the Chinese government for supporting basic research. Thanks to the close attention paid by the Chinese government and the strong support from scientists all over the country, the budget of NSFC, starting from 80 million yuan when it was first established, has been increasing at an annual rate of 26%, and reached 2.0 billion yuan in 2002 (Fig. 3: Annual Budget Increase). From 1986 to 2002, a total of more than 60,000 basic research projects were supported.

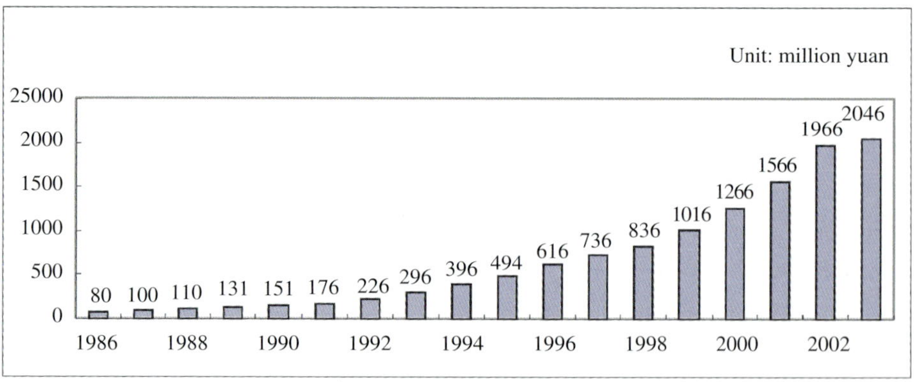

Fig. 3 Annual Budget Increase

3.1.1 The overall standard of Chinese basic research is significantly raised, with some disciplinary areas among the top in the world.

With the continuous support from NSFC and other related government agencies, the overall standard of Chinese basic research, including mathematics, physics, chemistry, astronomy, earth science, biology and material science, has been raised significantly. Among them the following are prominent:

In the International Congress of Mathematicians (ICM) held in 2002, China had one mathematician giving the One-hour Speech, and 11 domestic and 9 overseas Chinese mathematicians giving 45-minutes Invited Speech. The number of such speakers was far more than the total number of Chinese speakers in all of the past ICMs, and was among the top in all countries participated.

China has become one of the world important centers in paleontology and paleoenvironment research. In 2001, *Nature* published an anthology named *Rise of the Dragon* featuring 22 papers of paleontology studies on Chinese fossils published on *Nature* since 1997. These papers all received funding from NSFC. Henry Gee, the editor of the anthology made the following comments in the preface:

In less than a decade, paleontology in China has risen from relatively modest beginning to being a dominant force on the international scientific scene with a series of spectacular discoveries that have immediately enriched our understanding of key episodes in the history of life.

The research on fossil of dromaeosaurus with feathers by Dr. Ji Qiang of the Institute of Geology was ranked by the *Discovery* magazine number 19 on the top 100 important discoveries of the world in 2001. Recently, the study of Xu Xing and others on *gui Microraptor* fossils discovered in Liaoning was regarded as the most important work in the research of bird origins. Academician Liu Dongsheng was awarded the Taylor Award for Environmental Science, which was the highest international award in this area, due to his achievements in developing the study on Chinese loess and paleoenvironment.

3.1.2 A number of new disciplinary growing points have been created.

The advancement of science depends on the long-term accumulation of basic research and continuous intercrossing between different disciplines. NSFC maintains a wide range of support to encourage the development of interdisciplinary research and new disciplines and boundary disciplines by means of setting priority research areas, special fund, Major Research Plan, etc. Research in such areas as human genome, nano science and technology, quantum informatics, green production and environmental friendly chemistry, complexity science and bio-complexity, financial mathematics and management, global change and advanced manufacture, are in a large proportion received initial funding from NSFC and then start to develop widely in China. Some new branches of science, such as bio-informatics and proteomics, have been started from the scratch with the funding from NSFC.

3.1.3 Talented professionals and knowledge reserve have been provided for high-tech development of China.

Supported by NSFC and other agencies, Chinese scientists completed independently the "fine picture"of Chinese rice (hsien rice) genome. This is the first fine picture of a crop genome, meeting the international standard for precision genome pictures. Since NSFC initiated its support to human genome research at the beginning of the 90's, studies related to genome have been developing rapidly and a number of high quality research results have been obtained. Taking the year 2002 as an example, Prof. Kong Xiangyin and others proved for the first time in the world that HSFC gene mutation may result in genetic cataract of children, and proposed that this gene may have relations with the occurrence of cataract of the elderly. The group led by Prof. Zeng Yixin established the largest sample database of family tree of high incidence of nasal and throat cancers, and successfully located the sensible gene of nasal and throat cancers in the 4p15.1-q12 region of the number 4 chromosome.

NSFC has been supporting research on nano science and technology since the end of the 80's in the last century. Research on nano materials, carbon nanotube and nano-electronics has made large impact internationally. Academician Zhu Daoben, Prof. Liu Hanfan, Prof. Xie Sishen and their research teams made significant achievements in the research of several basic problems of C_{60} physics and chemistry, polymer and steady metal nano clusters, directional carbon nano tube preparation, structure and physical properties, etc. The research group led by Dr. Lu Ke made breakthroughs on solving the difficult problem of metal surface nitrification using surface nano technology of metal materials. Prof. Cheng Huiming and others obtained good results in the research on hydrogen carbon nano tube, which was highly regarded by Nobel Laureate for Chemistry Prof. Miles as "very impressive and exciting work".

3.1.4 A large number of research results have been granted national awards.

The winners of the National Supreme Award in Science and Technology Academician Wu Wenjun and Academician Huang Kun have all been strongly supported by NSFC since the mid-eighties of the last century. The winner of the first class prize of the National Natural Science Award Academician Jiang Xikui and key members in his research group have received support from more than 10 NSFC projects, and the funding received from NSFC is about 68% of their total research expenses. Among the research achievements which have been given the National Natural Science Award, those supported by NSFC has been increasing each year, with an average proportion of 91.6% since 1999.

3.2 The Talent Strategy of NSFC Has Achieved Significant Results

The key for the progress of science is talented professionals. Through the competitive mechanism, NSFC continuously supports a basic research contingent consisting of about 60 thousand people, and maintains a reasonable scale of basic research in China. In order to facilitate the growth of young scientists, NSFC sets up the Young Scientists' Fund for researchers under the age of 35. Up to now, nearly 10 thousand young scientists have received funding for research, many of them have become the key members of basic research, and over 300 have been awarded the National Science Fund for Distinguished Young Scholars. The National Science Fund for Distinguished Young Scholars was set up in 1994, which has funded a total of 1,015 outstanding young scholars under 45. Among them, 17 have been elected Academicians of the Chinese Academy of Sciences and the Chinese Academy of Engineering. Starting from 2000, NSFC is trying out the pilot program of Innovative Research Groups, and 55 such groups have been funded up to now.

The national natural science fund has also played a key role in the training of graduate students. The number of graduate students participating in NSFC funded projects is rising rapidly, which increased from 4,228 in 1998 to 16,163 in 2002, among them Ph.D. students increased from 793 to 7,710.

With the joint efforts of NSFC and other government agencies, Chinese basic research force has formed a reasonable age structure. Among the principal investigators of the General Programs of NSFC in 2002, 69.8% were under the age of 45. The peak on the age distribution of principal investigators has moved from the middle-aged and senior group in the 90's of last century to the 35-40 years-old group at present. This is a historic shift (Fig. 4 Age Distribution of PI's in NSFC General Program

Projects). It shows that young scientists have become the major force in basic research in China, and the overall age structure is approaching the ideal status with the optimum vitality and potential for innovation.

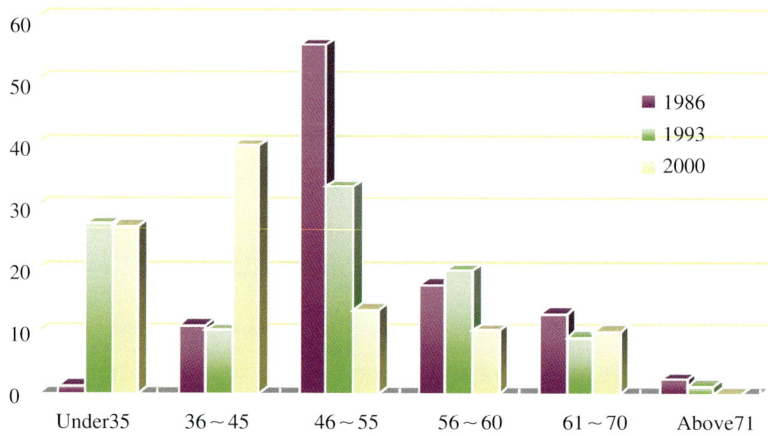

Fig. 4 Age Distribution of PI's in NSFC General Program Projects

NSFC is active in competing for international talents. NSFC is making efforts to make full use of the international human resources and taking various measures to bring in outstanding overseas Chinese scientists to conduct research in China. For example, the Joint Research Fund for Overseas Young Scholars has funded 264 overseas Chinese scholars to conduct joint research with their counterparts in China, and has supported 71 outstanding overseas Chinese scholars to set up joint research bases both in their countries of residence and with their counterparts in China.

3.3 Adhere to the Principle of Equality and Mutual Benefit and Create a Good International Environment for the Development of Basic Research

Basic research is for the human understanding of the universal laws of natural phenomena. International exchange and cooperation is an indispensable and important component of basic research activities.

NSFC attaches great importance to promoting international cooperation and exchanges in basic research. The budget of NSFC for international cooperation and exchanges is increasing annually and the channels for international cooperation are also expanding. NSFC will continue to develop its international cooperation in a more effective and diversified way. Up to now, NSFC has established cooperative and exchange relations with 58 science foundations or research organizations in 35 countries and regions (see Fig. 5: International Cooperation of NSFC) and NSFC and DFG of Germany have jointly set up the Sino-German Center for Research Promotion. A portfolio consisting of 5 categories of projects for international cooperation and exchange has been formed, and a total of nearly 10 thousand cooperative and exchange activities have been supported to date. NSFC has financed a number of major and substantial international joint research projects as well as the participation by Chinese researchers in major international research programs. It has also funded the establishment of virtual research centers in the areas of manufacturing, parallel computation and processing engineering, and has supported a number of outstanding foreign scientists to conduct long-term research in China.

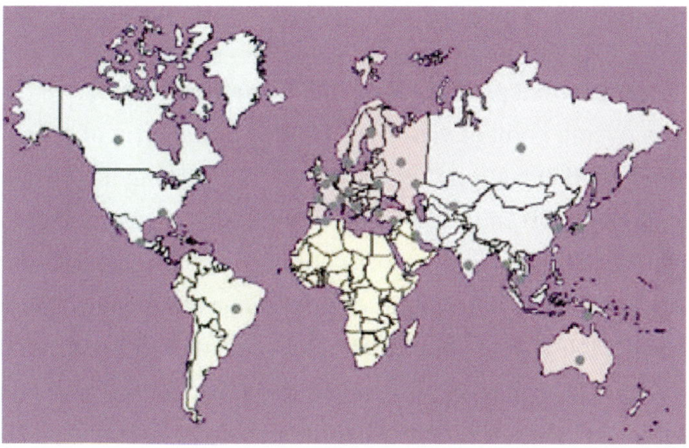

Fig. 5 International Cooperation of NSFC

4. FUTURE DEVELOPMENT OF NSFC

The 21st century marks an era of knowledge-based and globalized economy, as well as an era of fierce competition of comprehensive national strength based on science and technology. It is also a crucial moment for China to build a well-off society in an all-round way and revitalize grandly the Chinese nation.

Basic research is the fountainhead of science and the root of technology. Basic research and high-tech development are the driving forces of the Chinese modernization in the 21st century. Important achievements in basic research often start in the exploratory studies in the early stages. Without the tiny flows near the source there will be no surging tide in the river of science and technology. Practice in the past proves that the National Science Fund has played an indispensable role in promoting the innovation at the upper stream of the long river of science and technology. Therefore, in the future, we should be aware of the development of basic research in the world and the national strategic requirement, fully realize the strategic importance of basic research in maintaining the comprehensive national power and the fundamental role in gaining the leadership in the economy and science and technology, adhere to the mission of supporting basis research and fountainhead innovation, formulate the distinctive culture of national science fund, create a good environment for fountainhead innovation, and bring into full play the unique sensitivity and creative spirit of scientists. We will stick to the talent strategy, firmly adopt the idea of fostering talents, discerning talents, cherishing talents and gathering talents, continue to enrich and develop the talent supporting system, and keep talents by innovative projects and cultivate talents by creative practice. We should adopt an overall point of view, improve the mechanism of consultation and coordination with other national agencies of S&T management, and actively promote the joint support in the areas of major national strategic demands, and promote the building of the national innovation system. We shall further improve the management of the national natural science fund, and strengthen the integration, publication and demonstration of research achievements, informatization, legalization and management teams. We will open up continuously new prospects for science funding work, play an active role in strengthening the national capability of fountainhead innovation and science and technology competitiveness, and realize the transition from learning and tracking to independent innovation and the rapid development in science and technology.

Some Developments of Chinese Mathematics in the Computer Age

Wu Wenjun

Institute of Systems Science, AMSS, Chinese Academy of Sciences

Wu Wenjun *(Wen-Tsun Wu), a mathematician, is a research professor at the Academy of Mathematics and System Sciences and the honorary director of the Institute of Systems Science, Chinese Academy of Sciences (CAS). He is also a member of CAS and the Third World Academy of Sciences. He has served as the president of the Chinese Society of Mathematicians (1984-1987), director of the CAS Division of Mathematics & Physics (1992-1994), and a member of the national committee of Chinese People's Political Consultative Conference (CPPCC) and its standing committee member.*

Wu's research covers many aspect of mathematics, two of which deserve special attention. In the fifties, Wu made groundbreaking contribution to topology by discovering the Wu class and Wu formulas. After 50 years, these classic results are still used, e.g. by Fields Medal recipient E. Witten (1999). Since 1975 Wu devoted himself to the creation of a new discipline which he called Mathematics Mechanization. The Chinese ancient mathematics is constructive and computational with results mainly expressed as algorithm readily adapted to modern computers. Inspired by this, Wu developed a theory of zero-set structure of polynomial systems, known as Wu's method. The method had been applied with great success to Mechanical Geometry Theorem-Proving, which is considered as a landmark in the field of Automated Reasoning. Wu also applied his method to mechanism design, robot-

ics, CAGD, etc. Wu has received the following awards for his scientific contribution:First Prize Chinese National Natural Science Award, 1956; Tan Kah-kee (Chen Jia-geng) Prize in Math-Physical Sciences, 1993;Distinguished Scientist Award, Qiu Shi Science and Technology Foundation Hong Kong, 1994;Herbrand Award on Automated Deduction, 1997; National Supreme Award of Science and Technology, China, 2000.

Abstract: Different from the present-day mathematics which is governed by the deductive axiomatic treatment, the ancient Chinese mathematics had the characteristic features of being applications-oriented, being computational, constructive, and algorithmic. Influenced by the ancient Chinese mathematics, we introduced a successful method to mechanize the proving of geometry theorems. This method of geometry theorem-proving had evolved to a somewhat general movement of Mathematics-Mechanization with polynomial equations-solving as one of its main concerns.

1. SOME CHARACTERISTIC FEATURES OF CHINESE ANCIENT MATHEMATICS

The present-day mathematics is governed by the deductive axiomatic treatment with theorem-proving as its main concern which had its origin in the ancient Greek mathematics represented by the Euclid's "Elements of Geometry" of 3 B.C.. In contrast to this the ancient Chinese Mathematics paid little attention to theorem-proving and had even no such notions of axioms, theorems, and proofs. In fact, the Chinese ancient mathematics was rather applications-oriented, with problem-solving as its main concern, and the problems to be solved arose usually from practice, the rudimentary commerce of goods exchange, the area and capacity measuration, the reconstructions, the official administrations, etc.

As the known data of the problem to be solved and the resulting values to be sought for should be connected by some kind of equations, naturally and most frequently polynomial ones, so solving of polynomial equations became the main concern of Chinese scholars for thousands of years in ancient times.

The most important Chinese classics in mathematics which had been preserved up to the present day are universally recognized to be the "Nine Chapters of

Arithmetic" completed in 1c B.C. and its "Annotations" in year 263 A.D. due to Liu Hui in the Period of Three Kingdoms (220-265 A.D.). Below we shall denote these two classics by (NC) and(AN) respectively.

Already in (NC) there appeared methods for problems equivalent to the solving of simple linear equations and simultaneous linear systems of equations in the present day. There appeared also methods of square and cubic root-extraction equivalent to the present-day solving of simple quadratic and cubic equations. Since very remote times China had a perfect place-valued decimal system of positive integers, however large it may be. In solving the above-mentioned equations corresponding to the completion of some kinds of computations, the ancient Chinese scholars had successively enlarged the number system of positive integers to fractions, to negative numbers, and to (root-extraction)-irrational numbers. Liu Hui in (AN) even introduced the notion of infinite place-valued decimal numbers together with some limit concept and apparatus so that a complete real-number system was already arrived in that time. We have to point out that in Europe it was only in the later half of 19th century that the real number system was completed in diverse involved ways.

From the time of (NC) and (AN) onwards the solving of polynomial equations had been incessantly developed in China and cultivated in some classic (1247) of Song Dynasty (960-1279 A.D.) to general methods of numerical solutions of polynomial equations of arbitrary degree in numerical coefficients.

In Song and Yuan Dynasty (1271-1368 A.D.) there occurred a creation of utmost importance, viz. the introduction of the notion of Heaven's Element, Earth's Element, etc., corresponding to unknowns in the present-day terminology. These notions rendered the previous intricate task of turning a problem into equations almost a triviality. Moreover, in treating the Heaven's Element, etc. as some new kind of numbers added to the real number system so that arithmetic operations may be done in the usual manner, there will naturally be introduced the notions corresponding to present-day polynomials and rational functions, as well as their algebraic manipulations with elimination procedure in particular. All these imply the essence of modern algebraic geometry and modern (polynomial) algebra in some sense. Moreover, these developments led to ideas and methods for the solving of systems of polynomial equations actually in arbitrary number of unknown variables. The methods were clearly described in some classics (1299, 1303) due to the scholar Zhu Shijie of Yuan Dynasty. Though naturally there were many defects and incompleteness in these methods, they became the starting point of our general study in the present computer age.

It is very important to point out that all the methods of polynomial equations-solving occurred in our ancient mathematics were expressed in the form of Shu literally meaning rule or method which are actually equivalent to the present-day algorithm. Thus, the solving of simultaneous linear equations was expressed in the form of Rectangular-Array Shu and Positive-Negative Shu for transposition of terms. The solving of (simple) quadratic and cubic equations were expressed in the form of (Square and Cubic) Root-Extraction Shus, (eventually with Cong Shu). The numerical solving of polynomial equations with numerical coefficients in arbitrary degree was expressed in the form of Positive-Negative Root-Extraction Shu. The solving of (simultaneous) high-degree polynomial equations is expressed in the form of Tian-Yuan Shu, 4-Element Shu, etc.. There were in fact many Shus of diverse character in our past-time classics.

Besides polynomial equations the present-day so-called Diophantine Equations were also studied in our ancient mathematics. Thus, already in the classic (NC) there was a problem solved by some Shu which gave the complete set of exact integer-values of the three sides of a right-angled triangle (called Gou-Gu Form) in our ancient times. In (AN) Liu Hui gave even a logically rigorous proof of this Shu (of course in a style different from the usual Euclidean type). We remark in passing that general formulae about the integer-valued sides of right-angled triangles did not appear anywhere else in ancient times other than China until hundreds years later in Diophantine's work and without any indication of proof. In fact, most of historical works on mathematics gave false descriptions in this respect. The calender-making of China since quite remote times led to the problem of solving some kind of congruence equations in the present-day terminology. Successive developments cultivated to the Da-Yian 1-Seeking Shu in the 1247- classic mentioned above. When the above result was transmitted later to Europe in 18th century it was henceforth known as Chinese Remainder Theorem which played an important role in modern mathematics.

The present-day so-called binomial coefficients together with some associated Diagram equivalent to the later known Pascal Triangle were already discovered in Song Dynasty around 10c A.D.. It was created with the purpose of high-degree root-extraction equivalent to the present-day solving of equations of the form $x^n = a > 0$, where n is an arbitrary positive integer. Also in Song Dynasty the great politician and great scholar Shen Kuo (1031-1095 A.D.) had created some theory and Shus dealing with formulas alike to the present-day combinatoric identities. Shen's creation had been much extended in Song and Yuan Dynasties with for example an identity which was re-discovered in recent years and was called accordingly Formula of Zhu Shijie and Van der Monde in the literature.

In conclusion, the ancient Chinese mathematics had the characteristic features of being applications-oriented and even highly practical, being computational, constructive, and algorithmic. In fact, most of important results in our ancient mathematics were expressed in the form of Shus, corresponding in general to the present-day algorithms, which may readily be turned into programs to be run on the computers, if one likes. As pointed out by D.E.Knuth, computer science may be considered as a science of algorithms. In this sense our ancient mathematics may be considered as a kind of computer mathematics. Its intrinsic value and possible influence in the nowadays computer age are quite clear.

The brilliant development of our ancient mathematics declined unfortunately during the Ming Dynasty (1368-1644 A.D.). It was replaced henceforth by the western mathematics of entirely different characteristics. However, the spirit of our ancient mathematics had been revived in China in recent years. In fact, in Institute of Systems Science of our Academy of Sciences, it was established in 1990 a Research Center bearing the name of Mathematics Mechanization (abbr. MMRC). Through the efforts of members of MMRC and their collaborators spread over a vast part of China, important achievements along the line of thought and spirit of algorithmic method of our ancestors had been done as shown in the next section.

2. MATHEMATICS-MECHANIZATION STUDIES OF CHINA IN THE COMPUTER AGE

Being inspired by the ideas, methods, and achievements of our ancient mathematics, the present author had begun to apply computers to the study of mathematics toward the end of 1970 in the last century.

Everybody had learned since his or her study in primary schools how difficult and intricate for the proving of geometry theorems. Under the influence of ancient

Chinese mathematics the present author tried in the end of year 1976 to find a mechanical way of proving geometry theorems. For this purpose we first turned to algebraic forms the geometry theorems by means of coordinates. After several months of painstaking trials we ultimately succeeded in discovering some algorithmic way of proving some essential part of elementary geometry theorems by algebraic manipulations. Since that time hundreds of difficult geometry theorems had been trivially proved or even discovered on the computers by this method. Later on our algorithmic method of geometry theorem-proving had evolved to a somewhat general movement of Mathematics-Mechanization with polynomial equations-solving as one of its main concerns.

Following the line of thought of our ancestors with some techniques borrowed from modern mathematics we have been able to give a general algorithmic method of solving arbitrary system of polynomial equations in the case of characteristic zero. The results are expressed in the form of various Zero-Decomposition Theorems about Zero(PS), the Zero-Set of a polynomial system PS. These theorems permit to give a complete explicit determination of the solutions or zeros of an arbitrary polynomial system of equations PS=0 in case of characteristic zero. Our method of mechanical proving of geometry theorems is actually an application of the above general method of polynomial equations-solving. As a further application, we have shown how to determine in some mechanical way the explicit forms of unknown relations known to exist.

The above method of polynomial equations-solving had been extended to the differential case. Thus, for some systems of algebraic-differential equations DPS=0 there are also Zero-Decomposition Theorems which permit to give a complete set of solutions or zeros in certain sense of such systems. As in the algebraic case, the method had been applied to the mechanical proving of differential-geometry theorems and to the automatic discovery of unknown differential relations. In particular, this had been applied to the automatic discovery of Newton's Inverse-Square Law from the Kepler's Observational Laws. Besides, the method had been applied with great success for the determination of complete set of solutions of soliton-type of a large number of partial differential equations occurring in physics and other realms of natural sciences. All these were achieved by a unique method, while in the literature such solutions can only be found for each individual partial differential equation by some intricate method peculiar only to that equation alone, even often with incomplete set of possible solutions.

A (projective) complex algebraic variety is defined as the (homogeneous) zero-set of (homogeneous) polynomial system. For projective varieties with no singularities, there may be defined the celebrated Chern Classes or Chern numbers via the associated tangent bundle of the smooth non-singular variety in question. For a variety with singularities for which tangent bundle is not defined, then the known method of extending the notions of Chern classes and Chern numbers is a very intricate one and is actually impossible to determine them explicitly in even very simple concrete cases. However, our method of polynomial equations-solving had permitted us to define such generalized Chern Classes and Chern Numbers in quite simple and natural way which are even explicit and computational. Thus, the well-known Miyaoka-Yau Inequality between Chern Numbers which are known only for complex 2-dimensional nonsingular surfaces of certain type, had been generalized by our method through easy computations to a large number of equalities and inequalities for high dimensional hypersurfaces with arbitrary singularities. The above Miyaoka-Yau inequality is only a very particular extreme case and its truth does not require any limitation on singularities. This shows clearly the powerfulness of our general method.

Optimization problems, or min-max problems, are abound in most fields of science, engineering, and technology. Undoubtedly the solving of such problems are extremely important for our economic constructions. It is well known that such problems are usually solved by various kinds of convergent approximation methods of numerical mathematics. Such methods give usually only isolated and local optimal values. On the other hand when the objective function and the restricted conditions are all polynomial in form, which occur quite often in the nature, then our general method of polynomial equations-solving will give the global optimal values whenever they do exist. In particular, in the case of the important bilevel programming, we have shown that for certain particular examples in the literature, what the global optimal values found by some intricate numerical methods are actually not the global optimal values and are far behind the true global values determined by our method.

Our method of polynomial equations-solving in the characteristic zero case had been extended to the case of finite characteristic. This was applied to theorem-proving in finite geometries. It turns out that the results gave light to the interesting phenomena that the theorems of same hypothesis will give rise to quite different conclusions for even and odd characteristic.

In combinatorics we know that the earliest successful algorithm due to Sister Celine for the proving of combinatorial identities depends somewhat on the solving of some polynomial equations. On the other hand we have discovered some general algorithmic method of directly proving some class of combinatorial identities, including in particular the Zhu-Van der Monde identity mentioned above.

Our method of polynomial equations-solving had also been applied to the study of cryptology and some related problems.

As problems in science and technology naturally lead to polynomial equations-solving, so our general method have naturally diverse applications in science and technology. Thus, we have found new solutions of the well-known Yang-Mills Equations and Yang-Baxeter Equations in theoretical physics, not found before by other methods. Besides, we have applied our algebraic-geometry treatment to find whole set of definite-type solutions of problems in surface-fitting of CAGD. We have also dealt with various kinds of mechanisms involving four-bar linkages, manipulators, Stewart platforms, etc., eventually explicit solutions in some concrete cases. We have also shown how to apply our method to the study of computer vision, etc., with noticeable success.

There are also further contributions in diverse directions which we are unable to enumerate here. For these we refer to the writings due to members of MMRC and their collaborators.

A general software named MMP independent of any other alike ones has also been completed with the package about our method of polynomial equations-solving as one of the central parts.

3. FUTURE POSSIBLE DEVELOPMENTS OF MATHEMATICS-MECHANIZATION

Mankind is now entering a new era of information or computer age characterized in particular by the presence of the powerful tool of computers. In past eras from 18th century onwards the mankind had encountered various stages of industrial revolutions which may be characterized as mechanization of physical labor. In the coming era we will be faced with a new kind of industrial revolution characterized by the mechanization of mental labor. Now mathematics is a typical mental labor. It is universally recognized as the fundamental basis of all kinds of sciences and technologies. At the same time mathematics enjoy the widest applicability to actually all kinds of activities. Hence among all kinds of mental labors, mathematics should have the highest priority and utmost urgency to be mechanized. On the other hand mathematics has the peculiarity of being clear, precise, concise, and unambiguous in its exposition which are not possessed by any other kind of mental labor. Hence among all kinds of mental labors, mathematics seems to be the easiest one to be mechanizable. Our success in the mechanization of geometry theorem-proving shows that this is really the case.

Our Mathematics Mechanization is so proposed to meet the necessity of mechanization of mental labors in the present computer age. However, our development of Mathematics-Mechanization is yet in quite nascent stage. Thus, so far theorem-proving is concerned, what we have done successfully is restricted only to the very narrow and not so important domain of elementary geometry or (local) differential geometry. However, each domain of mathematics has its own problems of theorem-proving, to be solved in some way peculiar to that domain, not necessarily reducible to solving of polynomial equations. Furthermore, it is known from mathematical logic that if the domain of mathematics in consideration is too large in some sense, then the proving of theorems in that domain may be undecidable in logician's terminology, or non-mechanizable in our terminology. On the other hand if the domain in question is too small in some sense, then the theorems to be proved may be devoid of any mathematical interest, though the domain as a whole is mechanizable. In view of this we have launched the following program to be studied in years to come:

Cover as much as possible the whole of mathematics by domains each of which is sufficiently small to be mechanizable, at the same time also sufficiently large to contain lot of theorems or problems of high mathematical interest.

So far as polynomial equations-solving is concerned, we remark that our method is symbolic in character, quite different from the usual numerical methods. Such symbolic methods lead usually to polynomials with milliards of terms to be out of control of the computers. It seems that to render our method practically workable, the only way is to develop some hybrid method of combining both the superiority of the symbolic method and that of the numerical method, which should of course have a solid mathematical basis.

The present-day mathematics is somewhat governed by such subjects owing much to the development of calculus which is of somewhat infinite character. On the other hand computers can deal only with objects of finite type, and work only in a finite way. Hence combinatorics as a certain kind of mathematics of finite character seems to become more and more important in the present computer age, the more so in view of information safety closely connected with national defense. Therefore we have to pay more attention to the algorithmic study of combinatorics and its alike than any time before.

We have shown how to automatically derive the Newton's Square-Reciprocity Law from the Kepler's Observational Laws. It gives a concrete example of a general method in automatic discovering of theories in mathematical form from observational or experimental facts and should be tried in diverse sciences in years to come.

Furthermore, though our general method of polynomial equations-solving have wide applications in science and technology, it waits still to become real productive forces in our technology and industry.

Finally, our software MMP needs to be improved and extended to raise up its efficiency and widen its applicability. In short, it waits to become a universal powerful tool to be used by the whole world.

REFERENCES

1. Chou, S.C., Mechanical Geometry Theorem Proving, D. Reidel, (1988).
2. Gao, X.S. and Wang, D.M. (Eds.), Mathematics Mechanization and Applications, Acad. Press, (2000).
3. Mathematics Mechanization Preprints, No.1-21, Mathematics-Mechanization Research Center, Institute of Systems Science, CAS, (1987-2002).
4. Wu W.T., Basic Principles of Mechanical Theorem-Proving in Geometries, (Part on Elementary Geom etries), (in Chinese), Science Press, (1984). English translation by D.M. Wang and X. Jin, Springer, (1994).
5. Wu W.T., Mathematics Mechanization, Mechanical Geometry Theorem-Proving, Mechanical Geometry Problem-Solving, and Mechanical Polynomial Equations-Solving, Science Press/Kluwer Acad. Publisher, (2000).

Some Advances in Mathematics in China

Yang Le

Institute of Mathematics,Chinese Academy of Sciences

Yang Lo(Le), Mathematician, Professor, born on Nov.10th, 1939, Nan Tong of Jiangsu Province. Starting from 1956, he studied in Mathematics Department of Beijing University. In 1962, after graduation from Beijing University he became a graduate student of the Institute of Mathematics, Chinese Academy of Sciences and has been doing research work in the Institute since then. He served as the director of Institute of Mathematics, the president of Academy of Mathematics and System Sciences(CAS), General Secretary and Chairman of Chinese Mathematical Society. He is a member of presidium of Chinese Academy of Sciences. In 1980, he was elected Member of Chinese Academy of Sciences.

Abstract: Modern mathematics researches arose in China at the beginning of the 20th century and developed gradually. The founding of the People's Republic of China brought great changes into mathematical researches and educations. Since 1978, mathematics researches became the order of the day, while groups of young mathematical talents came out, and mathematics publications and academic exchanges boomed up. Examples of the most outstanding achievements are "The theory of functions of several complex variables in classical domains","Researches on characteristic classes and embedding classes","Symplectic geometric algorithm of Hamiltonian systems",

"Researches on Goldbach conjecture" and"The stabilities of differentiable dynamical system".

Mathematics is a subject in which Chinese are traditionally skilled. Enjoying a long history, Chinese classic mathematics developed gradually from Shang and Zhou periods to Song and Yuan dynasties, and reached its peak in the thirteenth century. Some world leading results at that time had been achieved, among them were Shangao theorem (Pythagoras theorem); Chinese remainder theorem; Zugeng principle (Cavalieri's principle); Liuhui's method for determining segment areas and calculating the value of p; Yanghui triangle (Pascal triangle) and Qinjiushao's algorithm for numerical solution of higher polynomial equations, etc. The close link with practical problems characterized Chinese mathematics in ancient and medieval times.

Since the late Ming Dynasty, development of Chinese mathematics was suspended somehow. While in Europe, along with the Renaissance, mathematics entered a new period of rapid development with the introduction of Descartes' coordinate geometry and Newton-Leibniz's calculus. By comparison, Chinese mathematics lagged behind, and modern mathematics did not arise in China until the 19th century.

1.ARISING OF MODERN MATHEMATICS IN CHINA

From the middle of the nineteenth century on, western modern mathematics came into China. Around the beginning of the last century, Chinese scholars started to go abroad and studied mathematics in universities in Europe, Japan and the United States. On their coming back, they devoted themselves to establishing departments of mathematics in Chinese universities, to training young qualified scientists and to advocating original research on mathematics. Till the early 1930s, however, the number of scholars who were engaged in modern mathematics teaching and research was very limited, among them Hu Dunfu, Feng Zuxun, Hiong King-lai, Chen Kien-Kong, Chiang Li-fu, Yang Ko-Chuen, Su Buchin, Kiang Tsai-Han,etc. were the leading persons.

1930s was a crucial period in the development of modern mathematics in China. In Tsinghua University, Zhejiang University and some other universities, there appeared a group of mathematical talents, among them excellent representatives were Hua Loo-keng, Chern Shiing-shen and Shu Pao-Lu. 1935 saw the founding of the Chinese Mathematical Society. In the next year, "Acta of Chinese Mathematical Society"(predecessor of "Acta Mathematica Sinica") and"Journal of Mathematics"(predecessor of "Mathematics Bulletin") were launched. Even during the anti-Japanese war, under the difficult conditions in the rear, Chinese mathematicians never stopped mathematical research and teaching. Later in 1947, the Institute of Mathematics of Central Academy was set up in Shanghai.

According to statistics, before 1949, less than 80 Chinese scholars published only about 300 mathematical papers in total. Moreover, there existed many gaps in research fields. In most universities the mathematical departments enrolled only one or two students each year. Though master degrees were offered in Tsinghua University, there were less than a dozen students who attended the program in more than ten years. Until 1949, the Acta of Chinese Mathematical Society had published simply three issues in two volumes.

2. DEVELOPMENT OF MATHEMATICS IN CHINA SINCE 1949

Since the People's Republic of China was founded in 1949, tremendous changes have been brought into mathematical research and education. In 1952, the Institute of Mathematics of Chinese Academy of Sciences was set up, it conducted mathematical research along with some famous universities. Since then, a number of excellent mathematical results appeared and a batch of young mathematical talents grew up. The gaps in research fields were gradually filled up, in particular some applied subjects, such as differential equations, mathematical statistics, operations research, control theory and computational mathematics, etc. developed even faster with the expansion of contingent of researchers. Besides, the number of students and teachers in universities increased rapidly due to the adjustment of colleges and departments. In many universities, a hundred or more students of mathematics enrolled each year, among them there were many excellent young scholars, who were determined to devote themselves to mathematics.

In 1956, graduate programs were introduced in the Chinese Academy of Sciences as well as in some leading universities, though the number of graduate students was quite limited. In 1950s, a great number of monographs and textbooks written by leading Russian mathematicians were translated into Chinese, which resulted in a renewal of contents of mathematical courses in China. Varied mathematical activities were organized constantly by Chinese Mathematical Society."Acta of Mathematica Sinica", "Advances in Mathematics" and"Mathematics Bulletin"were published regularly. Essential progress was made successfully in application, transmission and popularization of mathematics, as well as in organizing mathematical Olympiads competitions.

From 1949 to 1966, modern mathematics in China strode forward independently and became more mature. About 450 scholars published over 1800 mathematical articles, of which all the papers appeared in "Acta of Mathematica Sinica" were translated into English and republished in US. On the other hand, this period saw frequently disturbance of political events, and mathematical researches and education experienced many complications.

During the following years of so-called "Culture Revolution" from 1966 to 1976, the whole country went through a catastrophe, from which the mathematical researches and education failed to escape as well. All the research work conducted in institutes and universities were seriously interrupted, talents training were totally suspended, and the publication of academic journals halted. Pure mathematics was criticized as "pseudo-science being divorced from practice", Euclid, Newton, Gauss and Hilbert..., all were unjustifiably attacked. Many mathematicians suffered political persecution and were treated unfairly. It was really a bitter lesson for us that should never be forgotten.

Earlier in the fifties and the first half of sixties of the last century, mathematical researches in China were already in rather a closed situation with almost no academic contacts with the western countries. When Chinese mathematicians woke up from the nightmare of ten years disaster (1966-1976), world mathematics had taken on a totally new look with arising of a series of new fields and new directions with which Chinese mathematicians had not been familiar. Even the Atiyah-Singer index theorem appeared in the sixties was then a stranger to Chinese scholars, not to mention the developments afterwards, for example, solving of Calabi conjecture, results in topology of lower dimensions, new advances in algebra geometry, revealing close links between partial differential equations and differential geometry, emergence of geometry analysis, realizing the increasing importance of computational mathematics, and the applications of operational research and mathematical statistics in economics, finance and management and so on.

As to the mathematical education then, undergraduate courses were far from demands, while in the international community graduate education was believed to be a main approach to foster young talents. For instance, in US only, about one thousand Ph.D students major in mathematics each year. More mathematical journals were published and more mathematical conferences were held, which presented a flourishing and dynamic picture of the world mathematics. In contrast, having suffered the ten years disturbance, the gap in mathematics between China and the advanced countries was getting even greater than before the"Cultural Revolution".

3. SPRING OF CHINESE MATHEMATICS

Since the National Congress on Sciences in 1978, mathematics, like other subjects in China, greeted its bright springtime and saw great chances to develop. Although confused at first by the mathematical concepts and formulas from which they had been away for ten years, Chinese scholars, with the traditional characters of diligence and persistency, worked ever so hard with strengthened determination and soon galloped in respective fields of mathematics. During the twenty years from the end of seventies to nineties, in spite of the harsh conditions for working and living, everyone kept high spirits and was absorbed in research on mathematics.

From the end of the nineteen seventies on, more and more Chinese mathematicians went abroad and visited universities and institutes in North America and Western Europe as well as in other countries and areas, where they conducted advanced studies and cooperative research, got a better knowledge of the latest advances of related subjects of mathematics with the widened field of vision, and made their own original work deserving of commendation. At the same time, research institutes and universities in China invited overseas mathematicians, especially overseas Chinese scholars like Chern Shiing-Shen, Yau Shing-tong, etc., to come to give lectures, to hold seminars and conferences and to build closer international academic exchanges. With more than twenty years efforts, the situation of Chinese mathematics improved apparently, and the level of mathematical researches and education was raised steadily.

A number of mathematical research institutes were established, and the mathematical research became the order of the day. Within Chinese Academy of Sciences, apart from the Institute of Mathematics, other three mathematical institutes were set up in succession, they are the Institute of Applied Mathematics,

Institute of System Sciences and Institute of Computational Mathematics and Scientific/Engineering Computing. In 1998, by reforming, above-mentioned four institutes merged into one named the Academy of Mathematics and System Sciences, which became one of the first combined bases for the Knowledge Innovation Project. At the same time, there were the Morningside Center of Mathematics of Chinese Academy of Sciences, the Institute of Mathematics at Nankai University, United Research Center of Advanced Mathematics at Peking University and Research Center of Mathematical Sciences at Zhejiang University, of which all conducted mathematical researches and academic exchanges aimed at promoting nationwide development of mathematics. Besides, many leading universities, including Peking University, Fudan University, Science and Technology University of China, Zhejiang University, Nanjing University and so on established their own institutes of mathematics, and encouraged teachers to carry out researches. Some outstanding technical universities, such as Tsinghua University, Shanghai Jiaotong University, Xi'an Jiaotong University and Southeastern University, realized as well the importance of mathematics for training engineers and other technical personnel of high-standard and therefore re-established departments of mathematics or set up new mathematical institutes.

After the end of 1970s, graduate education has developed greatly. The academic degree system was officially established in 1982. Mathematics, as a subject of first grade under the system, is divided into five directions including pure mathematics, applied mathematics, computational mathematics, probability and mathematical statistics, operations research and control theory. Since then, many graduate programs (for Ph.D and master degree) were gradually introduced into various universities and institutes, and a large number of Ph.D supervisors were appointed. Later in the mid 1980s, the post-doctoral system was further initiated, by which young scholars obtained more opportunities for more advanced studies. Nowadays, there are normally several dozens or even hundreds of graduates and post doctor majors in mathematics in one university or institute only. Many talented graduates themselves have become professors and Ph.D supervisors of mathematics in both domestic universities and research institutes. Some young scholars went to the first class universities in the United States, Western Europe and other countries and areas for further studies. Supervised by mathematicians of international reputation and nurtured in the best academic circles, some of them have already become full professors in world famous universities.

At present, thousands of scholars in China are engaged in mathematics, their research cover almost all subjects including some of the newest frontiers. Mathematical publications flourished in an unprecedented manner. Every year, around five hundred papers written by Chinese scholars appear in the international mathematical journals. The major mathematical periodicals published domestically are "Sciences in China (Ser.A)", "Acta Mathematica Sinica(both English and Chinese version),Chinese Annals of mathematics (English and Chinese versions), Acta Mathematica Sientia (English and Chinese Version)","Acta of Applied Mathematics (English and Chinese version)", "Algebra Colloquium", "Journal of Computational Mathematics", and "Advances of Mathematics in China". An influential mathematical serials–"Pure and Applied Mathematics Monograph Serials" is published by the Science Press in Beijing, which reflect the most important advances in mathematics in China. Forty volumes of the serials have appeared to the present, and most of them have been translated into English and published by Springer-Verlag. According to the incomplete statistics, over the last two decades, about 300 monographs and proceedings written or edited in English by Chinese mathematicians have been published abroad. In the 1980s, "Mathematics" volume of "Great Encyclopedia of China" appeared with one thousand pages and two million Chinese characters, which is now one of the most important reference books in China comprised nearly 1000 items dealing with all branches of mathematics and distinguished mathematicians of all countries at all times. In the 1990s, "Russian Encyclopedia of Mathematics" was translated into Chinese and published in five volumes, of which all the items were written by academicians and corresponding academicians of the Academy of Sciences of former Soviet Union.

The National and International Academic Exchanges are booming up. In the General Assembly of International Mathematical Union (IMU) held in Oakland, California, USA from July 31st to August 1st 1986, China, in combination of the Chinese Mathematical Society and the Mathematical Society located in Taipei, China, became successfully an IMU member of the fifth group having the highest five votes. Since 1986, Chinese mathematicians have participated in all the International Congresses of Mathematicians and General Assemblies, some of them were invited to give lectures at related sessions. In August of 2002, China hosted successfully the ICM-2002 in Beijing, as well as more than 40 satellite conferences held around the Congress, which covered almost all subjects of mathematics and attracted a great number of mathematicians from all over the world.

Now it is quite common to see world famous mathematicians to visit China to give lectures and attend conferences. Many international conferences and workshops are held in different parts of China every year. On the international mathematical stage Chinese scholars are playing more and more active role. They appear at various international conferences often at a reasonable large ratio of the participants, make contributions by giving well-received lectures and pushing forward various cooperation programs. In addition, Chinese school students have been performing very well in competitions at the International Mathematical Olympiads, and the 31st International Mathematical Olympiads was held successfully in Beijing in 1990.

4. EXAMPLES OF OUTSTANDING RESULTS IN MATHEMATICS RESEARCHES

Chinese mathematicians have achieved a large amount of significant results, some of them have made notable impact on related fields and received high regard from the international mathematical community. Among the achievements which have received the Chinese National Science Awards, results in mathematics possess a large proportion. Some mathematicians were awarded the Science and Technology Advancement Awards of He-Liang-He-Li Fund, Hua Loo-keng Mathematics Prize, Chern Siing-Shen Mathematics Prize, Awards in Mathematics of the Third World Academy of Sciences, Chen Jiangeng Awards, Outstanding Scientist Awards of Qiushi Fund, and Young Scientist Awards of China.

Here we present only a few examples of the most outstanding results obtained by Chinese mathematicians.

4.1 The Theory of Functions of Several Complex Variables in Classical Domains

While the theory of functions of several complex variables was still in its infancy, Hua Luo-keng and his students studied the automorphic functions of several complex variables profoundly. He then developed systematically a deep theory of the harmonic analysis on functions of several complex variables, and made great contributions. By means of the theory of group representation, Hua solved explicitly some basic problems of analysis and geometry on classical domains. He constructed analytic automorphism group and complete orthogonal system, obtained Cauchy integral formula, and set up the theory of kernel function and harmonic function.

Hua's monograph on the theory of functions of several complex variables has been translated into Russian and English, which was widely received by the international mathematical community. The related results showed far reaching influence on the further development of harmonic analysis, theory of functions of several complex variables, even of such fields as partial differential equations, theory of group representation, theory of symmetrical space, theory of automorphic function and mathematical physics. This outstanding work brought on a series of concepts named after Hua, such as Hua operator, Hua equation, Hua measure, etc. Hua opened up indeed the entirely new direction of research on the theory of several complex variables in China, trained a group of young experts who continued working along the direction and obtained more results of high level. According to the international authoritative scholars, Hua's work on this field led his western colleagues by 10 years and still has its influence.

4.2 Research on Characteristic Classes and Embedding Classes

In the 1930s, the characteristic classes were introduced respectively by E.Stiefel (Swissland), H.Whitney (US), L.S.Pontrjagin (Russia) and S.S. Chern. Wu Wen-Tsun, by Grassmann manifold, systematized the theory of characteristic classes, revealed the links between the different approaches mentioned above. At the same time, Wu applied his theory to construction of manifolds. The cohomological class introduced by Wu is now named the Wu characteristic class, and the two formulae obtained by him, which depict the topological invariance and the homotopic invariance, are also called Wu formula.

Wu also did excellent work on the theory of imbedding. He developed a general constructive method with non-homotopic topological invariants, which was systematically applied to research on embedding problems, and put forward the concept of embedding classes of complexes. In a similar way, Wu probed the immersion problem and isotopic problem, and introduced the immersion classes and isotopic classes.

Above-mentioned research on characteristic classes and embedding classes constituted valuable contributions to the field of topology. The related results have been widely cited and used by some outstanding scholars.

We omit here Wu's researches on mechanical proof of geometry theorems and mathematics mechanization, which will be dealt with in a separate article in this book.

4.3 Symplectic Geometric Algorithm of Hamiltonian Systems

Solving numerically different kinds of equations appeared in mathematical physics as a main topic for scientific and engineering computing. Feng Kang et al. proposed and developed a symplectic geometric algorithm for evolution equations in Hamiltonian form, creating an important field with wide prospect for application.

Nearly all the traditional computational methods are not symplectic, consequently man-made dissipation can not be avoided. Symplectic geometry makes it possible to preserve the construction of systems, possesses therefore the unique advantage in stability and long time tracing, which solves the computational problem of long time prediction in dynamics.

The symplectic geometric algorithms proposed by Feng et al. have been applied successfully to astrophysics, molecular dynamics and fluid geo-mechanics. Their work has exerted great influence in theinternational computational community and inspired a great number of succeeding research.

4.4 Research on Goldbach Conjecture

Three Chinese mathematicians, Chen Jingrun, Wang Yuan and Pan Chendong, made distinguished contributions in this area.

In the middle of the eighteenth century, German scholar C.Goldbach proposed the famous conjecture: every even integer bigger than 4 can be represented as the sum of two primes. The conjecture, other than a number puzzle, shows close link with L-function theory and the studies on it led in fact to the development of the sieve methods and the circle methods, of which both are powerful weapons in mathematics. The results obtained by researches on the conjecture have produced impact on many fields in mathematics far beyond the number theory.

Based on the previous research, in the middle fifties of the last century, Wang Yuan proved that every sufficiently large even integer can be represent as the sum of a product of no more than three primes and a product of no more than four primes, which is simply denoted by 3+4. Wang soon improved his result and obtained 2+3. In the early sixties, Pan Chengdong proved further 1+5 and 1+4. Eventually in the middle of sixties and the early seventies, Chen Jingrun made a great stride forward and proved successfully 1+2. Chen Jingrun's work has drawn numerous attention from the world mathematical community, and is considered as the most excellent result achieved in research on Goldbach conjecture so far.

4.5 The Stabilities of Differentiable Dynamical System

Theory of differentiable dynamical systems is a subject in which the major concern are the laws of evolution of a system, i.e. the global structures and asymptotic properties of phase graph of paths under certain transformations. Liao Shantao started his work on this subject early in 1960, when the concerned investigations were just unfolding abroad. Liao introduced a series of basic concepts such as canonical differential equations, obstruction sets, minimal distortion set, etc., around which the research in the field were systematically advanced. He probed the stabilities and related problems and obtained many significant results.

Liao's work on the differentiable dynamical systems formed a distinctive system with unique methods, which threw much light on a series of essential problems about the global properties of ordinary differential systems, including applications of ergodicity, a proof of the C' closing lemma, and some conjectures regarding characters and stability of certain systems. His work was widely acclaimed in China and abroad.

As well as the theoretical pursuits, Chinese mathematicians have done much in applied research-undertaking various projects of practical backgrounds, popularizing mathematical methods, and introducing mathematics into other fields including information science, life science, energy science, atmospheric science and oceanography, environment and population sciences, and so on. They played important role in the development of social economy and national defense (in particular the nuclear program and the development of artificial satellites) and made valuable contributions to the modernization of China.

In summary, over the past half a century or more, mathematics in China developed unprecedentedly, and considerable progress has been made. It is no longer a hobby for only a few people, nor longer fragmentary branches transplanted from foreign countries. It is now a great endeavor in China with a huge contingent of researchers, covering almost all fields and looking forward to a brilliant future. Compared with advanced countries, however, we know that China still has a long way to go in catching up with and surpassing the most developed countries in mathematics. Chinese mathematicians must redouble the efforts to raise further our research level, to foster young talents, to enable China's mathematics to rank among the world's most advanced, and to make greater contributions to the prosperity of China and the progress of mankind.

Brief Introduction of Physics Researches in Chinese Academy of Sciences

Zhao Zhongxian

Institute of Physics, Chinese Academy of Sciences

Zhao Zhongxian, *Prof./Research Scientist. Director, Working Committee on Consultative and Evaluation of Chinese Academic of Sciences. Vice president, China Association of Science and Technology. Chairman, Academic Council of Institute of Physics, CAS. Awards: TWAS Award in Physics, for pioneering work on High Tc superconductivity, especially on* Y-Ba-Cu-O, *1986. Tan Kah-Kee (Chen Jia-geng) Prize in Material Sciences, 1988; First class of Chinese Natl. Award in Natural Sciences, (as a member of the group) 1990; Wang Dan-ping Prize, 1992; HLHL Science and Technology Progress Award, 1997. Membership, TWAS, 1987; Internatl. Acad. of Ceramics, 1989; Chinese Acad. of Sciences, 1991. Honorary Doctor of Science, Chinese Univ. of Hong Kong 1988. Honorary Fellow of India Materials Research Society, 1991.*

Abstract: This report will give the outline of the team, organizations, research examples and large-scale facilities of physics research in the Chinese Academy of Sciences. The synopsis is selected from the source materials provided by some institutes.

1. THE TEAM AND ORGANIZATION

Modern physics research in China was started approximately in 1920s. Before 1949 several Chinese physicists had made quite remarkable contributions in different fields. The independent research on physics in China in the strict sense has been developed since the founding of the People' s Republic of China.

China has a big team doing research on physics, which is distributed mainly over universities, Chinese Academy of Sciences (CAS), and some laboratories in industries. There are a few societies related to physics. The main one is Chinese Physical Society (CPS), which has nearly 40000 active members and 27 committees for the different fields of physics including physical term standards in Chinese and physics education. The CPS has 10 journals; three of them are published in English *(Chinese Physics monthly; Chinese Physics Letter monthly, Communications in Theoretical Physics monthly).* CPS also has a website for physics education and it is one of the most active societies in the world and a member of IUPAP & AAPPS.

CAS has more than 10 institutes involved in the physics research. They are:

1.1 Institute of Physics at Beijing, which is a multi-disciplinary comprehensive institution engaged in research on basic and applied physics, including condensed matter physics, optical physics, atomic and molecular physics,plasma physics and condensed matter theory.

1.2 Institute of High Energy Physics at Beijing, in which the main research areas cover high energy physics, theoretical physics, cosmic ray and high energyastrophysics, accelerator physics and technology, nuclear detectors and electronics, applications of synchrotron radiation, nuclear analysis technology and applications, free electron laser, radiation protection, applications of computer and network, etc.

1.3 Institute of Theoretical Physics at Beijing was established in 1978 with an aim of creating a center of the highest standards for fundamental research in theoretical physics. The institute encourages interdisciplinary research at the intersections of various related fields. The following fields are the research focus of the institute: theory of particle physics and quark structure of nuclei; field theory, string theory and cosmology; condensed matter and computational physics; statistical physics and theoretical biophysics; quantum physics and quantum information.

1.4 Institute of Semiconductors at Beijing focuses on the following main research fields: optoelectronics and their integrated technologies; bulk, thin film and microstructure semiconductor material science and technologies; fundamental research on low-dimensional quantum system, quantum engineering and quantum information; semiconductor artificial neuro-network and special microelectronics research.

1.5 Institute of Acoustics at Beijing was established in 1964, its main research areas are underwater acoustics and sonar technology, air-acoustics and noise control, ultrasonics and ultrasonic electronics, phonetics and speech signal processing, digital audio/video signal processing and system integration.

1.6 Shanghai Institute of Technical Physics is mainly engaged in the research field of infrared of physics and optoelectronics, infrared optoelectronic devices and material, optical thin film technology, miniature cooling technique, infrared and multi-spectral remote sensing technique, optoelectronic information processing technique and etc.

1.7 Shanghai Institute of Nuclear Research (SINR), founded in 1959, is a comprehensive institute of non-power civil nuclear technology, which mainly devotes to fundamental and applied research, and subsidiary to the hi-technique researches and its industrialization.

The research activities are concentrated on synchrotron radiation light source, beam technologies, applications of synchrotron radiation and related subjects, as well as nuclear science, nuclear techniques and its application, and inter-disciplines based on low energy accelerators. SINR has a large science project of Shanghai Synchrotron Radiation Facility, which is preparing to construct.

1.8 Shanghai Institute of Optics and Fine Mechanics, Chinese Academy of Sciences (CAS), was inaugurated in 1964. It is the earliest and one of the largest research institutes in China specializing in laser science and technology. As a multidisciplinary and comprehensive institute, it explores fundamental problems in laser science and develops laser technology and applications. The broad spectrum of activities at the institute involves high intensity optics, high power laser physics, information optics, quantum optics, laser techniques and applications, opto-electronic techniques, and crystalline and noncrystalline laser materials etc.

1.9 Institute of Plasma Physics (ASIPP) in Hefei was established in September 1978. The institute focuses on R&D of high temperature plasma physics and magnetically confined fusion science and other related advanced technologies. ASIPP is one of the most important fusion research bases in China and the Nuclear Fusion Research Center sponsored by the World Laboratory.

The main research projects are:

- HF-7U Superconducting Tokamak, a National grand project, is under design and construction, which will be completed around 2003
- HT-7 Superconducting Tokamak was constructed in 1994 and have been used to study high performance of the Tokamak under long pulse operation condition with radio frequency wave heating and microwave current drive as the main resorts
- Ion Beam Bioengineering study in ASIPP is directed at research in molecular biology, microbiology, origin of life, environment and health
- A 20 Tesla Steady-state Hybrid Magnet, the largest in China, was built together with other water-cooled, superconducting and pulse magnets
- Dye-sensitive solar cells, a new type of solar cell

1.10 Institute of Modern Physics (IMP) in Lanzhou was founded in 1957. It focuses on basic research in heavy ion physics and its related interdisciplinary sciences. The accelerator physics and technology are developed accordingly.

The main research facilities in IMP are the Heavy Ion Research Facility in Lanzhou (HIRFL), the Radioactive Ion Beam Line in Lanzhou (RIBLL) etc. The National Laboratory of Heavy Ion Accelerator in Lanzhou (NLHIAL) was established in IMP in 1991. The HIRFL Cooler Storage Ring (CSR) as one of the key national scientific projects is currently under construction and will be completed in 2005.

Main research fields are as follows:

- Nuclear physics with ion beams including radioactive ion beams in the energy from few keV to few GeV/u

- Exploration of the existing limits of nuclei
- High energy density in matter
- Highly charged heavy ion atomic physics and physics of swift heavy ion in matter
- Biology effects of the heavy ion irradiation and heavy ion therapy
- Physics and technology of accelerators including cyclotron, synchrotron, cooling storage ring, statistic accelerator and emphasizing on high current beams

1.11 Institute of Solid State Physics in Hefei was established in March 1982. Its main focus is to study

- The Design, Fabrication and Properties of Quasi-one-dimensional Nanomaterials and Assembly Systems of Nanostructures
- The physical Properties of Mesoporous Composite Systems
- The Structure and Optical Properties of Quantum Dots and Composite Films
- The mechanical spectroscopy or internal friction spectra in materials including solid materials, soft matter, and liquids
- Theoretical Condensed Matter Physics and Computational Material Science

1.12 Wuhan Institute of Physics and Mathematics (WIPM), which is a multidisciplinary comprehensive institution, engaged in basic and applied basic researches, and high-tech development.

The main research interests in WIPM include quantum physics and quantum computation, the new methods and application of liquid and solid state NMR, principles of magnetic resonance imaging (MRI) for biological and medical purpose, strong field effects of Rydberg atoms, quantum chaos, trapping and cooling of ions and atoms, physics on cold atoms, regional properties of ionosphere structure and disturbances in China, ionospheric dynamics and storm at mid and low latitude regions, nonlinear partial differential equations, and complex harmonic analysis and function space theory. Meanwhile, researches on new mechanism of atomic frequency standards, light detection and ranging (Lidar), and new methods of ultrasonic detecting and imaging are also active subjects in the institute.

1.13 Xinjiang Institute of Physics established in 1961 is a physical technical institution. Three research areas in the institute are described as follows:

• Radiation physics

The main research activities in this area are radiation effects, radiation damage mechanism, radiation hardening of semiconductor materials and MOS electronic devices, as well as energy deposition and microdosimetry of ionizing radiation in electronic materials and layers.

• Material Physics

The main research activities in this area are ceramic functional materials, sensitive materials and their characteristics

• Computer information technology

Software and information technology include management information system (MIS), computer control automation system, software and database construction, etc. These researches have made important contributions to the technical innovations of the enterprises in Xinjiang.

1.14 Changchun Institute of Optics, Fine Mechanics and Physics (CIOMP) was reorganized on the base of Changchun Institute of Optics & Fine Mechanics and Changchun Institute of Physics in July 1999.

The institute is mainly specialized in the fundamental research, applied fundamental research and engineering technology research in the field of luminescence of wide gap II-VI group semiconductor, microcavity laser, organic and inorganic films of electroluminescence. The applied optics focuses on luminescence, short wavelength optics, space optics and related fields and has achievement and innovative achievements with extensive applications which will be introduced in other papers of the book.

1.15 Institute of Metals Research (IMR) is located in Shenyang and was established in 1953.The main part is the Shengyang National Laboratory for Materials. The great contributions of this institute will be introduced in other papers of the book. Its main research projects are:

• Nano-crystalline materials technology
• Corrosion mechanism and coating protection of materials at elevated temperature
• Computer design, synthesis, and characterization of metallic functional films at the atomic scale

- The failure behavior and recovery of materials under service environment
- Materials corrosion and electrochemistry in natural environments
- High-temperature structural materials for advanced propellant systems
- Super steel for pipeline application
- Materials technology in extreme conditions
- Advanced techniques for materials synthesis and processing
- Development of energy-related new materials

2. RESEARCH EXAMPLES

2.1 Theoretical Physics

- Study on the Particle Physics (Particle Theory)

In 1958, Zhou G. Z. (Kuang-chao Chou) introduced, for the first time, the helicity amplitude for elementary particles and established the relevant mathematical method. In 1960, Chou derived in a simple and systematic way the theorem of partial conservation of axial current (PCAC) and he became one of the founders of this important theorem.

- Study on Straton Model

This work was accomplished by Peking group for theory of elementary particles, consisting of Institute of Atomic Energy, Institute of Mathematics of CAS, Peking University and University of Science & Technology of China during 1965-1966. The obtained results were presented at the 1966 Beijing International Symposium of Physics. Before a dynamical theory for hadrons was established,the theory of straton model was quite successful for explaining the structure of hadrons.

- Studies on global properties of quantum fields (Field Theory)

This work was accomplished during 1980s. The main contributions are:

A correct form of the gauge-invariant effective "anomalous" term was obtained; a deep connection between the 2n-dimensional non-Abelian anomalies, 2n+1-dimensional Chern-Simons (C-S) characteristic classes and 2n+2-dimensional Abelian anomalies was found; a global formulation for the 2n-dimensional non-Abelian anomalies was used to generalize the concepts of the secondary C-S characteristic classes and the well-known transgression formulas.

Generalized C-S characteristic classes were proposed for the first time, and the general form of the transgression formulas was obtained; based on these results the cohomology analysis of the gauge groups was carried out, and the relationship between the gauge group cohomology and Cech-de-Rham cohomology was pointed out.

This cycle of work was accomplished by Guang-zhao Zhou and his co-workers (ITP, CAS, North-Western Univ., Peking Univ., GSUST, CAS and IAM, CAS).

Applied symbolic dynamics and its application to the study of chaos in dissipative systems (Nonlinear Science)

This achievement was accomplished during the period of 1980-1990. The main contributions are:

a. The applied symbolic dynamics for the map on the interval and on the circle was systematically developed, and symbolic dynamics for two-dimensional maps by analyzing manifolds was further established, which is then applied to a detailed study of periodically driven and autonomous differential dynamical systems;

Based on rigorous qualitative method and combined with numerical technique, symbolic dynamics has become a practical tool for investigating real dynamical systems.

This contribution was made by professors Bai-lin Hao, Wei-mou Zheng and their co-workers at ITP (CAS).

- Theory of optical phonon modes in Semiconductor Superlattices (Solid State Theory)

In 1988, K. Huang and B. F. Zhu by making original use of a microscopic model compatible with the continuum model have succeeded in clarifying and resolving the serious discrepancy between the dielectric continuum model and Raman scattering results. Their work has further elucidated the interrelation between the bulk-like and the interfacial modes and demonstrated that the dispersion of bulk phonons causes a coupling between the bulk-like and interfacial modes. Moreover, they have established the optical phonon eigen modes and analytical expressions for their interaction with electrons. This theory, widely referred to as the Huang-Zhu model, is conceptually important for a proper systematic understanding of the optical modes in low-dimensional systems, which has stimulated a series of theoretical and experimental investigations on the subject and has now won wide acceptance.

- Theory of transport balance-equation in semiconductors

With proposition of a series of physical concepts, such as separation of the center-of-mass (c.m.) mechanical motion from the statistical relative motion of the electrons and selection of the optimal initial states, the nonlinear electronic transport balance-equation approach was developed and a balance-equation for describing nonlinear transport properties of hot carriers in semiconductors provides a reliable and practical technique for high-field transport calculation in three-dimensional semiconductors and has been widely used in the study of electromagnetic response of heterostructures, superlattices, quantum wires and other low-dimensional semiconductor systems. This work was accomplished by Xiao-lin Lei (SIM, CAS), in collaboration with C. S. Ting of University of Houston, USA.

• Study of liquid crystal model of biological membrane

This achievement was accomplished during the period of mid-80s to mid-90s. The main contributions are:

Using differential geometry and variation method, a general shape equation for lipid bilayer vesicles has been derived. It has been shown for the first time that under certain conditions there exists a stable torus shape cell with a ratio of the radii of the generating circles equal to square root of 2. This prediction has been confirmed by experiments reported by several laboratories. It opens a new direction in studying biological membrane surfaces of higher genus.

By presenting a tilted chiral lipid bilayers (TCBL) theory, a reasonable explanation has been given to the biological helical structures, which have been found since 1984.

The basic reason for three different shape equations for axisymmetric vesicles in the literature has been clarified. Some new exact analytical axisymmetric solutions were obtained, among which the remarkable one is the famous biconcave shape for red blood cells.

This theory was proposed and accomplished by Zhong-Can Ou-Yang (ITP, CAS) and his co-workers at Tsinghua Univ. and ITP, CAS.

• Studies on non-equilibrium, closed time-path quantum Green's functions (Statistical Physics)

The theoretical structure of this type of Green's functions was systematically analyzed, and an effective theoretical formulation was proposed, which has been applied to critical dynamics, nonlinear quantum transport, disordered systems, etc, clarifying a number of important theoretical issues and providing new results.

b. This series of works has received increasing attention of colleagues in China and overseas, and more and more physicists started to use this formalism. The principal investigators of this project were invited to present special talks at 5 international conferences. The review article on this subject has over 400 citations by other authors.

This theory is contributed by Guang-zhao Zhou (Kuang-chao Chou), Zhao-bin Su, Bai-lin Hao, Lu yu (ITP, CAS).

• Micromagnetic theory on the continuous-discontinuous change of the magnetization in ferromagnetism

Professor P. C. Pu and collaborators have established a unified theory of phase transitions for the single-parameter magnetizing process; they gave the criterion of both continuous and discontinuous phase transitions and clarified the initial behavior of continuous phase transitions.

2.2 High Energy Physics (particle physics)

In January 1979, Deng Xiaoping headed the Chinese government delegation for an official visit to the United States. During his visit, both sides signed "the Implementing Accord on Cooperation between the People's Republic of China and the United States of America in the Field of High Energy Physics" and established the PRC/US Joint Committee on High Energy Physics.

On October 16th, 1988, the first collision between electrons and positrons was realized. It symbolized that the dream of building our own research base of high-energy physics had come true.

BEPC is now the only accelerator operating in the energy region of 2 - 5 GeV in the world. Beijing Spectrometer (BES) installed on the BEPC is a large universal particle detector. In the 10-odd years after the completion of BEPC, high energy physics in China had made much headway as demonstrated in the obtainment of many important results, such as the precision measurement of τ mass, the precision measurement of R value in the energy region of 2 - 5 GeV, etc.

BEPC operates for two purposes, namely high-energy physics and synchrotron radiation application. Its synchrotron radiation facility has become a large frontier interdisciplinary research platform open to society, which has been playing a great role in developing science and technology in China. Now there are 12 beam lines and 14 experimental stations. Every year more than 100 users are provided with VUV to hard X ray for inter-disciplinary frontier study of 300-odd subjects involving life science, material science, nano science, condensed matter physics, resources and environment chemistry and chemical industry, etc.

IHEP has developed into a multi-disciplinary comprehensive research base for high energy physics, R&D of advanced accelerator technology, advanced radiation technology and radiation application. Now it has the key nuclear analysis laboratory, the key particle astrophysics laboratory, the Beijing free electron laser, the high power slow positron beam line, the Yangbajing international cosmic ray observatory, etc. It enjoys extensive international cooperation and exchange with all high energy physics laboratories and major universities in the world and has joined many important frontier high energy physics experiments.

2.3 Nuclear Physics Researches

◆ Nuclear physics researches at Institute of Modern Physics in Lanzhou

● Nuclear reaction

In 1974, the reaction system of ^{12}C+^{209}Bi has been studied systematically by using 72 Mev ^{12}C ion beam from the reformed classical cyclotron. It has been found that the massive (as many as eight nucleons) transfer reaction is of great cross section near the coulomb barrier. The variation of the cross section with incident energies shows two stages, in the lower part it increases slowly,which may correspond to pure transfer reaction, and in higher region the cross section has an exponential increase which may come from dissipative process. In 80s of 20 century, the deep inelastic scattering has been observed in light heavy ion reaction systems such as ^{16}O +^{27}Al. A concept of incomplete deep inelastic scattering has been proposed independently in the world, which says that in low energy heavy ion collisions a part of projectile transfers into target nucleus first, and then a deep inelastic scattering process between the remain of projectile and target-like nucleus happens. The evidence has been observed in experiment performed by themselves.

● Synthesis and studies of new nuclides

Synthesis of new nuclides is a very important approach to search for the existing limit of nuclei and the limit of nuclear mass.

Since 1990, synthesis and studies of the new nuclides in the heavy neutron-rich region with A >170 and near the drip line in rare earth region, where the challenge is the very low cross section and difficult to separate the new nuclides, has been started.

By 10 years effort, eight new nuclides: ^{175}Er, ^{185,186}Hf, ^{208,209}Hg, ^{237,238}Th and ^{239}Pa have been synthesized and studied at the Heavy Ion Research Facility in Lanzhou (HIRFL). Their half-lives have been measured and compared with several theoretical calculations. The longer half-lives of ^{208}Hg (T1/2=42 ± 23min) and ^{239}Pa (T1/2=106 ± 30min) means they are more active than predicted in r-process.

Along the line of A=4n +1 and Tz= − 3/2, the new nuclide ^{65}Se and the nuclide ^{69}Kr predicted to be on the proton drip line was synthesized and studied. Their decay schemes were also proposed.

Synthesis of super-heavy nuclide ^{259}Db and measurement of β -delayed fission probability for ^{230}Ac

A β -delayed fission island (βDF) was also predicted to occur in heavy neutron-rich region in the last 60's of 20 century, and pointed out that β-DF is a limitation of r-process path and so the synthesis of the heaviest element is limited by β-DF and there exists no super-heavy element in nature.

A new super-heavy nuclide ^{259}Db has been produced through bombarding ^{241}Am with ^{22}Ne ions delivered by HIRFL. The identification of this nuclide has been performed by measuring the alpha-particle emission of the mother and daughter nuclides by the detector array arranged with craft combined with He-jet transport system. ^{259}Db has a 0.51 ± 0.16 s half-life and decays by α -emission with the energy of 9.47 MeV. Its Q α value of 9.62 MeV fits well with the general trend in a "Q α vs. N-systematic" for isotopes with $Z = 105$.

- Nuclear physics research with radioactive ion beams

In the middle of the 90s of last century the first PF type radioactive beam line (RIBLL) that consists of two antisymmetric double-achromatic sections, was build up based on HIRFL and hundred radioactive ion beams were delivered by it.

It was found that ^{8}B, ^{9}C, ^{12}N, ^{23}Al, ^{27}P etc are of larger nuclear radii and proposed to be the candidate of proton halo nucleus. At the same time, it was also found that the first excited state of ^{13}C could be of a proton halo structure.

S. Hofmann emphasized the importance for the study of proton radioactivity from highly deformed ground states of extremely proton-rich nuclei along the Z=0.743N+11.6 line in rare earth region.

The coincidence measurements between β -delayed proton and the γ ray of the transition from 2+ to 0+ in the daughter nucleus by using He-jet combined with fast tape carriage system have been proposed and performed. This method increased the sensitivity by a factor of more than 50. 10 new nuclides, ^{121}Ce, ^{125}Nd, ^{128}Pm, ^{129}Sm, ^{135,137}Gd, ^{139}Tb, ^{139}Dy, ^{142}Ho, and ^{149}Yb near and on the drip line mentioned above have been synthesized and identified since then. The half-lives of all of them have been measured and the parities and spins have been assigned for most of them. The results indicate that except for ^{139}Tb, all the other new nuclides were found to be delayed proton precursors.

◆ Progress and achievements in nuclear physics research at Shanghai Institute of Nuclear Research

● In the studies of the EMC effect and relativistic many body theory, a constituent quark model in nucleus (CQMN), which was referred as "Zhu-Shen Model" in the literature, was developed to describe the quark structure of nucleon in nucleus. It is the first time in the world to predict quantitatively the behavior of broken scaling in small Bjorken light-cone variable region and sea-quark invariant in nucleus, and predict the existence of a spin-isospin zero sound phonon collective excitation mode in low energy pion absorption in nuclear matter, all of which were proved by experimental results afterward. These achievements won the second grade of Natural Science Award of CAS in 1991.

● The new nuclide ^{202}Pt was first synthesized in this institute in 1992. Together with the new nuclide syntheses by the Institute of Modern Physics, CAS, it won the second grade of the National Natural Science Award. A series of theoretical studies on the nuclear behavior under extreme conditions have been carried out. The predicted limiting temperature of hot nuclei was cited four years later by experimental nuclear physicist to explain the newest experimental data and get the best fit. This is the first time in the world to prove the convergence of BBG theory.

● A series of studies on the exotic nuclei produced in projectile fragmentation, collective effect of heavy ion reaction, equation of state and halo or skin structures of exotic nuclei have been carried out experimentally as well as theoretically. The project "Study of collective effect of heavy ion reaction, production cross sections and properties of exotic nuclei" has made the following contributions: the best empirical formula, named Shen-Fomula, has been suggested to study the total reaction cross section of stable nuclei; the nuclear transport theory has been introduced to study the total reaction cross section; it solved the problem in the widely used Glauber Model, which underestimates the experimental total reaction cross section at intermediate energy; it can be used to extract nuclear equation of state and in-medium nucleon-nucleon cross section from experimental total reaction cross section.

• Relativistic Heavy Ion Collider (RHIC) is the running facility with the highest energy in the world. It can be used to study the evolution of universe and new matter form. The STAR-China collaboration will start to study the STAR physics.

• The progress and achievements of SMCAMS

SMCAMS (Shanghai Minicyclotron Accelerator Mass Spectrometer) designed by and built in the Institute is the first minicyclotron-based AMS facility and the first cyclotron capable of accelerating negative heavy ions in the world.

The successful technological, theoretical and methodological attempts to analyze radiocarbon on SMCAMS have all represented the success in the fields of cyclotron and AMS.

SMCAMS gives itself an advantage with its good analysis precision, compactness, low cost and free radioactive contamination etc. Tens of unknown radiocarbon samples including several archaeology samples from the Mawangdui archaeology site and some ancient geology samples have been analyzed on SMCAMS with a 1%precision and an optimum ^{14}C counting rate of 25 s^{-1}. Through this limited number of dating results SMCAMS has preliminarily exhibited a good performance that can be competed with that acquired from routine measurements on tandem AMS facility. Therefore, SMCAMS has basically been able to meet the requirements arising from AMS applications in biomedicine, environment, archaeology and geosciences etc.

• The interdisciplinary research between physics and biosciences, material sciences.

The interdisciplinary research work between theory physics and molecular biology (DNA and protein), cell biology and medicine is in progress.

2.4 Progress in Magnetic Confined Fusion Research

Magnetic Confined Fusion (MCF) research started in China in the middle of the last century. In early 1970s the first Chinese tokamak CT-6 (circular cross section, iron-core transformer, R=0.45m, a=0.12m, Bt=1Tesla) was built up and operated at the

Institute of Physics, Chinese Academy of Sciences (ASIOP). In 1970s the Southwest Institute of Physics, Ministry of Nuclear Industry (SWIP, 1970) and the Institute of Plasma Physics, Chinese Academy of Sciences (ASIPP, 1978) were founded. Since then several tokamaks were built up and operated. They are HT-6B (ASIPP, circular cross section, air-core transformer, R=0.45m, a=0.12m, Bt=1Tesla), HT-1 (SWIP, circular cross section, iron-core transformer, R=1.22m, a=0.20m, Bt=3Tesla), HT-6M (ASIPP, circular cross section, air-core transformer, R=0.65m, a=0.20m, Bt=1.5Tesla), KT-5 (USTC - University of Science and Technology of China, circular cross section, air-core transformer, R=0.32m, a=0.09m, Bt=1Tesla). Physical experiments, such as effect of the externally applied magnetic field on both magneto hydrodynamic (MHD) instabilities and confinement of plasma, feature of edge turbulence in plasma, ultra-low density operation, surface heating of plasma, radio-frequency (RF) wave heating and current drive and others, have been carried out on these tokamaks. The theoretical research in this period of time is mainly on the gyro-kinetic theory, effect of the energetic particle on MHD instabilities, MHD instabilities and disruption, RF wave-plasma interaction etc. A set of computer codes for plasma physics studies in tokamak were developed and numerical simulations were conducted.

At the same time for developing MCF research, significant progress in the application of MCF technology to other research fields, industry and agriculture, was made. The nonlinear resister technology, based upon the achievements in the electric engineering and material science research, has been successfully applied to industry in both the failure protection of the power stations and the surge arresters. A 20-Tesla hybrid magnet, based upon the achievements on magnet (including the superconducting magnet) technology studies, was built up and a serial experiments on physical, chemical, material, biological and medical sciences have been carried out on it. High magnet science and technology is becoming a crossing point of many disciplines' development. The application of ion beam technology onto the agriculture and industrial microbiology got a very notable achievement. A new branch called low energy ion beam biological engineering, based upon the development in the research of the effect of low energy ions on biological species, is being formed.

Since 1990s, Chinese MCF program has been brought into a new stage. It is characterized by the re-buildup of the formal Soviet Union's T-7 (a superconducting engineering test tokamak) to HT-7 (a superconducting physically-experimental tokamak). Its aim is the research in quasi-steady state high performance plasma physics and the related material and control techniques on HT-7 (ASIPP, circular cross section, iron-core transformer, superconducting toroidal field coils, R=1.22m, a=0.28m, Bt=3Tesla). In the recent years the results on the quasi-steady state high temperature plasma experiments indicate that an essential and encouraging step towards this aim has been made. The contribution to the MCF research were also made by the experiments on HT-6M, CT-6B (ASIOP, circular cross section, iron-core transformer, R=0.45m, a=0.12m, Bt=1Tesla) and KT-5 in the mechanism of the high performance mode formation in tokamak plasma and the features of the edge turbulence in plasma; the experiment on HL-1M in plasma fueling; the experiments on CT-6B in a.c. discharge and non-inductive start-up. The theoretical as well as the computational fusion researches in this period were mainly on several aspects related to the high performance plasma physics, such as plasma transport, MHD instabilities and their control, micro-instabilities and boundary layer physics.

In the middle of 1990s two engineering projects, HT-7U and HL-2A, were launched. HL-2A (SWIP, circular cross section, iron-core transformer, R=1.64m, a=0.40m, Bt=2.8Tesla) was completed in its built-up in the December 2002. HT-7U (ASIPP, elongated D-shape, air-core transformer, both superconducting toroidal field and poloidal field coils, R=1.8-1.94m, a=0.40-0.50m, Bt=3.5Tesla, $\kappa_{separatrix}$=1.6 - 2.0,$\delta_{separatrix}$ = 0.3 - 0.7) project is one of the giant scientific engineering projects of the China's 9th "Five years National Economic Development Plan". Now it is in construction. Its completion and operation will bring China into its significance in the world MCF research program. Its forthcoming experiments should provide the database in the steady-state advanced tokamak studies.

Started from 1980s the international cooperation in MCF research in China has been developed. Since then the fruitful collaborations with IAEA (International Atomic Energy Agency), EC (European Community), Russia, Japan and United States have been carried out. They include undertaking of the research projects, personnel and equipment exchange, cooperative research, co-hosting international meetings, etc. The Fusion Research Center, World Laboratory (in ASIPP) and the Training Center in Plasma Physics, Third World Academy (in

ASIOP) made their contributions in training the young scientists from the third world countries.

2.5 Condensed Matter Physics

In the earlier years Chinese physicists in CAS had made some pioneering contribution on the rare earth permanent magnets especially for Nd-Fe-B, quasi-crystals especially about 5-fold symmetry quasicrystal in Ti-Ni-V, new non-linear optical crystal especially the BBO crystal and high temperature superconductors especially the liquid nitrogen superconducting Ba-Y-Cu-O, which had made great impact in the related field.

In the recent years some new progress has been made, for example,

- The research group on Methods of Solving Crystal Structures, Institute of Physics,in Beijing is dedicated in developing new methods for diffraction analysis from proteins to incommensurate crystals and for image processing in high resolution electron microscopy.
- The program VEC (Visual computing in Electron Crystallography) has been released since 2000. It contains unique methods and algorithms developed in the group for image deconvolution, resolution enhancement and for *ab initio* determination of incommensurate modulated structures without relying on pre-assumed structure models. Up to now the program has more than 300 academic users from 51 countries/areas.
- A direct method of breaking the phase ambiguity in protein crystallography has been proposed and extensively tested. The program OASIS (One-wavelength Anomalous Scattering and Single Isomorphous Substitution) based on the method has been incorporated in the worldwide-distributed protein-structure-solving program suite CCP4 (Collaborative Computational Project No. 4, Daresbury Synchrotron Laboratories, UK). OASIS has been applied in laboratories of China, UK, USA and France. It is proved powerful in phasing single-wavelength anomalous diffraction data from proteins. There is a page on British Royal Society's website describing the work of protein crystallography in this group under the title:

International > News and events > Science stories from China > Third world Academy of Sciences Physics Prize.

The group's web page has been included in Google's Worldwide Open Web Directory under the category:

Science > Physics > Crystallography > Laboratories. It is the 10th among the total 25 entries in the above category according to Google's ranking order.

• Prof. Li F. H . has recently obtained L'Oreal - UNESCO Awards for Women in Science (2003) For quasicrystal and crystal structure study by electron microscopy.

The main contributions she made are

a. The almost continuous transformation from quasicrystals to crystals was, for the first time, observed and interpreted so that a new research field was created. The theoretical relationship between quasicrystals and crystals was derived. A new approach to quasicrystal structure determination was proposed and applied to two quasicrystals.

b. A new technique to crystal structure determination has been set up by introducing diffraction crystallography into high-resolution electron microscopy (HREM) and applied to superconductors and related compounds. The advantages are: no preliminary structure information is needed; available to crystals weak under the electron beam irradiation and the resolution of determined structures has been enhanced from 0.2 to 0.1 nm.

c. A new technique of image processing for field emission HREM has been set up and applied to crystal defect determination for semiconductors. The resolution of determined defects has been enhanced from 0.2 to 0.14 nm.

d. A new image contrast theory in HREM was derived. It plays an important role in interpreting images and obtaining additional important experimental results. With its instruction the atoms as light as lithium was observed. It is the theoretical basis of two techniques mentioned in 2 and 3.

• Concerning with nano physics the physicists at Institute of Physics in recent years also made important contribution especially for following aspects:

a. Nanometer-scale carbon structures

Nanometer-scale carbon structures are presently among the fundamental elements for mechanical, electrical, and optoelectronic nanodevices. Well-aligned carbon nanotubes have been synthesized by using a catalytic template on the substrates, and also the very long and narrowest carbon nanotubes. A new carbon nanocones (tubular graphitic cones) with a base size of several nanometers and a tip size of several micrometers were reported very recently. In addition, nitrogen-doped carbon nanotubes show polymerized nanobell structures.

b. Large-scale magic nanocluster array

The first large arrays of metallic nanoclusters that are both well ordered and uniform in size have been obtained. The process can be more precisely controlled by using the atoms of a silicon surface as a kind of template. The precise pattern depends upon the mixture of elements from indium, manganese, silver, aluminium, gallium, and cobalt. So far more than 16 different nanocluster arrays are reported.

c. Formation and Decay of Surface-based Nanostructures

The understanding of the kinetics involved in the formation and stability of surface-based nanostructures durig expitaxy is of importance for the fast-growing area of nanotechnology. A reaction limited aggregation (RLA) theory is developed to study a counter-intuitive fractal-to-compact island shape transition induced by increasing deposition flux or decreasing growth temperature in the presence of surfactant. A generic scaling law for decay characteristics of two-dimensional islands on both isotropic and anisotropic surfaces is proposed. The stability of three-dimensional nanostructures after their creation is discussed based on the fundamental mechanisms of atomic scale mass transport on surfaces.

Of course there are more beautiful works related to this field they will be reported in other specialized papers in the book.

Infrared Optoelectronic Physics and Technique at Shanghai Institute of Technique Physics (SITP)

• The Shanghai Institute of Technique Physics (SITP), Chinese Academy of Sciences, began to make foundation research of IR detective materials and detectors in the sixties. The Ⅲ - Ⅴ group and the Ⅱ - Ⅵ group compound semiconductor materials have been investigated to develop new IR detectors. The keystone in foundation research in the SITP is the basic physical properties of HgCdTe materials. It's included optical properties, electrical properties, energy band parameter and crystal lattice librations in HgCdTe materials. The complete absorption spectrum in HgCdTe has been obtained. Furthermore, the optical constant and dielectric function spectrum in different bands have been measured.

In SLQW(super lattice quantum well) field ,the SLQW system in high pressure and very low temperature conditions has been comprehensively investigated by optical spectral measurements. The basic data about the transitions of electron state, sub-band state, and coupling between electron state and photoelectron have been obtained.

2.6 Research on the Optics at Institute of Physics, CAS

They have done important step towards understanding the generation and propagation of high-energy hot electrons in fast ignition process. In fast ignition concept, it is very critical to understand the physics of generation and propagation process of high energy hot electrons, because they are the main energy carrier for the ignition process. A group at the Institute Physics made a breakthrough in this forefront. They observed the directional emission of hot electron beam from specially tailored plasmas for the first time. They also discovered important effects of laser polarization on hot electron emission and reached a deep understanding of the physics. Their achievements mark an important step in the forefront of this field. Their results were published in well-known international journals and attracted the focus of international community.

The light localization is demonstrated in a disordered one-dimensional photonic crystal. In the study of a periodic two-dimensional photonic crystal we find that the gap of quasiperiodic photonic crystals is independent of incident direction. Furthermore, the existence of photonic band gap in amorphous photonic materials with just short-range dielectric orders is demonstrated both experimentally and theoretically. It is verified that the periodicity is not the necessary condition for photonic band gap. The three dimensional FCC photonic crystals with the ultraviolet photonic gaps are formed by self-assembly method.

Four-wave mixing spectroscopy and its application have been investigated. A method by use of Rayleigh-enhanced nondegenerate four-wave mixing was proposed and employed to study the ultra fast processes in the frequency domain. The time resolution of this method is independent of the laser pulse width. This method can be used to observe very fast relaxation processes (less then 10 fs) by use of nanosecond laser pulses.

Some new organic and inorganic photorefractive materials and their properities have been studied. Some new phenomena have been found and the mechanisms have been understood.

The nonlinear properties for newly developed crystals (such as LCBO and KBBOF) were investigated. Several all-solid-state laser sources were developed, such as diode pumped Nd:YVO_4 and Nd:YAG lasers, from which highly affected red, green and blue lights were achieved using intracavity frequency doubling with nonlinear crystals LiB_3O_5. Also, a novel all-solid-state cascaded optical parametric oscillator by quasi-phase-matched technique in periodically poled $LiNbO_3$ was fabricated. The violet to mid-infrared multiwavelengths can be obtained simultaneously, which is a wide waveband tunable source.

The optical chaos in hybrid optical bistability system has been investigated systematically. The dynamic behaviors of the system under the condition of long delay feedback and short delay feedback have been studied.

A method used for amplitude-phase-retrieval in the optical system of general linear transformation has been proposed for the first time. In this method the inverse source problems can be classified into three types: pure phase retrieval, pure amplitude retrieval, and hybrid amplitude-phase retrieval. The set of related equations and the effective iteration algorithm have been presented. This method has been successfully used to design various diffractive phase elements (DPEs) that can implement multiple optical functions simultaneously. Some DPEs have been fabricated in practice, their performances have been measured and are in good agreement with the theoretical designing results.

2.7 Atomic, Molecular and Optical Physics Research at Shanghai Institute of Optics and Fine Mechanics

- Compact ultra-short-pulse high-intensity laser

SIOM is the pioneer in China in the research field of high-power laser and large energy laser since its establishment in 1964. Since late 1980s, SIOM has made a series of achievements in the research and development of compact ultra-short-pulse high-intensity laser.

In 1991, a compact high-power ultra-short pulse Nd:glass laser with variable pulse-width (from 20 picoseconds to several nanoseconds) was developed, which was awarded by CAS in 1993. The invention of several novel techniques for the ultra-short-pulse high-power laser system was awarded the second class National Invention Prize in 1999.

In 1996, a 2.8TW/43fs ultra-short-pulse high-intensity laser system based on the scheme of chirped pulse amplification was developed, which was then upgraded to

5.4TW in 1998, 15TW in 2001, and 23TW/34fs in 2002, this laser facility is presently the highest peak power laser system in China, and ranks the world advanced level.

SIOM is also engaged in exploring the new schemes for developing ultra-short-pulse and ultra-high-intensity laser. For example, SIOM is among the first several institutes studying the optical parametric chirped pulse amplification (OPCPA) scheme in the world. In 2002, SIOM demonstrated a 16.7TW/120fs laser system based on the OPCPA scheme. This laser system is presently the highest power OPCPA laser in the world to the best of our knowledge.

• Soft-X-Ray laser and high-order harmonics physics

Since early 1980s, SIOM has been studying soft-X-ray laser physics. The population inversion in He-like Al plasmas was observed and a new inversion region was found in 1981. Amplification in soft-X-ray region was demonstrated at 10.57 nm and 15.47nm by Li-like Al ions in 1988. Thereafter eight new wavelengths of X-ray laser were obtained for the first time in the world at very low pumping laser power by using recombination schemes of Li-like and Na-like ions, and the shortest wavelength was down to 4.68nm, which is near the"water window" spectral region of great importance for significant potential application. In 1998, a new longitudinal pumping scheme was proposed to realize highly efficient transient collisional excitation X-ray laser in Ni-like Mo ions. In 1997 and 1998, very high order harmonics and the double-peak splitting in the harmonic spectra were observed in high-intensity laser irradiated argon atoms, and the direct evidence of the effect of ionization on the intensity of high-order harmonics generated in neon atoms was found. Some of the achievements were awarded the second class National Natural Science Prize in 2001.

• Laser plasma interaction and laser fusion

Laser plasma interaction and laser fusion research have been carried out since mid-1960s in SIOM. The early work in 1960s and 1970s, such as the generation of high temperature high density plasma (1972), the realization of neutron emission (1973) and pellet compression (1977), the development of computer simulation codes (1976) and the establishment of a six-beam nanosecond high power laser system (1976) etc, led laser fusion research in China into the advanced level of the world.

The experimental and theoretical study of laser plasma interaction and laser fusion was then extended to the frontier of high power laser interaction with matter, in particular with the plasmas using the home-made high power laser systems. Creative and systematic achievements on the study of nonlinearity and instability processes in laser-plasma interaction are impressive, such as the first observation of the correlation of fast ion emission with the inhibition of electron thermal conduction, the first systematic clarification of the effect of parametric decay instability on second harmonic generation, the discovery of new mechanism of the self-generated small size magnetic field driven filamentation instability, the first discovery of large scale jet structures in the line-focused laser plasma and their dynamic process etc. Some of the achievements were awarded the second class National Natural Science Prize (1995).

● High-field ultra-fast laser physics

Since late 1980s, in particular with the advent of high-intensity femtosecond laser in SIOM in mid-1990s, both theoretical and experimental progress in the field of high-field ultra-fast laser physics has been achieved. SIOM is the core institute in the research of ultra-short ultra-intense laser science in China and is now responsible for the national key basic research project of some problems in the frontiers of ultra-short ultra-intense laser science. Significant progress has been made in the study of ultra-short ultra-intense laser interactions with electrons, atoms, molecules, clusters and plasmas.

Original achievements are evidenced by a large number of impressive publications in well-known international journals. For example, a series of important results were achieved in the forefront of laser matter interaction at relativistic intensity. The acceleration of electrons in the low-density plasma in front of a solid target by a propagating short relativistic laser pulse was studied for the first time and new acceleration scheme for electrons and ions was found in 1999. A novel method was proposed in 2002 to enhance the laser field by two orders of magnitude between two thin foils driven by counter-propagating short relativistic laser pulse. A new scheme to generate high-intensity correlated atoms beam was proposed in 2001, and a way to control coherent distant atoms by polarization interference effect was put forward recently.

The research on sub-femtosecond and attosecond extremely ultra-fast phenomena was also pursued. The role of the absolute phase of the driving laser in the generation of attosecond high-order harmonic in atoms was investigated. A timing jitter of the order of tens of attoseconds between the attosecond pulse and the driving laser field was found. A novel method for in-situ detection of the carrier-envelope phase of the few-cycle short-pulse laser was proposed.

Time-resolved diagnosis of the laser absorption and electron density of the laser-irradiated atomic clusters were performed with a chirped-pulse spectral interferometry, Energetic ions with the maximum energy up to 1.3 MeV generated in the explosion of large Xe clusters were measured. A modified hydrodynamic expansion model for studying cluster explosion was developed by taking the effective dielectric constant into account.

● Quantum optics

The research on laser cooling and trapping of atoms has been carried out since 1979. The low temperature of 60μK was achieved in laser cooled Na atom in 1988. The theoretical and experimental research on Bose-Einstein Condensation (BEC) system was pursued since 1999. The realization of BEC in dilute rubidium gas was observed in early 2002. The BEC was achieved in a quadrupole and Ioffe configuration trap. The number of condensed atoms is around 4 × 104 in total 5 × 105, and the transition temperature is 250nK. Further work on atomic clock and quantum information technology by using cooled atoms is on the way.

● Shenguang-II laser facility

Shenguang-II laser facility (SG-II) is the largest high power laser fusion system built up in China on the basis of the original Shenguang-I laser facility (SG-I), with key technological innovation and integration. It is ranked among the few giant laser systems in the world. Its operation symbolizes that a new period for laser fusion research is coming. SG-II has provided an unreplaceable experimental means for the researches of laser and target interaction, such as inertial confinement fusion, X-ray laser, high accuracy measurement of materials at extreme high pressure.

The main parameters of SG-II are as follows:

* Beam number: 8
* Beam size: φ 240mm/beam
* Total laser power: 8 × 1012W

* Laser energy and pulse width: 800J/1ns, 100J/100ps for each beam
* Beam quality: 3.5 × diffraction limit

2.8 Atomic and Molecular Physics Research at Wuhan Institute of Physics and Mathematics

The research work on the frequency standards, the Nuclear Magnetic Resonance and ionospheric physics began in China at early 1960s. Under the direction of Professor Wang Tian-juan, Chinese physicists had successfully set up many kinds of atomic frequency standards, such as, NH3 maser, Rb maser, Rb clock, active and passive H clocks.

In recent years, much important progress has been achieved. For example. In recent years, study on highly stable frequency standard for space is under the way. The intracavity microwave frequency doubling has been gotten and a new patent microwave resonator had been designed. Thus the size and Q value of this type of Rb frequency standard has closely reached to the top in the world.

In NMR works, Subharmonic Resonance Phenomena in the pulsed FT-NMR was first reported . Orientation dependence of longitudinal relaxation time in solid was discovered. New methods for fast magic angle setting and for direct measurement of multiple-quantum relaxation time were proposed. The basic physical properties of Radiation Damping Effect in NMR experiments were revealed. A Dynamic Nuclear Polarization (DNP) spectrometer was constructed with the highest frequency and the largest microwave power in the world so far. 129Xe NMR signal of dilute gas with enhancement of polarization-transfer mechanism on a high-filed NMR instrument was first observed and successfully applied it to Magnetic Resonance Imaging (MRI).

Many new methods (B-spline method. potential model method and modified FCPC method) have been developed and used successfully to calculate the accurate energies and oscillator strengths of atoms in strong magnetic or electric fields. Under the collaboration with the the Imperial College of London and Queen's University of Belfast, the diamagnetic Stark spectra of Ba and Sr in magnetic fields of 2.89T and 2.48T were measured respectively, and the electric field strength of 0- −250V/cm. The two polarization separated spectral components are found to differ obviously, which has been identified to be from the quantum defects by comparing with a successful theoretical calculation.

In Quantum information, a new method was proposed that gives multi-order transformation and entanglement swapping based on the couplings of only adjacent qubits . Modified 7-qubit (the highest qubit number realizable in the world) D-J quantum algorithm has been experimentally demonstrated using NMR method. And we realized the 7-order coupling transformation for the first time. Moreover, with the NMR technique we demonstrated many multi-qubit quantum algorithms.

Now, Laser Cooling and Trapping of Neutral Atoms and ions are carried out. The unique feature of the second stability region of a Paul trap and the magnetron-motion-free mode in a combined trap and the origin of the correlated motion and nonlinearity of ions in the Paul trap have been revealed. Experiments of cluster and multi-charged ions were successfully conducted. The Doppler cooling and laser trapping of Rb was successfully realized in 1998 . Multi-dark resonances and remarkable reduction of light speed have been observed in these experiments. A new all-quantum analytical method to solve Bose-Einstein Condensation system was also developed .The research works on the frequency standards and its application under the direction of Professor Wang Tian-juan began in our institute in early 1960s. For 40 years, we successfully set up many kinds of atomic frequency standards, such as, NH_3 maser, Rb maser, Rb clock, active and passive H clocks.

3. LARGE-SCALE FACILITIES

Physics Research needs some large-scale facilities. The government has paid more attention to those facilities

3.1 Beijing Electron-Positron Collider (BEPC)

Besides BEPC, China have an SRS in Hefei (China University of Science and Technology). The energy of the electron beam is 800MeV. Now it has been upgraded with more beam lines. It has been used for physics, chemistry, life science and micro electronics.

There are plans to build a new SRS in Pudong of Shanghai-SSRF (Shanghai Synchrotron Radiation Facility)

Energy: 3.5 Gev

Average Current: 200-300mA

Number of Beam line: 7 Beam lines

Circumference: 396 meter

The State Science and Technology Leading Group has approved in principle the report submitted by CAS on the developmental target of high energy physics and advanced accelerator in China and gave green light to the upgrade project of BEPC, known as BEPC- II. It is expected that this project will be put in operation at the end of 2006. Once completed, the luminosity of the accelerator will be increased by a factor of two and the performance of the detector improved by a big margin. Thus our country will continue to have a high-energy physics experimental facility with the best performance in this energy region in the world. This will lay a foundation for China to maintain its leading position in the study of τ charm physics in the world and achieve important physics results for a considerably long period of time. China's high-energy physics faces another important historical opportunity of great development. And it will surely move towards an even more magnificent era of development.

- Heavy Ion Research Facility (HIRFL)

It is composed of a powerful 14.5GHz ECR ion source, an injector which is a sector focused cyclotron with K=69, a main one which is a separate sector cyclotron (4 sectors) with K=450, a radioactive ion beam line (RIBLL) and several experimental devices. All of them, except the injector, were designed and constructed by China. The accelerator can deliver heavy ion beams with energies from 100MeV/u for 12C to 15MeV/u for 129Xe and several hundred species of radioactive ion beams. The radioactive ion beam line, which consists of two antisymmetric double-achromatic sections, has been completed based on HIRFL. It is not only able to separate the products reaction induced by primary beam with high resolution, but also has a function of zero-degree spectrometer. The facility put a strong support to our nuclear physics researches including nuclear reaction, nuclear structure and synthesis and study of new nuclides, and the applications of nuclear physics to other research field.

- Cooling Storage Ring (CSR)

Scientific Purpose

Nuclear physics with Ion Beams including Radioactive Ion Beams

Exploration of super-heavy elements

High Energy Density in Matter

Highly charged heavy ion atomic physics and physics of swift heavy ion in matter;

Biology effects of the heavy ion irradiation and heavy ion therapy;

Physics and technology of accelerators including synchrotron, cooling storage ring and emphasizing on high current beams.

In order to meet the requirements of nuclear researches and related applications to other fields, the heavy ion Cooler Storage Ring (CSR) in Lanzhou has been under-constructed since 2000. CSR consists of two rings, the main ring for accumulating, accelerating beam, and the experimental ring for experiments with cooled beams. The ions of 12C to 238U can be accelerated in CSR to 900MeV/u to 420MeV/u, respectively, with as good as 1π mm mrad of emittance and 10-5 of momentum resolution. After completion of the project, the nuclear researches in China will possess a good base.

Characterization

- CSRm(accumulating, cooling, accelerating) Circumference: 161.20m
- Maximum energy: 2350MeV(p), 900MeV/u (12C6+), 420MeV/u(238U72)
- Max. magnetic rigidity: 10.64T-m
- CSRe(high sensitive and high accuracy spectrometer) Circumference: 128.9m
- Maximum energy: 2000MeV(p), 620MeV/u (12C6+), 400MeV/u (238U72)
- Max. magnetic rigidity: 8.4T-m

3.2 HT-7 and HT-7U Tokamak in Hefei

Institute of Plasma Physics

HT-7

- Major Radius R=1.22m
- Minor Radius r=0.30m Toroidal Field BT=2.5T
- Plasma Current I p =100-250kA Discharge Duration DT =10s

HT-7U Superconducting Tokamak

HT-7U Superconducting Tokamak, a non-circular advanced steady-state plasma experimental device, was rectified by the Leading Team of Science and Technology, the State Council on June 3, 1997 as a National Mega-Project of Science Research. Its scientific mission is to realize stable operation and carry out experiments on heating and confinement improvement under advanced operation of the tokamak and finally realize the tokamak advanced operational mode.

HT-7U will have a long pulse (60-1000s) capability, a flexible PF system, and auxiliary heating and current drive systems, and will be able to accommodate divertor heat loads that make it an attractive test facility for the development of advanced tokamak operating modes. HT-7U consists of nine sub-systems, namely superconducting tokamak device, cryogenic and refrigerator system, power supply system, control and data acquisition system, lower hybrid current drive (LHCD) system, ion cyclotron resonance heating (ICRH) system, vacuum pumping and gas puffing system, diagnostic system, and water cooling system. The main parameters are:

Table 1 Main Parameters of HT-7U Device

Toroidal Field, Bo	3.5 T
Plasma Current, IP	0.5 MA
Major Radius, Ro	1.7 m
Minor Radius, a	0.4 m
Aspect Ratio, R/a	4.25
Elongation, Kx	1.6 - 2
Triangularity, d x	0.6 - 0.8
Heating and Driving:	
ICRH	3 MW
LHCD	4 MW
ECRH	0.5 MW
NBI	Non(maybe will has)
Pulse length	1-1000 s
Configuration	Double-null divertor Pump limiter Single null divertor

At the beginning of 2003 China proposed for being one of possible partner countries to participate in the International Thermonuclear Experimental Reactor (ITER) project. This may be an opportunity for Chinese MCF research to join the international cooperation,which will deeply influence the development in the Chinese MCF research and the related engineering and technology.

Acknowledgements: For drafting this report, some institutes of Chinese academy of sciences have provided related materials. These include the Institute of Physics, Institute of Theoretical Physics, Institute of High Energy Physics, Institute of Semiconductor, Changchun Institute of Optics, Fine Mechanics and Physics, Shanghai Institute of Technical Physics, Shanghai Institute of Nuclear Research,

Institute of Plasma Physics, Wuhan Institute of Physics and Mathematics, Institute of Modern Physics, Shanghai Institute of Optics and Fine Mechanics.

Prof. Zhou Guang Zhao's report at IUPAP is the frame of the draft. Some colleagues in the Institute of Physics, in International Cooperation Bureau of CAS have done a lot of work for the report. Here the draftsman would like to thank all of them for their help.

Looking Back at a Lifetime of Original Research

Huang Kun

State Key Laboratory for Superlattices and Microstructures, Institute of Semiconductors, Chinese Academy of Sciences

***Huang Kun**, male, born on 2nd September 1919 in Beijing, graduated at Yenching University in 1941. 1945-1948, graduate student at the University of Bristol, PhD in 1948. 1949-1951 post-doctorate research work at the University of Liverpool. 1951-1977 a professor of physics at Peking University. From 1977-1983 was the Director of the Institute of Semiconductors, Chinese Academy of Sciences; since 1983 has been Honorary Director of the Institute. Was elected a member of the Chinese Academy of Sciences in 1955, was elected a foreign member of the Royal Swedish Academy of Sciences in 1980, was elected a fellow of the Third World Academy of Sciences in 1985. Huang Kun theoretically predicted, in the late forties, diffuse X-ray scattering associated with the impurities in crystal lattices, which was experimentally confirmed in the sixties, and later named "Huang Scattering", and has already developed into an effective and direct method for studying micro-defects in solids. His multiphonon transition theory through its "Huang-Rhys factor" has become widely known. With a pair of equations, proposed by him, relating optical displacement, macroscopic electric field and electric polarization (Huang Equations) he was led to discover for the first time coupled vibratory modes between optical vibration and the electromagnetic field which has come to be called*

"polariton". He is widely known for his collaboration with Max Born in writing the monograph "Dynamical Theory of Crystal Lattices".

Abstract: This paper introduces several important pioneering contributions of Kun Huang in solid state physics, including: 1. X-ray scattering of dilute solid solutions; 2. the writing of "Dynamical Theory of Crystal Lattices" with Max Born; 3. Huang equations and polariton; 4. theory of multiphonon transitions; 5. nonradiative multiphonon transitions; 6. research on semiconductor superlattices and the Huang-Zhu model.

In the course of my research career I have made some notable contributions to be described below:

1. X-RAY SCATTERING OF DILUTE SOLID SOLUTIONS

X-rays incident upon matter will be scattered by the atomic electrons. As X-ray wave lengths are close to the lattice constants of solid matter, they are widely used in the analysis of material structures. In the case of perfect crystals when the path difference of X-rays scattered by neighboring atomic planes differs by an integral number of X-ray wavelengths, they show sharp diffraction peaks or, expressed alternatively, when incident and scattered rays differ in their wave number by the reciprocal lattice vector, the diffraction shows a peak. In the case of real metals, the atoms deviate randomly from their strictly periodic positions. The reason might be the thermal motion of the lattice atoms. This produces additional X-ray scattering around the Bragg peaks known as X-ray diffuse scattering. Another source of diffuse scattering could be foreign impurities or crystalline defects. I took up the problem of diffuse scattering due to foreign atoms in dilute solid solution by introducing a simple effective theoretical model for this study with the following assumptions[1]:

The impurity atoms have a different volume from the lattice atoms, but all atoms scatter X-rays equally.

The randomly distributed impurity atoms are considered to be independent isolated deformation centres so that their scattering contributions add linearly.

The elastic displacements due to each centre are taken to be isotropic and inversely proportional to the square of the radial distances.

With the above approximate model, the theoretical calculation of the resulting X-ray scattering proved to be straight-forward. The main results can be summarized as follows:

The presence of the foreign atoms in the alloy lattice causes a reduction in the intensity of the Bragg peaks by a factor depending on the square of the reciprocal lattice vector and the concentration of impurity atoms.

Around the Bragg peaks, the impurity induces diffuse scattering with a specific intensity distribution. In directions perpendicular to the reciprocal lattice vector, the diffuse scattering has zero intensity, in the direction parallel to the reciprocal lattice vector the intensity of the diffuse scattering falls away rapidly with the Bragg angle.

The impurity induced diffuse scattering appears similar in form to the thermal diffuse scattering. However, their microscopic mechanisms and quantitative values are very different. At room temperatures the impurity induced diffuse scattering is concealed by the stronger thermal diffuse scattering.

It was 20 years after my theoretical prediction of X-ray diffuse scattering that the German physicists Peisl and Spalt, in their experimental investigation on X-ray scattering from γ radiation on LiF, accidentally observed the diffuse scattering caused by the long range elastic field of point defects, thereby confirming the theoretically predicted diffuse scattering[2]. Subsequently, many scientists within China and abroad observed this diffuse scattering using, for example, fast neutron irradiated copper crystals and electron irradiated aluminum crystals. In order to distinguish from thermal diffuse scattering this scattering is called Huang's Diffuse Scattering (HDS) or just Huang scattering.

In 1972 Trinkaus, a German physicist, worked out a typical Huang scattering formula for point defects such as interstitial atoms and holes. For isotropic conditions the distribution of Huang scattering in reciprocal space forms two families of spheres tangential at the reciprocal lattice point; they are usually referred to as Huang spheres. Trinkaus and Dederichs carried out averaging over the diffuse scattering from various anisotropic elastic fields. Their analysis showed that using the position and shape of the zero intensity surface (line) and equal intensity surface (line) of Huang scattering one could deduce the structure of the point defects, namely, their type, symmetry and size. Thus by measuring the intensity distribution of X-ray scattering in reciprocal space the concentration and structure of crystal defects can be determined[3]. Huang scattering is now an effective means for directly studying defects in crystals.

2. WRITING "DYNAMICAL THEORY OF CRYSTAL LATTICES" WITH MAX BORN

Small vibrations of the atoms around their equilibrium sites in crystalline solids are the basic form of their motion. The great scientist A. Einstein was the first to realize the importance of this type of motion and initiated this theoretical research. Around 1910 Born, von Karman and Debye put forward the concept of lattice waves. Thereafter, Born took up systematic research work on the dynamical theory of crystal lattices and the progressive development of the subject had been mainly due to Born himself or his students and colleagues. So when I first went to see him he was the undisputed authority. I had already read a little book he had written, early in 1915 in German, and as at that time, not many physicists in England could read German, Born was apparently, pleased to find a young scientist who had already read his early work and had a basic knowledge of lattice dynamics. There-upon he gave me an unfinished manuscript on lattice dynamics to read. This unfinished manuscript was what he had already written of a new book on the dynamical theory of crystal lattices. In this book he planned to derive, completely deductively, on the basis of quantum mechanics, a general dynamical theory of crystal lattices. He suggested I collaborate with him to complete writing the new book, and I agreed.

But I did not completely agree with Born's idea of writing the book solely by deduction from general quantum mechanic principles, so I suggested including an elementary part which would gradually lead to the "general theory" based on deduction from quantum mechanics. Born apparently considered only the general theory part of interest. However, as I rather insisted on my idea, he finally agreed. Thus the finished book[4] consists of two parts, the elementary theory and the general theory. The three chapters on elementary theories are: *Atomic Forces, Lattice Vibrations*, and *Elasticity and Stability* which includes all the basic theories and experimental applications of crystal lattices. The four chapters on general theories are: *Quantum Mechanical Foundation, The Method of Long Waves, The Free Energy*, and *The Optical Effects*. The general theories lead to more complicated theories and experimental work including theoretical predictions of some new light phenomenon such as the fine structure of the infra-red absorption in crystals (residual lines), and the Raman effect. In the sixties, after the discovery of laser, these theoretical results were all experimentally confirmed. The wide use of this book is due to the elementary theories while the general theories are very important for resolving important physics problems because the theories represent a deeper level of truth.

3. HUANG EQUATIONS AND "POLARITON"

Periodicity is basic to crystal lattices and the smallest unit of periodicity is known as an elementary cell. The structures of all elementary cells are similar. Once we know the vibratory mode of a single cell, that of another differs only by a phase factor. Lattice vibrations can be seen as wave propagation with a phase factor. Suppose the elementary cell contains two atoms each with 3 degrees of freedom. In such a case, the lattice vibrations take the form of 3 branches of acoustic waves and 3 branches of optical waves. In the limit of long lattice waves (i.e. with wave lengths much larger than the lattice constant), in an acoustic wave the two atoms in the unit cell move in unison as a rigid unit, while in an optical wave the two atoms have opposite phase thus keeping their centre of mass at rest.

The quantization of lattice waves is known as a phonon, so we can also say for a crystal with 2 atoms in an elementary cell that its vibrational spectra are constituted from 3 branches of acoustic phonons and 3 branches of optical phonons. For polar crystals (also known as ionic crystals) the 2 atoms are charged differently, being positive and negative ions. For example, in common crystals such as gallium arsenide (GaAs) the gallium ions are positively charged and the arsenic ions are negatively charged. In optical vibrations the positive and negative ions will be accompanied by a polarizing electric field. The macroscopic electric field thus produced will affect the frequency of the optical mode of vibration giving rise to longitudinal optical vibrations (the direction of vibration is parallel to the vibrating wave) and transverse optical vibrations (the direction of vibration is perpendicular to the vibrating wave) with different frequencies. In investigating the interaction between polar crystals and infra-red light and the interaction between electrons and the lattice vibrations, the long wave optical vibrations are of particular importance. The deep going research carried out between 1930 and 1940 shows that the microscopic mechanism of these polarizing effects are especially complex. The polarization of the ions includes not only polarization of the electron cloud; it is also accompanied by deformation of the cloud. The hard ion model adopted, at that time, in the investigation of the optical vibrational modes was totally inadequate to account for the polarization effect. Moreover, the treatment of the effect of the long range Coulomb force on the dynamical processes by way of a microscopic model is by itself a complex problem. In actual fact, the lack of proper understanding of the effect of long range Coulomb force led to some erroneous conclusions in the earlier research. To overcome these difficulties I was led to consider a phenomenological approach.

On the one hand the dynamics of the optical vibrations includes the long range Coulomb force and the restoring force in the vibration depends not only on the displacement but also on the macroscopic electric field. On the other hand, one must realize that the polarization which determines the electric field also depends on both the optical displacement and the macroscopic field. Thus in 1950, based on the above two points, I put forward a pair of phenomenological equations[5] to resolve the problem of optical vibrations in polar crystals. I introduced the macroscopic field as a new variable to designate the effect of the Coulomb field on the optical vibrations of ions, and the macroscopic electric field and the displacement of the ions both contributed to the electric polarization. These pair of equations gives the relation between the optical displacement W (directly proportional to the displacement between the positive and negative ions), the macroscopic electric field E and the polarization P as follows:

$$-\omega^2 W = b_{11} W + b_{12} E,$$

$$P = b_{21} W + b_{22} E.$$

These equations have 4 coefficients, but according to the conservation of energy it can be proved that b_{12} equals b_{21}. So effectively, there are three independent parameters b_{11}, b_{12}, and b_{22} which can be independently determined from the experimental values of $\varepsilon(0)$, and $\varepsilon(\infty)$ that is; these phenomenological equations are completely determinable. These equations became known as "Huang Equations".

The most natural and easiest application of the Huang equations is to obtain the plane wave solution of the equations in conjunction with the equation of electrostatics (the use of electrostatics is tantamount to the use of a Coulomb force). The solutions are found to give the frequency of the transverse and longitudinal waves and the ratio of these frequencies agrees exactly with the Lyddane-Sachs-Teller equation:

$$\omega^2_{LO}/\omega^2_{TO}=\varepsilon(0)/\varepsilon(\infty).$$

This agreement strongly proved the validity of the two phenomenological equations.

At that time I read a well-known paper showing the microscopic calculations of the lattice vibrations to be divergent. The author conjectured this might be because he had not considered the retardation effect of the electromagnetic waves in his calculations. I knew this to be wrong, but this very fact led me to be curious about how the solution would come out if the radiative retardation were properly taken into account. Then I noted this would be another ideal problem for applying the two phenomenological equations. Moreover, to solve this problem just means combining the equations with all the Maxwell equations instead of just the equations of electrostatics. Thus I obtained solutions of great interest[6]. They are not to be interpreted like the passage of conventional electromagnetic waves. The result introduced a mode of motion involving the electromagnetic waves and the lattice phonons of polar crystals which shows many new features, and has come to be known as a polariton.

Phonon polarities were first observed, by C. H. Henry and J. J. Hopfield in 1965[7] in gallium phosphate (GaP) using Raman scattering. At present polarities, as a basic form of motion, are studied in many fields of solid state optics, and been widely made use of.

4. THEORY OF MULTIPHONON TRANSITIONS

This multiphonon transition theory has general importance in the investigation of the transition processes in localized electronic states. But this discovery came from investigating a specific problem: the spectrum of colour F-centres in crystal lattices.

An F-centre is just the vacancy site left over by a missing negative ion; this is the most intensively studied crystal defect in ionic lattices. Thus an F-centre plays the role of a positive ion which can bind an electron as a localized state. An optical spectrum is a very important means for the study of F-centres. As F-centres are localized electronic states that can bind an electron, an electron can absorb a photon and jump to a higher electronic state, which can be observed through an absorption spectrum. In an emission spectrum an excited electron can release a photon and return a lower state. Thus the optical spectrum is a very basic means for the investigation of F-centres. In general, a bound electron in a localized state making a transition to another electronic state corresponds to definite transition energy and is observed as a line spectrum that is a spectrum with a definite frequency. However, this is not the case with F-centres, here the observed spectrum is of considerable width. Since in molecular physics a wide band spectrum is commonly observed, people generally know that this is due to the thermal motion of the ions. These wide bands are equivalent to the energy of tens of phonons and if treated by the perturbation theory would involve perturbation of incredibly high orders, thus it was thought practically impossible to achieve a quantitative theory for the process at that time. Traditionally, the optical transition in solids is determined by the dipole matrix element between the initial and final electronic states, and the contribution to this process due to phonons comes from the perturbed electronic state through the electron-phonon interaction. In the first order perturbation the phonon number can only change by one i.e. increase or decrease by one phonon. For the case of a change of 2 phonons the order of perturbation has to be raised by one order i.e. second order perturbation. Thus for a spectrum with tens of phonons the perturbation treatment would have to be carried to an unthinkably high order, so a different approach apparently, had to be found.

As the writing of the book with Max Born proceeded I thought more about the problem of atom vibrations and began to note that the usual perturbation treatment when applied to the F-centre problem is misleading. The appropriate scheme for dealing with such a case is the well-known adiabatic approximation. With the scheme of adiabatic approximation when an electronic state makes a transition to another state the equilibrium positions of the nearby ions are more or less shifted. This effect is known as lattice relaxation and is a basic characteristic leading to many important effects.

With the above understanding of the lattice relaxation effect I saw that this effect could play a decisive role in causing the apparent anomalous behavior observed in F-centre spectra. Thus if there were no displacement of the equilibrium positions of the ions, or there were no interaction between the electrons and phonons there would not be any change in the number of phonons during an electron transition (orthogonally of the phonon wave function). However, the presence of relaxation can cause a change in the equilibrium position of the surrounding atoms breaking down the orthogonality in the vibrational wave function. The larger the displacement of the atoms and as the coupling between the localized states and the neighboring lattice ions increases, the greater the change in the number of the phonons. This understanding led to the formulation of a systematic theory of multiphonon transitions[8] based on quantum mechanics as follows:

(1) Within the framework of the adiabatic approximation and the Condon approximation, i.e., the dipole transition matrix element between two electronic wave functions is independent of the lattice coordinate, the intensity of a multiphonon optical transition is proportional to the square of the overlap integral between the relaxed vibrational wave functions of the two states that are involved in the transition.

(2) Under the harmonic approximation for the dispersionless optical phonons in linear electron-phonon interactions, at low temperature, a p-phonon transition event evolved in the overlap integral consists of p transitions of emitting one phonon, thus the p-phonon transition probability is proportional to

$$|M|^2 \exp(-S)\, S^p/p!,$$

where M is the matrix element of the dipole electronic transition, and the parameter S, which is proportional to the square of the shift of the equilibrium positions of the ions, represents the strength of the lattice relaxation and is now widely referred to as the Huang-Rhys factor. Hence this theory predicted that at low temperatures the line spectrum is replaced by a series of phonon-bands, which exhibits a Poisson distribution with a peak associated to S-phonon transitions. This prediction was verified by J. J. Hopfield in an experiment on the emission spectrum of CdS in 1959.

(3) At finite temperature, a p-phonon transition event consists of p+q transitions of emitting one phonon as well as q transitions of absorbing one phonon, thus the p-phonon transition probability is proportional to

$$\exp\left[-S(2n+1)\right]\left[(n+1)/n\right]^{p/2}I_p\left(2S\sqrt{n(n+1)}\right),$$

where I_p is the Bessel function with imaginary argument, and n is the average phonon occupation number at that temperature. This theoretical result agrees well with the experimental data for F-centre absorption curves in KBr at various temperatures.

Afterwards Kubo-Toyozawa and Lax developed a more general theory on general phonon distribution. In 1959 Hopfield[9] was the first to verify, in the bandedge emission from cadmium sulphide (CdS), the theoretically predicted multiphonon spectrum. By the sixties, solid state optical spectroscopy had received very important development including the widely observed multiphonon transitions.

5. NON RADIATIVE MULTIPHONON TRANSITIONS

Both in the papers by S. I. Pekar and that by A. Rhys and me on multiphonon transitions, similar systematic theories on the optical transitions are developed. But in our paper a theory on non radiative multiphonon transitions is given, based on the mechanism that the energy change involved in electronic transitions is compensated by phonon emission and absorption. Theories of non radiative transitions are closely related to important practical problems such as efficiency of light emission centres and the dynamics of deep-level centres; but, compared with radiative transitions, have met with more difficulties in their theoretical development. This is due partly to the fact that experimental check on theory relies more on the numerically calculated results of transition probabilities, and it happens that such calculations in the case of nonradiative transitions are particularly complex and difficult. As regards the basic confusions developed in the numerical calculations, different approximations led to conflicting schemes of theoretical calculations. Under such conditions we have been able to analyse and clarify the origin of their differences, thereby unifying the scheme of calculation. This clarification has been followed by more theoretical work on the efficiency of light emission centres and the development of deep-level centres, and furthermore, has caused attention to the theory of crystal relaxation and multiphonon transitions.

6. RESEARCH ON SEMICONDUCTOR SUPERLATTICES AND THE HUANG-ZHU MODEL

Below is an account of more recent research work which centres around the physics of semiconductor superlattices. What we mean by superlattices are periodic structures formed of alternating thin layers of semiconductors with different energy bands. Superlattices led to a new field of research in the physics of semiconductors.

The electronic structures serve as the basis for the theoretical investigation of solids. At the time when I and my colleagues started our research on semiconductor superlattices, the structure of the complicated hole sub-band and the quantum well exciton represented two of the most competitive research subjects. To catch up with this development Tang Hui and I[10] achieved our purpose by successfully developing a method of plane wave expansion within the framework of Luttinger-Kohn effective mass Hamiltonian.

Excitons are the basis of optical processes within quantum wells. In 1985 I particularly noted the importance of the effect of the two dimensional nature of quantum well excitons on optical processes within quantum wells. Taking proper account of the complicated whole structure Zhu and I worked on the wave function of the quasi two dimensional exciton in terms of the four component spinor[11]. We found that each of the four components of the spinor represents a corresponding angular momentum and accordingly deduced the proper selection rules in the exciton transition[12].

Lattice vibrations are one of the basic physical processes underlying solid state physics. Optical vibrations in quasi two-dimensional quantum structures are usually divided into bulk like modes and interface modes. Such classification and the mode characteristics are all basically derived on the basis of the dielectric continuum model. Since 1965, invariably in use for LO bulk like modes in quasi two-dimensional quantum structures are the macroscopically determined solutions for a dielectric slab. Thus follows the electrostatic boundary conditions, i.e., the electrostatic potential is equal to zero at the interface. If the width of the quantum well is d, and the origin set in the middle of the quantum well, then the electrostatic potential Φ is given by:

$$\Phi(z)=\cos(m\pi z/d), \quad m=1,3,5,\cdots$$

$$\Phi(z)=\sin(m\pi z/d), \quad m=2,4,6,\cdots$$

Since the optical displacement is the derivative of the electrostatic potential and the slab model is taken to be a standing wave, the optical displacement at the interface is a maximum.

In 1985 a small group under Cardona[13] at the Max Planck institute made an unexpected discovery in their experiments on Raman scattering of short period superlattices. They found that the polarization of various orders of bulk like modes is just the reverse to what was predicted on the basis of the dielectric continuum model. This seemingly simple fact led to various attempts for its explanation, but it seemed to defy a satisfactory solution. Then in 1987, we saw that if we wanted to make any further progress, it was absolutely necessary to take account of the microscopic nature in any model they employ. It just happened that I had, in the fifties, for an altogether different purpose, introduced a model consisting of a lattice of dipole oscillators to simulate a polar crystal. On this basis, we introduced a clear and simple microscopic dipole oscillator superlattice model which is yet completely compatible with the macroscopic dielectric continuum model. Accordingly, they were able to clarify the difficulties in previous various conflicting models and led to a unified systematic theory which has been named "Huang-Zhu Model ". The main points are[14, 15]:

The optical displacement and the electrostatic potential at the interface are both zero, which is different from the continuum dielectric model and the guide model where the optical displacement and the electrostatic potential at the interface are out of phase.

The interface mode and the bulklike mode are interconnected. The interface modes are anisotropic. When the phonon wave vector is parallel to the growth direction of the superlattice the interface mode is reduced to a standing wave whose half wavelength is equal to the width of the quantum well.

Phonon dispersion induces interface modes to mix with bulk like modes of close frequencies.

The shortcoming of the continuum dielectric model (it neglects phonon dispersion) means that the solutions of connections with bulk like modes are arbitrary, and the assumed sinusoidal standing waveform leads to wrong symmetry. However, there is no arbitrariness in the mixing of longitudinal and transverse optical phonons corresponding to the interface mode. Thus is solved the puzzle why the wrong bulk like modes and correct interface modes are obtained from the dielectric continuum model.

There are some grounds for the guide model; it gives the correct optical displacement equations for the case where the phonon wave vector is parallel to the growth direction of the superlattice.

After analyzing the discrepancies, and based on theoretical calculations we arrived at equations to describe approximately the mode of optical vibrations and electrostatic potential in superlattices:

$$\Phi(z)=\cos(m\pi z/d)-(-1)^{m/2}, \qquad m=2,4,6,\cdots$$

$$\Phi(z)=\sin(\mu_m \pi z/d)+C_m(z/d), \quad m=3,5,7,\cdots$$

where μ_m and C_m are determined by the condition that Φ and its derivative both vanish near the interface.

Recently, our model has been widely used to study low dimension semiconductor materials and many physics problems related to phonons including, transport, Raman scattering, the relaxation of thermal carriers, plasma-polariton, free carrier absorption and the dynamics of bandedge picoseconds light emission. These subjects are directly related to the light emission and transport characteristics of low dimension semiconductor devices which have important scientific implications and application.

REFERENCES

1 K. Huang, X-ray Reflexions from Dilute Solid Solutions, Proc. Roy. Soc. (London), 1947(A 190): 102-117.

2 H. Peisl, H. Spalt, and W. Waidelich, X-Ray Diffuse Scattering in γ -Irradiated LiF, Phys. Stat. Solidi, 1967(23): K75-78.

3 H. Trinkaus, On Determination of the Double-force Tensor of Point Defects in Cubic Crystals by Diffuse X-ray Scattering. Phys. Stat. Solidi (b), 1972(51): 307-319.

4 M. Born and K. Huang, Dynamical Theory of Crystal Lattices, Oxford University Press, 1954.

5 K. Huang, Phenomenological Equations of Motion for Simple Ionic Lattices. E. R. A. Report Ref. L/T, 1950(239): 3-8.

6 K. Huang, On the Interaction between the Radiation Field and Ionic Crystals. Proc. Roy. Soc. (London), 1951(A 208): 352-365.

7 C.H. Henry and J. J. Hopfield, Raman Scattering by Polaritons, Phys. Rev. Lett., 1965(15): 964-966.

8 K. Huang and A. Rhys, Theory of Light Absorption and Non-radiative Transitions in F-centers. Proc. Roy. Soc. (London), 1950(A 204): 406-423.

9 J. J. Hopfield, A Theory of edge-emission phenomena in CdS, ZnS and ZnO, J. Phys. Chem. Solids, 1959(10): 110-119.

10 Tang Hui, Huang Kun, Hole Subbands in GaAs-$Al_xGa_{1-x}As$ Superlattice, Chinese J. Semicond., 1987(8): 1-10.

11 B.F. Zhu and K. Huang, Effect of Valence-Band Hybridization on the Exciton Spectra in GaAs-GaAlAs Quantum Wells, Phys. Rev. B, 1987(36): 8102-8108.

12 B.F. Zhu, Oscillator Strength and Optical Selection Rule of Excitons in Quantum Wells, Phys. Rev. B, 1988(37): 4689-4693.

13 A. K. Sood, J. Menendez, M. Cardona, et al., Resonance Raman Scattering by Confined LO and TO Phonons in GaAs-AlAs Superlattices, Phys. Rev. Lett., 1985(54): 2111-2114.

14 K. Huang, B. F. Zhu, Long Wavelength Optic Vibrations in a Superlattice, Phys. Rev. B, 1988(38): 2183-2186.

15 K. Huang, B. F. Zhu, Dielectric Continuum Model and Fr ö hlich Interaction in Superlattices, Phys. Rev B, 1988(38): 13 377-13 386.

The Advancement of Nanoscience and Nanotechnology in China

Bai Chunli

Nanoscience and Nanotechnology Center of the Chinese Academy of Sciences

***Bai Chunli,** Vice President of the Chinese Academy of Sciences (CAS), graduated from Peking University in 1978 and received his Ph.D. degree from CAS in 1985. During 1985-1987, he was at Caltech, conducting research work on scanning tunneling microscopy as a visiting scholar. Upon his return in 1987, he continued his research at the Institute of Chemistry, CAS. From 1991 to 1992, he was a visiting professor at Tohoku University in Japan. His work in the nano-field includes the successful development of several new types of scanning probe microscopes in China for nanostructure characterizations. Using scanning probe microscopes and other techniques, Bai and his laboratory have studied a wide variety of materials with an emphasis on the organic and biomaterials structures, which opened new horizons in nanotechnology. As a project leader, Bai has won 19 prizes and awards including SCI International Medal and TWAS Medal Lecture. He has more than 300 research papers to his credit, and has authored 12 monographs and several book chapters in English and Chinese. He is Member of CAS and TWAS. Bai also serves as IUPAC Bureau member, the president of Chinese Chemical Society; the chief scientist of State Steering Committee for Nanoscience and Nanotechnology and Chairman of the Academic Committee of CAS Research Center for Nanotechnology. In addition, he is the President of the Graduate School of CAS and holds concurrent professorship at a dozen of universities.*

Abstract: Nanoscience and nanotechnology are important emerging fields of research and development focusing on the artificially fabricated structures in the nanometer range (1~100 nm). Chinese scientists have followed with the main stream in the development of the nanoscience and nanotechnology since its initial stage. In the present paper, the achievements and present status of China in relative researches such as nanomaterials, nanodevices, and construction and characterization of nanostructures are described.

1. INTRODUCTION

Nanoscience and nanotechnology are the creation and utilization of materials, devices, and systems through the control of matter on the nanometer-scale. Its essence is the ability to work at these levels to generate larger structures exhibiting novel physical, chemical, and biological properties and phenomena. The aim of nanoscience and nanotechnology is to learn to exploit these properties and efficiently manufacture and employ the structures. The control of matter on nanoscale has already played an important role in scientific disciplines as diverse as physics, chemistry, materials science, biology, medicine, engineering, and computer simulation. For example, it has been shown that carbon nanotubes are ten times as strong as steel with one sixth of the weight, and that semiconductor nanoparticles are used in biolabeling.

Chinese scientists have followed with the main stream in the development of the nanoscience and nanotechnology since its initial stage. Tens of national conferences have been held in China since 1990 covering a wide range of topics in the related fields. In Beijing, the Chinese Academy of Sciences (CAS) sponsored the 7th International Conference on Scanning Tunneling Microscopy (STM'93) and the 4th International Conference on Nanoscale Science and Technology (Nano IV)[1-2]. These conferences have served academic exchanges and collaborations both nationally and internationally. To date, more than 20 institutes of CAS, 50 universities and 300 enterprises have engaged in the research and development of nanoscience and nanotechnology. Several centers for research and development of nanoscience and technology have been established in CAS, Tsinghua University, Peking University, Nanjing University, East China University of Science and Technology, etc.

Among these research centers, CAS pioneered the investigation on nanoscience and technology in China. A series of significant research projects were carried out in the late 1980s. The principal fields supported by CAS are as follows: bond-selective chemistry under the control of laser and the manipulation of single atoms with scanning probe microscopy (SPM); molecular electronics research on molecular materials and molecular devices; giant-magneto resistance materials and related physics; photo catalytic and photoelectronic chemistry study of nanosemiconductor; SPM studies on surface and interface; study on carbon nanotubes and other nanomaterials; study on the structure and physical properties of artificial "super-atom", and others. In the year 2000, CAS organized 11 institutes of CAS to take joint efforts in a major research project of "Nanoscience and Nanotechnology", with the main target to improve or to invent new synthetic methods and techniques for nanostructures, to produce new nanomaterials and nanodevices with important significance. In the recent years, the leading achievements have been made in China.

2. NANOMATERIALS

Synthesis and processing of nanostructures will employ diverse materials – organic, inorganic, and biological-well beyond examples already realized. The driving forces will be creativity in broad areas of science, technology and economics. Increasing emphasis will be placed on synthesis and assembly at a very high degree of precision, achieved through innovative processing. The result will be control of the size, shape, structure, morphology, and connectivity of nanostructured materials. Chinese scientists have devoted much attention to the preparation of nanomaterials. A great many of nanostructures, including nanoparticles, nanowires, nanorods, nanotubes, nanomembranes and nanocrystals, have been successfully fabricated through various chemical and physical methods.

In respect of the preparation of nanoparticles, the Institute of Solid State Physics of CAS manufactured the silicon-based nano-oxide (SiO_{2-x}) with high specific surface area (~640 m^2/g), and established a one-hundred-ton-scale production line with enterprises. East China University of Science and Technology is establishing a 150,000 tons/year super-fine $CaCO_3$ production line based on the industrial test of 3,000 tons/year. Peking University has achieved good results in the production of nanoscale powder of Ni and applied the material with the largest Ni-H battery company in China. Based on the super-gravity synthesis methods, the Beijing University of Chemical Engineering developed a production line of 3,000 tons/year nanoscale powder, whose scale and techniques ranked first class in the world in 1994. The successful preparation of nanoscale iron powder by Tianjin University made China the second country in which nanoscale metal powder can be industrially produced. Qingdao University of Chemical Engineering has accumulated a wealth of experience in the research and development of nanoscale Cu catalyst. To date, there are more that 20 production lines with ton-scale capacity to prepare nanoscale powder materials. The great varieties include: nano-oxides (ZnO, TiO_2, SiO_2, ZrO, MgO, Co_2O_3, NiO, Cr_2O_3, MnO_2, Fe_2O_3, etc.), nano-metal and nano-alloy (Ag, Pd, Cu, Fe, Co, Ni, Ti, Al, Ta, Ag-Cu alloy, Ag-Sn In-Sn, Ni-Al, Ni-Fe and Ni-Co, etc.), nano-carbonate (W_2C_3, C powder, SiC, TiC, ZrC, NbC, B_4C_3, etc.), nano-nitronate (Si_3N_4, AlN, Ti_3N_4, BN, etc.).

One-dimensional (1D) nanostructures have great potential for testing and understanding fundamental concepts about quantum confinement effect and for applications in nanoelectronics, nanophotonics, high density recording, and scanning probe microscopy. Concerning the preparation of 1D nanostructures, great achievements have been obtained in nanowires, nanorods and especially in carbon nanotubes[3-5]. A research group at the Institute of Physics of CAS invented a template method based on chemical vapor deposition catalyzed by iron nanoparticles embedded in mesoporous silica in 1996[6]. The obtained nanotubes are approximately perpendicular to the surface of the silica and form an aligned array of isolated tubes with spacings between the tubes of about 100 nm. This approach avoids the possible problems such as entanglements compared with other methods. In 1998, this group produced very long, multi-walled carbon nanotubes that reach about 3 mm in length through the pyrolysis of acetylene over iron/silica substrates, which is an order of magnitude longer than that described in most previous reports[7]. In 2000, the thinnest carbon nanotube with a diameter of 0.5 nm was first produced[8]. Then, the Department of Physics in Hong Kong University of Science and Technology prepared the thinnest single-wall carbon nanotube (0.4 nm) arrays using zeolite as template[9]. Afterwards, the existence of small single wall carbon nanotubes with diameters of 0.5 and 0.33 nm was demonstrated by high resolution transmission electron microscopy (HRTEM). The 0.33 nm carbon tube observed is likely a (4,0) tube[10]. In 1999, masses of single-walled carbon nanotubes with a large mean diameter of about 1.85 nanometers were synthesized by a semicontinuous hydrogen arc discharge method in the Institute of Metal Research of CAS. They also investigated the characteristics of hydrogen-storage, and reported that the mass capacity of hydrogen storage in carbon nanotubes can reach 4.2%[11].

In respect of the synthesis of nanoscale inorganic material, scientists in the University of Science and Technology of China developed the hydrothermal synthetic route to prepare GaN microcrystal[12]. They manufactured, for the first time, the GaN microcrystal with a size of 30 nm at about 300℃. The research team also applied a reduction-pyrolysis-catalysis method to prepare the diamond powder, and therefore developed a technological route with high economic values[13]. Through a carbon nanotube-confined reaction, a research group in Tsinghua University prepared 1D GaN nanocrystals, which have a diameter of 4 to 50 nm and a length of up to 25 μm[14]. This study demonstrated the possibility to synthesize other nitride nanorods through similar carbon nanotube-confined reactions.

A research group in the Institute of Metal Research of CAS synthesized a bulk nanocrystalline pure copper with high purity and high density by electrodeposition[15]. For the first time, an extreme extensibility (elongation exceeds 5,000%) without a strain harden effect was observed when the nanocrystalline copper specimen was rolled at room temperature. This behavior demonstrates new possibilities for scientific and technological advancements with nanocrystalline materials. This discovery was considered as of "a breakthrough in this field". Recently, they also successfully realized nitriding iron at lower temperatures by means of surface nanocrystallization[16]. The microstructure in the surface layer of a pure iron plate was refined at the nanometer scale by means of a surface mechanical attrition treatment that generates repetitive severe plastic deformation of the surface layer. The subsequent nitriding kinetics of the treated iron with the nanostructured surface layer were greatly enhanced, so that the nitriding temperature could be as low as 300℃, which is much lower than conventional nitriding temperatures (above 500℃). This enhanced processing method demonstrates the technological significance of nanomaterials in improving traditional processing techniques and provides a new approach for selective surface reactions in solids.

Recently, the surfaces of small diameter (1-7 nanometers) silicon nanowires (SiNW) were well studied by scanning tunneling microscopy (STM) in the City University of Hong Kong[17]. STM images of SiNWs, performed both in air and in ultra high vacuum, revealed atomically resolved images that can be interpreted as hydrogen-terminated Si (111)-(1×1) and Si (001)-(1×1) surfaces corresponding to SiH_3 on Si(111) and SiH_2 on Si(001) respectively. These H-terminated SiNW surfaces seem to be more oxidation resistant than regular silicon wafer surfaces, since atomically resolved STM images of SiNWs were obtained in air after several days exposure to ambient environment. Scanning tunneling spectroscopy measurements were performed on the oxide-removed SiNWs, and used to evaluate the electronic energy gaps. The energy gaps were found to increase with decreasing SiNW diameter from 1.1 eV for 7 nanometers to 3.5 eV for 1.3 nanometers (Fig.1). Among the possible exciting applications anticipated from the findings are the UV light-emitting diodes and lasers from SiNWs, which may become feasible if the wide band gap is a direct one, as predicted for small diameter SiNWs.

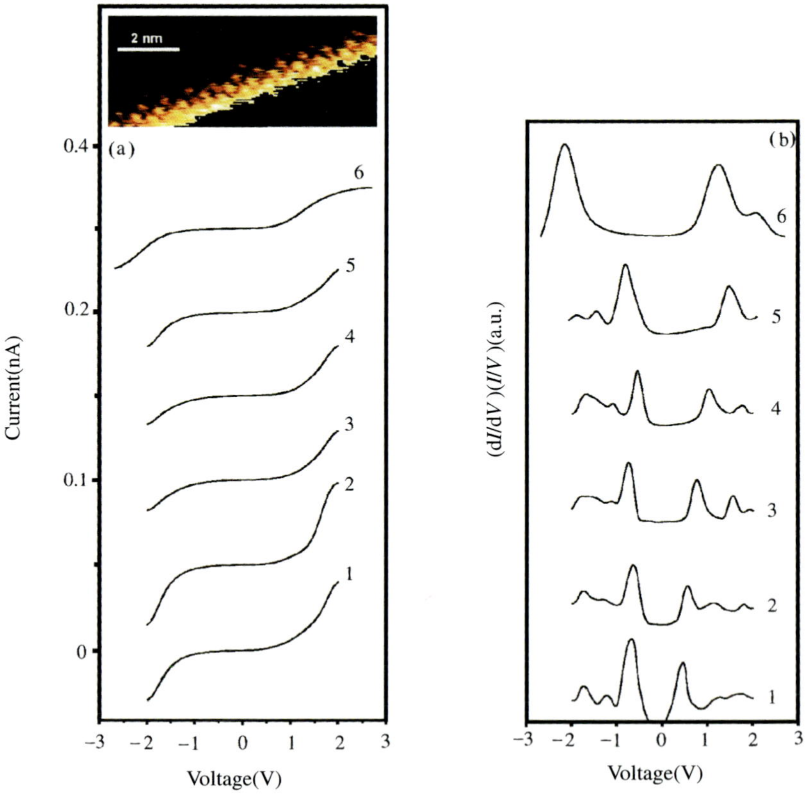

Fig. 1 Electronic properties of the SiNWs surfaces: (a) Current (I)-voltage (V) curves obtained by STS on six individual SiNWs; the diameter of wires 1 to 6 being 7, 5, 3, 2.5, 2, and 1.3 nm respectively; The inset shows the atomically resolved STM images of wire 6. (b) The corresponding normalized tunneling conductances, $(dI/dV)/(I/V)$; the curves are offset vertically for clarity

3. NANODEVICES

In the broadest sense, nanodevices are the critical enablers that will allow mankind to exploit the ultimate technological capabilities of electronic, magnetic, mechanical, and biological systems. While the best examples of nanodevices at present are clearly associated with the information technology industry, the potential for such devices is much broader. Nanodevices will ultimately have an enormous impact on our ability to enhance energy conversion, control pollution, produce food, and improve human health and longevity. In the past decade, our ability to manipulate matter from the top down, combined with advances and in some cases unexpected discoveries in the synthesis and assembly of nanometer-scale structures, has resulted in advances in a number of areas. Particularly striking examples include the following: quantum electronics devices, ultra-high density data storage etc.

Concerning quantum electronics devices, several groups have studied the single electron tunneling effect at room temperature, single electron tunneling of single atom junction, Coulomb blockade effect at room temperature using UHV-STM, high performance opto-electrical detectors. The Tsinghua University has achieved 100 nm (0.1 μm) level MOS devices, and a series of silicon-based integrated sensors, microphones, micro-motors and micro-pumps etc. They also developed new techniques and Microsystems using 3-D microscopic lithography methods. The Institute of Semiconductors of CAS has developed sensors based on infrared (13-15 μm) adsorbing quantum dots, and semiconductor quantum dot laser (with the wavelength range of (0.7-2.0 μm)[18]. A prototype single electron device has been achieved in Institute of Physics of CAS[19-20]. A model device of field emission display panel made of nanotube material has been fabricated in the Xi'an University of Transportation, and has already been under continuous test for over 3,800 hours.

The research on the ultra-high density data storage using organic materials was carried out in the laboratory of Vacuum Physics of CAS cooperated with Peking University. The diameter of arrays of dots on NBPDA organic thin films reached 1.3 nm in 1997, 0.7 nm in 1998 and finally 0.6 nm in year 2000[21]. This data is an order of magnitude smaller than the reported one by groups in other countries, and could lead to an increase of nearly one million times in storage if commercialized CD. Groups in Peking University adapted binary composite material TEA/TCNQ as ultra-high density data storage material and obtained large area data arrays of 3 × 3 μm, with 8 nm dot diameter. Fudan University successfully fabricated high-speed, high-density memory unit by using bi-stable thin films, and synthesized several patented organic monomers as basic materials organic integrated circuits.

A research group in Tsinghua University has found new approaches to pull carbon nanotubes into continuous yarns up to 30 cm in length by being drawn out from superaligned arrays of carbon nanotubes[22]. A light-bulb filament can be constructed by winding a carbon nanotube yarn between two metal leads. The filament emits incandescent light when a DC voltage is applied (Fig. 2). A carbon nanotubes polarizer can also be constructed by parallel alignment of carbon nanotubes yarns. When a beam of light passes through the polarizer, photons having a polarization direction parallel to the axis of the carbon nanotubes are absorbed. The polarizer can work in the ultraviolet region, and could even be used for ultraviolet light of wavelengths in the region of tens of nanometers. The findings should translate the remarkable mechanical and electrical properties of CNTs to a macroscopic scale. Another interesting finding is the conductivity and strength of the yarns can be enhanced by heating them at high temperatures. Hence, the yarns, after appropriate heat treatment, should be woven into a variety of macroscopic objects that can be used to create high-quality bullet-proof vests or materials that can block electromagnetic waves.

The well-defined nanostructure can be constructed at solid/liquid interface under potential control with inorganic ions, organic molecules or other objects. The molecular orientation and structure could be controlled by applying an electrode potential in electrolyte solution. The formation process of nanostructure could be monitored by electrochemical STM. The so-constructed nanostructure is potentially useful in the future for nano-electronic devices [23].

It is noteworthy that Chinese scientists still remain at the level of the preparation and selection of materials for nanoscale devices, as well as at the level of the investigations for new physical phenomena. The investigation on the mechanism and structure of nanoscale devices are relatively weak and lack originality. To achieve a significant progress in the research of nanoscale devices, China is ready to throw much more funds into this field to alter the present experimental equipment and research conditions. Also, the collaborations between research units are encouraged.

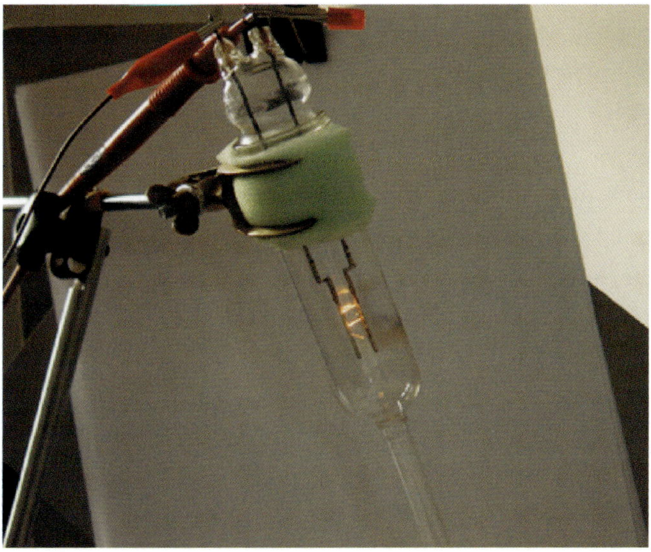

Fig. 2 A light-bulb filament by winding a carbon nanotubes yarn between two metal leads. The filament emits incandescent light when a DC voltage is applied

4. CONSTRUCTION AND CHARACTERIZATION OF NANO-STRUCTURES

The recent rapid advances in nanoscience and nanotechnology are due in part to our newly acquired ability to measure and manipulate individual structures on nanoscale. Whether it should be scanning probes, optical tweezers, high-resolution electron microscopes, or other new tools, instruments available to research workers in science and technology now permit them to create new structures, measure new phenomena, and explore new applications. There are limitations for various properties, such as the chemical composition of a single nanostructures and local electronic and thermal characteristics. Our research group has independently developed a series of instrumentations such as STM, AFM, BEEM, LT-STM, UHV-STM, and SNOM which have been extensively used in the characterizations on the nanoscale structures[24]. Peking University has designed and constructed UHV-SEM-STM-EELS system and LT-SNOM system. They also set up a set of complete systems consisting of SNOM-Near field spectrum, and classical optical methods, by which the structure of cancer cells was investigated.

With the development of these instrumentations, CAS such as our lab and lab for Vacuum Physics and Peking University began the surface lithography experiments by STM on the nanometer scale and even on atomic scale as early as 1990[24-25]. The characters of "CAS", the map of China, and Chinese characters of "China" have been successfully fabricated. These progresses in nanoscale fabrication helped to stimulate the awareness of the research of nanoscience and nanotechnology.

Great efforts have been made in achieving assembling organic molecules on solid surfaces in a controllable way. This practice requires the fine-tuning of the thermodynamic equilibrium of the molecular assembles, with various interactions, ranging from van der Waals, hydrogen bonds, electro-static, steric interaction, etc. The assembled structure could lead to the possibility of artificially designing of molecular nanostructures, possessing rich physical and chemical functions. Our lab has made good progress in the ordered self-assembly of organic molecules, aiming at the exploratory research on the nanoscale devices[26]. Various molecular nanostructures including single molecular wires, molecular superlattices and adlayer structures of atomic and molecules have been successfully constructed. For example, single molecular wires of phthalocyanine (Pc) have been assembled with nanometer sized alkane templates [27]. It was found that a new phase of homogenous molecular assembly of Pc with C18SH and C18X (X=Cl, Br, I, CN) formed in addition to the typical segregated domains when Pc and alkane derivatives were coadsorbed from solution onto HOPG (Fig. 3). By adjusting the molar ratio of Pc to alkane derivatives, homogenous intercalation of Pc in length scale of submicrometers can become the main form of Pc on graphite surface. Similar assembling behavior was also observed in other alkane derivatives, with model systems of C18X.

The self-assembly of 2,3,6,7,10,11-alkoxy-substituted triphenylenes with n-carbon side chains (Tn, n=12,14,16) on HOPG have been extensively studied by using STM [28]. When n equals twelve, two kinds of structures in one unit were formed. One is typical six-fold symmetry, the other is three-fold symmetry. With two additional methylene units of the alkyl compound of hexakistertradecyloxytriphenylene (T14), the whole arrangement of the molecule changes drastically. A characteristic dimer structure is persistent in a large area. Part of the alkyl chains form tightly packed structure. When the length of alkyl increases once more by two methylene units to form hexakishexadecyloxytriphenylene (T16), all the alkyl parts

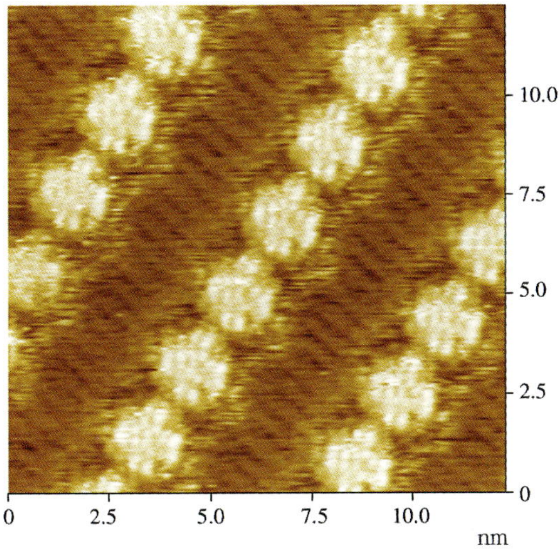

Fig. 3 Intercalated assembling structure of copper(II) 1,4,8,11,15,18,22,25-octabutoxy-29H,31H-phthalocyanine formed with 1-iodooctadecane on HOPG surface

are parallel to each other and form close-packing structure. The origin is considered in association with the interplay of molecular stearic interaction and the 2D crystallization of alkyl parts.

Atomic structures of adsorbed sulfur on Cu(111) surface was investigated by using in suit electrochemical scanning tunneling microscope (ECSTM), which has been proved to be a powerful tool to study solid/liquid interface with atomic resolution, in perchloric acid solution [29]. A well-defined S adlayers with a $(\sqrt{7} \times \sqrt{7})$ R19.1° symmetry was resolved at the potential range between −0.35 and −0.22 V (Fig. 4). The adlayer structures of pyrene, perylene and diaza-15-crown-5 on Cu(111), benzene and pyridine molecules on Cu(100), L-cysteine, calix[4]arene and calix[4]arene diquinone disulfide on Au(111) and dendritic molecules on HOPG have been also extensively studied [30-35]. The novel structures may open a new area of SAM in both science and industrial applications.

In addition, a group in the University of Science and Technology of China observed a single C_{60} array consisting of domains of two different orientations, in which the boundary separates two distinguishable domains by their internal features, providing the basic data for developing nanoscale devices [36-37].

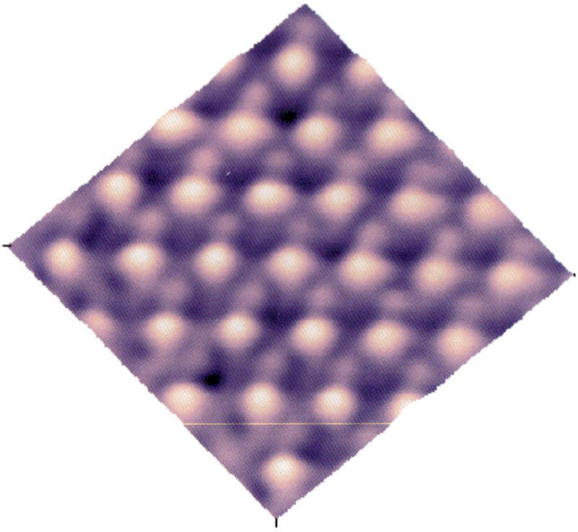

Fig. 4 A well-defined S adlayers with a $(\sqrt{7} \times \sqrt{7})$ R19.1° symmetry at the potential range between − 0.35 and − 0.22 V by ECSTM.

5. CONCLUSION

In summary, considerable achievements have been obtained in the field of nanoscience and nanotechnology in China, in particular the research of nanoscale materials, represented by carbon nanotube that is ranked among the front in the world. But the researches on nanoscale devices have just begun because of the limited research conditions. Most research still remain at the level of preparation of nanomaterials and investigation of the new physical phenomena. Few attentions have been focused on the designation and fabrication of nanodevices. A national technological platform will be established to strengthen the ability of surface processing. Good coordination should be made in order to organize different branches of learning the breakthrough of key techniques for nanoscale devices. The original and applied researches will be reinforced concerning nanoscale materials. It is expected that development of nanoscience and nanotechnology will be significantly increased in the near future.

REFERENCES

1 Bai C L, Colton R, Kuk Y (eds.). Papers from the 7th International Conference on Scanning Tunneling

Microscopy/Spectroscopy, The American Institute of Physics. New York: The American Institute of Physics, 1994

2 Bai C L. Progress of nanoscience and nanotechnology in China. Journal of Nanoparticle Research, 2001 (3):251~256

3 Yu D P, Bai Z G, Feng S Q et al. Synthesis of nano-scale silicon wires by excimer laser ablation at high temperature, Solid State Commun., 1998(105):403~407

4 Guo Y G, Wan L J, Gong J R et al. Preparation and dispersion of Ni-Cu composite nanoparticles. Phys. Chem. Chem. Phys., 2002(4):3422~3424

5 Guo Y G, Wan L J, Zhu C F et al. Ordered Ni-Cu nanowire array with enhanced coercivity, Chem. Mater., 2003 (15): 664~667

6 Li W Z, Xie S S, Qian L X et al. Large-scale synthesis of aligned carbon nanotubes. Science,1996(274): 1701~1703

7 Pan Z W, Xie S S, Chang B H et al. Very long carbon nanotubes. Nature, 1998(394):631

8 Sun L F, Xie S S, Liu W et al. Materials: Carbon nanotubes with a diameter of 0.5 nm. Nature, 2000(403):384

9 Wang N, Tang Z K, Li G D et al. Single walled 4Å carbon nanotube arrays. Nature, 2000(408):50~51

10 Peng L M, Zhang Z L, Xue Z Q et al. Stability of carbon nanotubes: How small can they be? Phys. Rev. Lett., 2000(85):3249~3252

11 Liu C, Fan Y Y, Liu H et al. Hydrogen storage in single-walled carbon nanotubes at room temperature. Science, 1999(286):1127~1128

12 Xie Y, Qian Y T, Wang W Z et al. A benzene-thermal synthetic route to nanocrystalline GaN. Science, 1996(272):1926

13 Li Y D, Qian Y T, Liao H W et al. A reproduction-pyrolysis-catalysis synthesis diamond. Science, 1998 (281):246~247

14 Han W, Fan S, Li Q et al. Synthesis of gallium nitride nanorods through a carbon nanotubes-confined reaction. Science, 1997(277):1287~1289

15 Lu L, Sui M L, Lu K. Superplastic extensibility of nanocrystalline copper at room temperature. Science, 2000(287):1463~1465

16 Tong W P, Tao N R, Wang Z B et al. Nitriding iron at lower temperatures. Science, 2003(299):686~688

17 Ma D D D, Lee C S, Au F C K et al. Small diameter silicon nanowire surfaces. Science, 2003, in press

18 Pan D, Zeng Y P, Kong M Y et al. Normal incident infrared absorption from InGaAs/GaAs quantum dot superlattice. Electronics Letters, 1996(32):1726~1728

19 Wang T H, Li H W, Zhou J M. Si single electron transistor with in-plane point-contact metal gates. Appl. Phys. Lett., 2001(78):2160~2162

20 Wang T H, Aoyagi Y. Single electron charge in parallel dot structure. Appl. Phys. Lett., 2001(78):634~636

21 Gao H J, Sohlberg K, Xue Z Q et al. Reversible nanometer-scale conductance transitions in an organic complex. Phys. Rev. Lett., 2000(84):1780~1783

22 Jiang K L, Li Q Q, Fan S S. Spinning continuous carbon nanotubes yarns. Nature, 2002(419):801

23 Wan L J, Noda H, Wang C et al. Controlled orientation of individual molecules by electrode potentials. Chem.Phys.Chem., 2001(2):617~619

24 Bai C L. Scanning tunneling microscopy and its applications. Springer-Verlag, Second Edn., Heidelberg, 2000

25 Wang C, Bai C L. Evidence of diffusion characteristics of field emission electrons in nanostructuring process on graphite surface. Appl. Phys. Lett., 1996(69):348~350

26 Qiu X, Wang C, Zeng Q D et al. Alkane-assisted adsorption and assembly of phthalocyanines and porphyrins. J. Am. Chem. Soc., 2000(122):5550~5556

27 Lei S B, Wang C, Yin S X et al. Single molecular arrays of phthalocyanine assembled with nanometer sized alkane templates. J. Phys. Chem. B, 2001(105):12272~12277

28 Wu P, Zeng Q D, Xu S D et al. Molecular superlattices induced by alkyl substitutions in self-assembled triphenylene monolayers. Chem.Phys.Chem., 2001(2):750~754

29 Wang D, Xu Q M, Wan L J et al. Atomic structures of adsorbed sulfur on Cu(111) in perchloric acid solution by in situ ECSTM. Surf. Sci., 2002(499):L159~L163

30 Wang D, Wan L J, Xu Q M et al. Adlayer structures of pyrene and perylene on Cu(111): an in situ STM study. Surf. Sci., 2001(478):L320~L326

31 Wang D, Xu Q M, Wan L J et al. In situ scanning tunneling Microscopy study of adsorption of diaza-15-crown-5 on Cu(111). Surf. Sci., 2001(489):L568~L572

32 Wan L J, Wang C, Bai C L et al. Adlayer structures of benzene and pyridine molecules on Cu(100) in solution by ECSTM. J. Phys. Chem. B, 2001(105):8399~8402

33 Xu Q M, Wan L J, Wang C et al. New structure of L-cysteine self-assembled monolayer on Au(111): studies by in situ scanning tunneling microscopy. Langmuir, 2001(17):6203~6206

34 Pan G B, Wan L J, Zheng Q Y et al. Self-organized arrays of calix[4]arene and calix[4]arene diquinone disulfide on Au(1 1 1). Chem. Phys. Lett., 2002(359):83~88

35 Wu P, Fan Q H, Zeng Q D et al. Observation of functional group mediated assembling of dendritic molecules by STM. Chem.Phys.Chem., 2002(3):633~637

36 Hou J G, Yang J L, Wang H Q et al. Topology of two-dimensional C_{60} domains. Nature, 2001(409):304~394

37 Hou J G, Yang J L, Wang H Q et al. Identifying molecular orientation of individual C_{60} on a Si(111)-(7 × 7) surface. Phys. Rev. Lett., 1999(83):3001~3004

Some Recent Progress of Chemical Studies in China

Zhu Qingshi

University of Science and Technology of China

Zhu Qingshi, *Professor of Chemistry, born in 1946 in Chengdu, Sichuan Province, graduated from the University of Science and Technology of China (USTC) in 1968. Before moving to USTC in 1994, he worked in the Chinese Academy of Sciences (CAS) Qinghai Institute of Saline Lakes and Dalian Institute of Chemical Physics. He was elected as an academician of CAS in 1991. He founded the CAS Key Laboratory of Bond-Selective Chemistry and USTC Biomass Clean Energy Laboratory. He was assigned as Vice-President of USTC in August 1996, and has been the President of USTC since June 1998. In 2001 he was elected as an academician of the Third World Academy of Sciences.*

Abstract: This review mainly focuses on the progress of chemical research in China in the past decade. Most of the cited papers were published in journals *Science, Nature*, *J. Am. Chem. Soc. and Angew. Chem. Int. Ed.* It was organized as inorganic and analytical chemistry, organic chemistry and chemical biology, physical and theoretical chemistry, and polymer chemistry and physics. For limitation of the author's knowledge, it was hard to cover all significant research work in this review. My colleagues, Dr. Lifeng Yan and Professor Qing-Xiang Guo are acknowledged for their valuable help in the preparing this review article.

"Science and Technology is the First Productive Force" is a famous statement of the Chinese Government, which has made a huge development of Science in China with plenty of funds pouring into the research goals of various fields. More and more research groups work in the frontier of modern science, and many important results have been achieved. At the same time, original research increases quickly. For Chemistry, we would have an outlook on the following subjects.

1. INORGANIC CHEMISTRY AND ANALYTICAL CHEMISTRY

In the second half of the 20th century, there were many achievements in Inorganic and Analytical Chemistry, such as Debang Hou's improvement of alkali manufacture, separation and analyses of B, Li and U isotopes, the theory of silica acid polymerization, which clears some errors in past years, and the precise measurements of the atomic mass of antimony, europium and cerium, which were accepted as a new standard for IUPAC. Good results have also been obtained in crystal chemistry field.

Bottom-up synthesis method provides great changes in solid inorganic chemistry to prepare nano-scale materials. Recently, some exciting research work has been completed in this field in Tsinghua University[1]. Gallium nitride nanorods were prepared through a carbon nanotube-confined reaction. Ga_2O vapor reacted with NH_3 gas in the presence of carbon nanotubes to form wurtzite gallium nitride nanorods. The nanorods have a diameter of 4 - 50 nanometers and a length of up to 25 micrometers. It is proposed that the carbon nanotube acts as a template to confine the reaction, which results in the gallium nitride nanorods having a diameter similar to that of the original nanotubes. The results suggested that it might be possible to synthesize other nitride nanorods through similar carbon nanotube-confined reactions. This research was selected as one of the ten achievements in 1997 by *Science*. Researchers in CAS Institute of Physics[2] reported large-scale synthesis of aligned carbon nanotubes by using a method based on chemical vapor deposition catalyzed by iron nanoparticles embedded in mesoporous silica. Scanning electron microscope images show that the nanotubes are approximately perpendicular to the surface of the silica and form an aligned array of isolated tubes with spacing between the tubes of about 100 nanometers. The tubes are about 50 micrometers long and well graphitized (Fig.1). They also reported[3] that simple improvements in the electric arc technique can create a carbon nanotube with a diameter of 0.5 nm – the same as a C36 molecule, and a method to prepare carbon tube with the length of about 2-3 mm[4].

Fig. 1 (A) Low-magnification SEM image of a film composed of aligned carbon nanotubes. This film with a thickness of 50 μm was obtained by growing for 2 hours. (B) Tip structure of the aligned tubes

Researchers in USTC[5] synthesized diamond powder through a metallic reduction-pyrolysis-catalysis route with the reaction of carbon tetrachloride and sodium at 700℃,in which the sodium was used as reductant and flux. The temperature is much lower than that of traditional methods. Although the yield was only 2 percent, this method is a simple means of forming diamond. A Benzene-Thermal Synthetic Route to GaN was reported by scientists in USTC [6]. Recently researchers in Tsinghua University[7] have successfully developed a method to produce single-walled nanotubes (electric arc, laser ablation, and chemical vapor deposition). The typical lengths of tangled nanotube bundles reach several tens of micrometers. They reported that long nanotube strands, up to several centimeters in length, consisting of aligned single-walled nanotubes can be synthesized by the catalytic pyrolysis of n-hexane with an enhanced vertical floating technique. The long strands of nanotubes assemble continuously from arrays of nanotubes, which are intrinsically long.

The SWNT strands, with lengths of 20 and 10 cm, generally consist of thinner ropes and have a diameter of the order of 0.3 to 0.5 mm. The inset shows one straightened strand and another tied in a knot, demonstrating the high flexibility of the nanotube strand (Fig. 2).

Although hydrothermal synthesis has been extensively used to prepare polymolybdate-containing inorganic/organic hybrid materials, the synthesis of heteropolymolybdate-containing lanthanides by a hydrothermal synthetic method

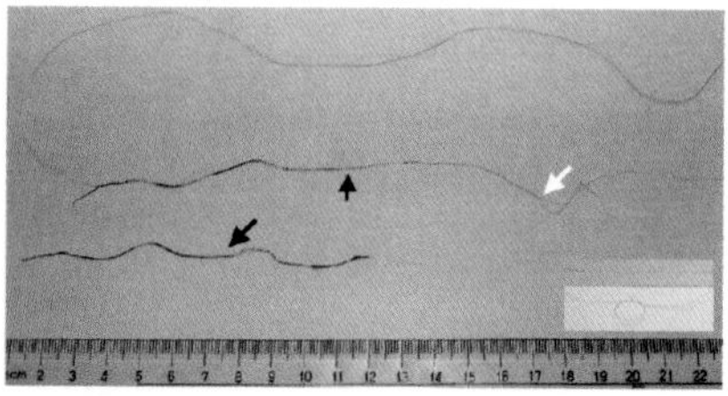

Fig. 2 Optical image showing a human hair and two as-grown SWNT strands (indicated by black arrows)

has fallen far behind. The successful synthesis of a novel three-dimensional framework was achieved in CAS Fujian Institute of Research on the Structure of Matter[8] (Fig.3). It suggests that the hydrothermal synthesis is an effective method to construct some higher dimensional frameworks for novel heteropolymolybdates-containing lanthanide elements.

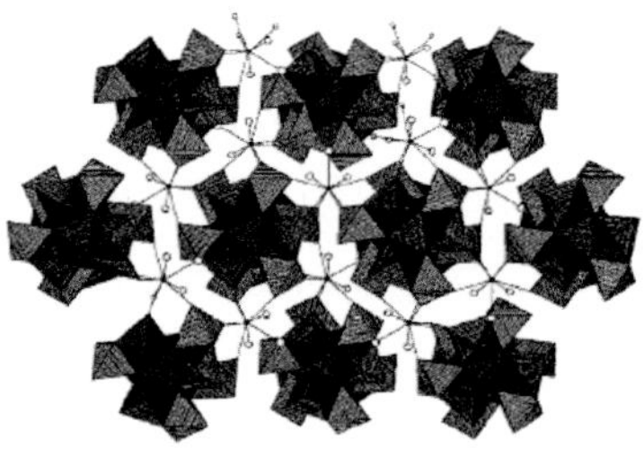

Fig. 3 Polyhedral representation of the $[GdMo_{12}O_{42}]_{9}$- anion and balland-stick representation of the nine-coordinated Gd^{III} cations down along the *c* axis

Nanowire, nanotube and nanorod are attractive nanostructures to be synthesized. Researchers in Tsinghua University[9] reported the synthesis of Bismuth nanotubes with diameters of about 5 nm and lengths ranging of 0.5-5 μm. A low-temperature hydrothermal reduction method with bismuth nitrate $[Bi(NO_3)_3]$ and aqueous hydrazine solution $(N_2H_4 \cdot H_2O)$ at 120℃ has been successfully used to synthesize large quantities of nested bismuth nanotubes. Researchers in Fudan University[10] reported a simple and mild solution-growth procedure, free of any templates and substrates, creating a unique self-supported pattern of radial semiconductor Ag_2S nanorod arrays at room temperature. It may provide a new method for direct growth of nanorods and related materials. Chemists in USTC [11] successfully prepared CdS nanorods in n-butylamine, which demonstrates that chalcogenide nanorods can be prepared by using a monodentate ligand as solvent. This result implies that one anchor atom in a ligand is necessary and sufficient for the formation of 1D nanocrystals, even though more anchor atoms may be present in a ligand. A facile hydrothermal synthetic method to prepare $Ln(OH)_3$ nanowires that were expected to exhibit some novel properties was developed in Tsinghua University[12]. In Fudan University[13], 3D bicontinuous cubic Ia3d-type mesoporous silica has been synthesized for the first time in acidic medium at room temperature by using a nonionic triblock copolymer as template. The resulting FDU-5 materials have extra-large pores with narrow size distributions. Such materials could have promising applications for the sorption, transport, and separation of large molecules. Researchers in Tsinghua University [14] also synthesized α- and β-MnO_2 single crystal nanowires by selected-control hydrothermal synthesis. Scientists in USTC [15] also developed solvothermal co-reduction (SCTR) route to the nanocrystalline III-V semiconductor InAs at a greatly decreased temperature, which is the lowest temperature in synthesis of nanocrystalline InAs up to now.

Isotope labeling of carbon nanotubes and observed ^{12}C-^{13}C nanotube junctions was studied in Tsinghua University[16]. The results provided a clear picture of the growth process of multiwalled carbon nanotubes in CVD and represent a step forward toward fully understanding the growth mechanism of nanotubes. The method can be naturally extended to study the growth mechanism of single-walled carbon nanotubes. Further, the formation of ^{12}C-^{13}C nanotube junctions suggests that nanotube-based intermolecular devices can be obtained by controlled chemical growth techniques. The isotope-labeling method also provide a powerful tool in investigating the catalytic growth mechanism for other forms of nanotubes and nanowires.

((DABCO)· $ZnGe(HPO_4)_3$ was synthesized in USTC[17], it was the first Zinco-Germanophosphate with a unique asymmetric cage by heating a mixture of GeO_2, $ZnSO_4$, 70%H_3PO_4, DABCO, ethylene glycol, and pyridine at 170℃ for 10 days in a Teflon-coated steel autoclave, to give colorless crystals insoluble in common solvents. The largest crystal size was about 2 mm long.

Researchers in Fudan University[18] obtained mesotunnels on the silica wall of ordered SBA-15 to generate three-dimensional large-pore mesoporous networks by high temperature hydrothermal treatment; the mesotunnels with the size of 2～3 nm randomly distributed on the silica wall of the 3D SBA-15.

A novel mild route for synthesis of nanocrystalline selenides, such as ZnSe, CdSe, Bi_2Se, SnSe and $Cu_{2-x}Se$, at room temperature was reported by the researchers in USTC [19].

Chemists in Jilin University[20] reported that complex porous structures could be fabricated by taking the advantage of the ethanol-assisted self-assembly of zeolite nanocrystals. The high concentration and reactivity of external surface silanol groups of the zeolite nanocrystals are critical for hydrogen bonding formation and for subsequent cross-linking (Si-O-Si covalent bond) during calcination. The simple processing scheme separates crystallization of zeolite nanocrystals from the morphological construction, allowing preparation of complex structures that are otherwise difficult to obtain by direct hydrothermal synthesis.

Researchers in Fudan University[21] demonstrated that strongly acidic and high-temperature hydrothermally stable mesoporous aluminosilicates with ordered hexagonal structure could be synthesized by self-assembly of preformed alumino-silicate nanoclusters with templating micella.

Recently, the iodine chemisorption on gold nanoparticles (NPs) was studied in CAS Changchun Institute of Applied Chemistry[22]. Compared with the previous aggregating studies, both aggregation and fusion/fragmentation of NPs were observed by TEM upon KI addition. They proposed that the physical origin of the fusion/fragmentation of gold NPs is energetic injection, which fits with the previous studies. The driving force of aggregation should be van der Waals attractive force. It is found that I-linked surface-bound-NP nanostructure can be fabricated by a layer-by-layer deposition technique.

Researchers in Tsinghua University[23] reported a simple and versatile vacuum-controlled pyrolysis procedure for the removal of the surfactant species from lamellar mesostructured composites of tungsten oxide with cetyltrimethylammonium surfactant cations (WO-L).

In CAS Institute of Chemistry[24], a new method to build nanostructured spherical aggregates from individual magnetic nanoparticles has been developed. The results demonstrate a strategy directed by p ± p interactions for the self-assembly of individual magnetic nanoparticles into structured ensembles. They also studied the aligned structure of ACNT films and very large fraction of air on the films determining the superhydrophobic properties, while ACNT films with a fluoroalkylsilane coating show super-amphiphobic properties[25].

2. ORGANIC CHEMISTRY AND CHEMICAL BIOLOGY

There have many important achievements in Organic Chemistry, such as total synthesis of cattle insulin, the development of elemental organic chemistry and its application to eliminate schistosomiasis, the total synthesis of yeast lactamic acid RNA, the synthesis of steroid and their reaction, the relationship between molecular structure and properties; the prcparation of butadiene from butylene by oxidation dehydrogen reaction, organic phosphorus chemistry, primary, secondary and stereo structure of Trichosanthin, the total synthesis of Artemisinin and its application, subsulfonating dehalogen, organometallic conductor, electron transfer reactions in fluorin chemistry, the synthesis of organic arsenic and antimony.

Recently, studies on the reaction mechanism, efficient synthesis and total syntheses of some special organic compounds were carried out. Structure-property relationships is a hot topic in Physical Organic Chemistry. The scientists in CAS Shanghai Institute of Organic Chemistry [26] have worked on a project of setting up a new σ^* scale (later designated as σ_{JJ}^*). They have demonstrated its trustworthiness by its successful applications to kinetic data derived from a rigorous methodology developed for the present study.

Scientists in CAS Shanghai Institute of Organic Chemistry and Beijing University[27] presented two different kinds of strong interactions between α,ω-diiodoperfluoroalkane and oxygen that pack in an analogous fashion. Meanwhile, a useful electron donor, hexamethylphosphoramide (HMPA), which is involved in two electron donor-acceptor bonds to halogen. The results may offer an alternative for researchers who concentrate their study on charge-transfer interactions, molecular recognition, reaction mechanism, supramolecular and many other prospective fields.

Palladium-catalyzed allylic substitution reaction is one of the brightest focuses in asymmetric synthesis during the past decades because of its great potential in organic synthesis. Various ligands have been synthesized and used in this reaction, especially with 1,3-symmetrically disubstituted substrates, and high ee% is realized. Researchers in CAS Shanghai Institute of Organic Chemistry [28] designed and synthesized a series of novel ferrocene ligands. A new chiral center was formed on the P atom during the reaction with BINOL, and all diastereoisomers were easily isolated as orange solids by column chromatography. High regio- and enantioselectivity were realized in Pd-catalyzed allylic alkylation and amination of monosubstituted allylic acetates employing these chiral ligands.

The reaction of bare atoms/clusters with compounds has been a very active research field. As an important class of reactions, it was interesting to insert bare atoms into molecules such as CH_4, CH_3-X, NH_3, H_2O, C_2H_2, and C_2H_4 for many years. However, there has been less attention paid to the insertion of atom dimers/clusters, and there is little knowledge about the reactivities of atom dimers/clusters vs. atoms. The research in USTC[29] showed direct theoretical evidence for the higher reactivity of B_2 compared to the single B atom.

Some superior catalysts that are cheap, easy to access, air-stable, and water-tolerant were developed in the CAS Shanghai Institute of Organic Chemistry[30]. They considered a sidearm approach, focusing on the improvement of bisoxazoline into pseudo-C_3-symmetric trisoxazoline. The results demonstrated that pseudo-C_3-symmetric trisoxazoline has the encouraging ability to achieve high face selectivity in asymmetric Michael reaction of indoles to alkylidene malonates.

In the transition metal chemistry, researchers in Nankai University[31] recently reported a class of macrocycles, which contain tetrahedral Mo_2FeS cluster cores and are formed by a self-assembly cyclization reaction. Macrocycles are of great importance in theory and practical applications. Furthermore, the macrocycles containing metal cluster cores along with their synthetic methods are little known up to now. So, they studied the new synthetic methodology by which such macrocycles could be synthesized.

Assembly of supramolecular architectures with inner cavities is a subject of current interest because of their ability for selective inclusion of ions and molecules, molecular recognition, and catalysis for specific chemical transformations. The potential use of such species largely depends on their cavity size and type, which are mainly controlled by the ligands *via* appropriate metal ions. Researchers in CAS Fujian Institute of Research on the Structure of Matter[32] reported a stable neutral nanometer-sized metallosupramolecular cube cage $[Ni_6(tpst)_8Cl_{12}]$ (tpst = 2,4,6-tri[(4-pyridyl)sulfanylmethyl]-1,3,5-triazine) with O_h symmetry, which was prepared from assembly reaction of $NiCl_2$ and the tpst ligand in DMF. In structure, the six Ni atoms are located very close to the faces of the cube, 7.9 Å from the cavity center. The centroids of the eight C_3N_3 rings in the ligands are each located at the 3-fold inversion axes and very close to the corners of the cube.

The use of transition metal catalysts in the carbocyclization alkenes and alkynes offers the unique means to construct a variety of synthetically important carbo- and heterocycles with high efficiency not normally accessible by traditional methods.

Scientists in CAS Shanghai Institute of Organic Chemistry[33] reported a new type of Pd(II)-catalyzed cyclization of enyne esters utilizing bidentate nitrogen-containing ligands for the synthesis of γ-butyrolactones initiated by acetoxypalladation with high efficiency as well as its catalytic asymmetric version.

Interest in metal bis-dithiolenes has increased over the past decades since they have been intensively studied as materials for electrical conductors. These compounds are also found to exhibit unusual optical properties. They are known to be very stable to intense irradiation in the near-infrared region and have been used as Q-switch dyes for Nd:YAG lasers. Researchers in Suzhou University[34] reported a new method to synthesize unsymmetrical dithiolene metal complexes such as $(Me_4N)_2[Cd(dmit)(Sph)_2]$ and $(Me_4N)_2[Zn(dmit)(Sph)_2]$ (Sph = thiophenolate group). Both complexes showed modest third-order NLO responses.

Based on the study of semiconductor and metal nanoparticles, many new physical phenomena, such as quantum confinement effects, and finite size effects, have been discovered, and many novel device concepts have been developed. Researchers in CAS Institute of Chemistry [35] reported that nanoparticles of 1-phenyl-3-((dimethylamino)styryl)-5-((dimethylamino)phenyl)-2-pyrazoline (PDDP) ranging from tens to hundreds of nanometers were prepared by using the reprecipitation method.

Chemists in CAS Shanghai Institute of Organic Chemistry [36] described highly efficient enantioselective catalysts for hetero-Diels-Alder reaction by high-throughput screening of a combinatorial library of chiral titanium complexes. The reactions of dienes with a variety of aldehydes producing dihydropyrone heterocycles with quantitative yield and 99.8% ee were carried out. An efficient method for one-step enantioselective synthesis of 1,3-disubstituted-2-silylvinylcyclopropanes was also developed in this institute[37]. As a result, both diastereoselectivity (cis/trans) and enantioselectivity are excellent in most cases studied.

The addition of organolithium compounds to nitriles predominately forms *N*-lithioketimines, which afford ketones or imines upon hydrolysis, and may be used as reactive intermediates for synthesis of *N*-substituted imines. Researchers in Beijing University[38] reported two novel and synthetically useful reaction patterns of organolithium compounds with nitriles affording pyridine derivatives as the final products.

The possibility of separating axial and equatorial conformers of comparable stabilities of monosubstituted chair cyclohexanes has intrigued chemists. The first completely stable axial conformers at room temperature of monosubstituted cyclohexanes which result from steric hindrance was discovered by the chemists in Beijing Normal University[39].

Recently, researchers in Xiamen University[40] have performed the arc discharge experiments in a different way where the discharge reaction is allowed to proceed in liquid chlorine rather than in the gas atmosphere. The products including perchlorinated aromatic compounds were obtained. The results in this work are of significance in understanding the formation mechanism of fullerenes.

A three-component tandem double-addition ± cyclization reaction that provided an efficient route to polysubstituted cis-pyrrolidine derivatives with matched relay and excellent regio- and stereoselectivity was developed in CAS Shanghai Institute of Organic Chemistry[41-42]. The enantioselective catalytic oxidative coupling of 2-naphthols by a series of novel achiral biphenol-derived diastereomeric oxovanadium(iv) complexes was studied in CAS Chengdu Institute of Organic Chemistry[43]. This discovery not only provides practical catalysts for the asymmetric synthesis of 1,1-binaphthols but is also of significant fundamental interest. Researchers in Shanghai Institute of Organic Chemistry [44] have developed a Pd(II)-catalyzed cyclization reaction of alkynes with carbon ± heteroatom multiplebonds under mild conditions. This reaction with high atomic economy, did not require organometallic reagents, additives, or redox systems.

Researchers in Zhejiang University[45] have shown that spontaneous and quantitative encapsulation of a drug could occur by deposition within the interior of a macromolecule, which specifically binds to it and initiates its precipitation. This process might be useful in controlling the sustained release of water-soluble agents.

Scientists in CAS Shanghai Institute of Organic Chemistry[46] have clearly demonstrated that two kinds of reducing systems ($NaBH_4/Me_3SiCl$ or $NaBH_4/BF_3 \cdot OEt_2$) in combination with the new, chiral and polymer-supported Nsulfonamide, are effective in the reduction of prochiral ketones. The best ee% values are obtained when the reduction is carried out in refluxing THF. The chiral polymer can be recovered easily and reused.

The reactions of organolithium compounds with carbon monoxide are most straightforward for introducing carbonyl groups into organic molecules and have attracted much attention for decades. A convenient and versatile one-pot preparation of 3-cyclopentenone derivatives with excellent regio- and stereoselectivities the fundamental protocol of RLi/CO systems was developed in Beijing University[47].

Nitric oxide (NO), known as the simplest intra and intercellular signaling molecule, plays the key role in regulating many important physiological functions in living bodies. Chemists in Nankai University[48] reported that the establishment of the first Y-NO bond energy was scaled by direct calorimetric measurements combined with relevant electrochemical data for three types of *N*-nitroso compounds, in order to facilitate the understanding of the driving force for NO release and capture. By using a modified eletrochemical method they also investigated of the methylene C-H bond dissociation energies (BDEs) of the onium-substituted toluene series and of the adjacent electron-pulling group-substituted acetophenone and fluorene series. Two series of bond cleavage energies of the S-NO bond in RSNOs (S-nitrosothiols) were obtained in acetonitrile with direct calorimetry measurements through a new thermodynamic approach. The first two series of Co-NO bond dissociation enthalpies in benzonitrile solution were determined by titration calorimetry with suitable thermodynamic cycles. The energetic and structural information disclosed in the work is believed to furnish hints to the understanding of cobalt nitrosyl porphyrins' biological functions *in vivo*.

Medicinal study is one of the most important fields in Organic Chemistry. Design, synthesis and bioconjugate studies of nucleic acid derivatives and analogues were carried out in Beijing University[49], and some interesting results have been obtained. Excitatory neuronal transmission within the central nervous system (CNS) is mediated predominantly by L-glutamate, which plays a role of utmost importance in many physiological processes such as neural plasticity, memory, and learning. Researchers in CAS Shanghai Institute of Organic Chemistry [50] synthesized 2,5-disubstituted pyrrolidines A, which were envisioned to be ideal building blocks for synthesizing the target molecule because its 2-position could be lithiated and then coupled with (*R*)-Garner aldehyde toassemble the right-hand part of kaitocephalin, and its C-C double bond could be converted into the left-hand moiety through Sharpless asymmetric dihydroxylation. They completed the total synthesis of kaitocephalin, the first naturally occurring AMPA/KA receptor antagonist.

3. PHYSICAL CHEMISTRY AND THEORETICAL CHEMISTRY

Many achievements in the field of Physical and Theoretical Chemistry were made before 1990s, e.g. Coordination Field Theory, Graphics of Molecular Orbital Theory and its application, new concept of atomic value and nxcπ rule, electronic structures of rare earth elements compounds, modified Huckel MO Theory, reaction dynamics of molecular beam, application of Group Theory in quantum chemistry, structures of Mo, Fe and S atomic clusters, studies on SPM, etc.

Scientists in Xiamen University[51] reported that a block-localized wave function was used to examine the stereoelectronic effects on the origin of the structural difference between trisilylamine and trimethylamine. The pyramidal geometry of trimethylamine along with its high basicity is consistent with the traditional VSEPR (valence shell electronpair repulsion) model for σ bonding. On the other hand, in trisilylamine, the d orbitals of silicon make modest contribution to the electronic delocalization, although the key factor in charge delocalization is still the $n_N \rightarrow \sigma_{SiH}$ negative hyperconjugation.

Electron transport in supramolecular assemblies containing redox centers has been a subject of great interest. Researchers in CAS Institute of Chemistry[52] described a new interfacial architecture that combined the high redox concentration in 3-D polymers and controllable structure and functionality of the 2-D monolayer systems. The new interface utilizes structurally defined redox liposomes engineered with biomolecular recognition capability by incorporating cell surface receptor GM1 into the bilayer membrane.

Methylnitrene, the CH_3N diradical, is one of the most interesting of all organic free radicals, and it has drawn a lot of attention in both experiments and theory. The CAS Institute of Chemistry presented the experimental and theoretical determinations of the ionization energies of different ionic states for the CH_3N diradical, this is the first report on ionization energy for the nitrenes[53].

Scientists in CAS Institute of Photographic Chemistry[54] studied the photosensitized oxidations of the olefins, *trans*-1,2-dimethoxystilbene (DMOS), *trans*-stilbene (TS), and *trans,trans*-1,4-diphenyl-1,3-butadiene (DPB) as well as the amine 2,2,6,6-tetramethylpiperidine (TMP) in mixed surfactant vesicles. All of the observations suggest that singlet oxygen generated in the bilayer or the inner water pool of one vesicle is able to diffuse out and enter into the bilayer of another vesicle through the aqueous dispersion and react with the target molecules. They extended the study of photosensitized oxidation of alkenes to Nafion membranes to establish the scope of the microreactor controlled selectivity approach.

A series of new stable fullerides A_3C_{70} (A = Sm, Ba) were synthesized by solid-state reactions in USTC[55]. The structure of Sm_3C_{70} has been identified to be monoclinic by simulation of synchrotron powder diffraction data. The structure of Ba_3C_{70} is derived from the A15 structure adopted by Ba_3C_{60}. The low symmetry relative to C_{60} fullerides could be an important factor in the absence of superconductivity in C_{70} fullerides.

Scientists in CAS Dalian Institute of Chemical Physics[56] established the double resonance spectroscopy (OODR-MPI) and molecular dynamics method, and found evidence for quantum interference in collision-induced intramolecular energy transfer within CO singlet-triplet mixed states.

It is a dream for chemistry to directly form or break a selective chemical bond. Modern laser and single molecular techniques make it possible. However, it is necessary to complete two basic research: one is search method to let the energy focus on one bond and stay a period of time without energy transfer, the other is directly observe the chemical bond inner molecules. Scientists in USTC studied the local mode vibrational states in multi-atom molecules[57], and established a new field of Bond-selective Chemistry (Fig. 4).

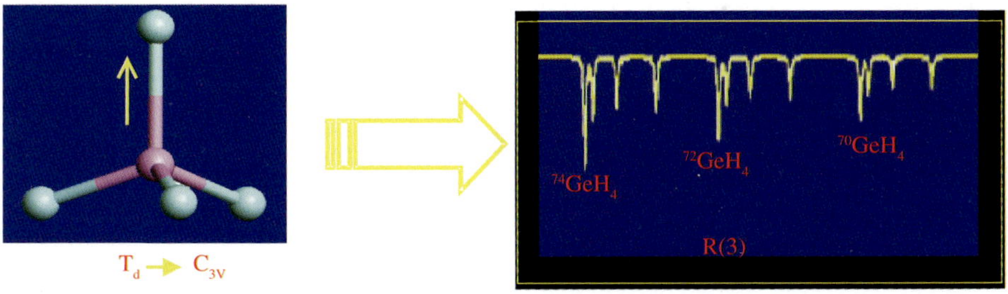

Fig.4 Symmetries decrease, change of vibrational spectrum of Germane in local mode vibration

Two-dimensional systems possess a unique topological ordering that is not found in either three- or one-dimensional systems. Using high-resolution scanning tunnelling microscopy, the results from USTC[58] showed that a C_{60} array on a self-assembled monolayer of an alkylthiol forme an ideal two-dimensional system which has another novel topological order originating from the orientational degrees of freedom. At a temperature of 5 K, the two-dimensional C_{60} forms a domain structure in which the correlation function of the molecular orientation within a domain is constant anywhere (so every C_{60} has the same orientation) but

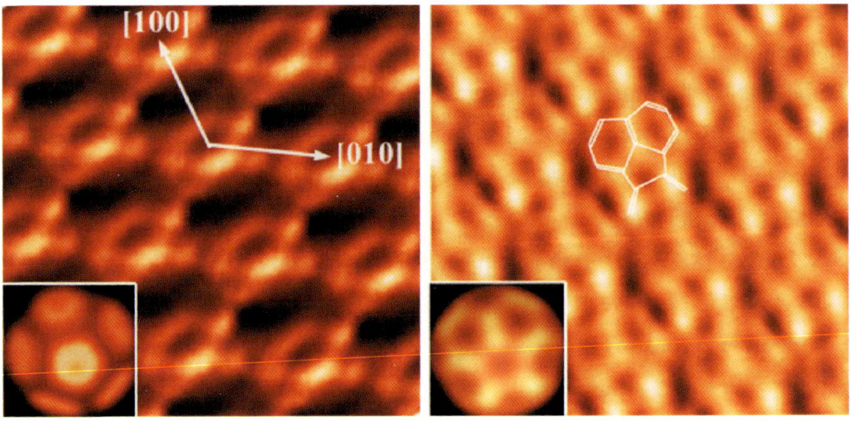

Fig.5 STM image of C_{60} molecule with clear chemical bonds

changes abruptly at domain boundaries (Fig. 5).

The epoxidation of olefins with H_2O_2 was performed with a tungsten-containing catalyst. This insoluble catalyst forms soluble active species by the action of H_2O_2, and when the H_2O_2 is used up, the catalyst precipitates for easy recycling. Thus, the advantages of both homogeneous and heterogeneous catalysts are combined in one system through reaction-controlled phase transfer of the catalyst. When coupled with the 2-ethylanthraquinone/2-ethylanthrahydroquinone redox process for H_2O_2 production, O_2 can be used for the epoxidation of propylene to propylene oxide with 85% yield based on 2-ethylanthrahydroquinone without any co-products. This approach avoids the problematic co-products which are normally associated with the industrial production of propylene oxide[59].

Masses of single-walled carbon nanotubes (SWNTs) with a large mean diameter of about 1.85 nm, synthesized by a semicontinuous hydrogen arc discharge method in CAS Shenyang Institute of Metal Research[60], were employed for hydrogen adsorption experiments in their as-prepared and pretreated states. A hydrogen storage capacity of 4.2 weight percent, 78.3 percent of the adsorbed hydrogen could be released under ambient pressure at room temperature.

Asymmetric adsorption of chiral molecules on chiral crystals may occur, and it has been applied to liquid chromatography for chiral separations. Moreover, in terms of the asymmetric adsorption of amino acids on chiral quartz, scientists tried to explain the dominance of L-amino acids in the living organisms. Researchers in Beijing University[61] studied the adsorption of amino acids on Cu(001), and unexpectedly found the chiral restructuring phenomenon of steps induced by adsorption of L-lysine [(2*S*)-2,6-diaminocaproic acid, NH_2-$(CH_2)_4$-$CH(NH_2)$-COOH]. This experiment indirectly proves that chiral transition metal surfaces are able to exhibit enantioselectivity.

Cisplatin was widely used as anticancer drug in the world. The hydrolysis of it was believed to be the key activation step before the drug reaching its intracellular target DNA. The hydrolysis of cisplatin and the hydrolysis of $[Pt(dien)Cl]^+$(dien = diethylenetriamine), were studied by Hartree-Fock (HF) *ab initio*, electron correlation methods, pure density functional theory (DFT) methods, and hybrid HF-DFT methods with several basis sets in Nanjing University[62]. The results provided theoretical guidance on detailed understanding on the mechanism of the hydrolysis of cisplatin, which could be useful in the design of novel Pt-based anticancer agents.

Phosphodiesterases are clinical targets for a variety of biological disorders. Understanding the structure and mechanism of a phosphodiesterase will provide a solid basis for rational design of the more efficient therapeutics. Although a three-dimensional X-ray crystal structure of the catalytic domain of human phosphodiesterase 4B2B is recently reported, it is still uncertain that a critical bridging ligand in the active site is a water molecule or a hydroxide ion. The identity of this bridging ligand was theoretically determined by the researchers in Huazhong Normal University with quantum chemical calculations on models of the active site[63]. The results indicate that this critical bridging ligand in the active site of the reported X-ray crystal structure is a hydroxide ion, rather than a water molecule, expected to serve as the nucleophile to initialize the catalytic degradation of the intracellular second messengers.

Phycobiliproteins, including phycoerythrin, phycoerythrocyanin, phycocyanin and allophycocyanin, are brilliantly colored, highly fluorescent components of the photosynthetic light-harvesting antenna complexes of cyanobacteria, red algae, and cryptomonads. In the past decades, studies of phycobiliproteins focused mainly on the photophysical processes of light capture and energy transfer of the different chromophores of these proteins' communication. Researchers in CAS Institute of Photographic Chemistry[64] reported that the detection of reactive oxygen species, i.e. superoxide radical anion ($O_2^{\cdot -}$), hydrogen peroxide (H_2O_2) and hydroxyl radical (· OH), generated from the photosensitization of the three phycobiliproteins, *R*-phycoerythrin (*R*-PE), *C*-phycocyanin (*C*-PC), and allophycocyanin (APC). They have demonstrated the photosensitized formation of reactive oxygen species (·OH, H_2O_2, and $O_2^{\cdot -}$) upon irradiation of the three phycobiliproteins (*C*-PC, *R*-PE, and APC) with light >470 nm, which may provide an understanding of both their biological functions and their photodynamic action mechanisms.

The gas phase He photoelectron (PE) spectrum of $BrONO_2$, an important atmospheric trace species, was reported by the researchers in Lanzhou University[65]. The assignment for the PE spectral bands of $BrONO_2$ was made on the basis of band shape, fine vibrational structures, and *ab initio* calculations.

A recent effort in nanomaterial research has led to the emergence of a new class of exotic materials that are composed of complex building blocks such as stable large molecules and clusters. In these materials, there exists an important parameter that is absent in more conventional materials: the orientational degrees of freedom of the building blocks. Researchers in USTC [66] investigated the orientationally ordered structures of two-dimensional (2D) C_{60} at low temperature theoretically and experimentally. A majority of these theoretical findings are confirmed by the low-temperature STM study of a 2D C_{60} array supported on a self-assembled monolayer.

The chirality of (R)-PPA and (S)-PPA molecules has been observed on Cu(111) in solution by using ECSTM in Beijing University[67]. Although the two molecules adsorbed on Cu(111) are in the same symmetry, the molecular arrangements are different from each other. The chirality of the two molecules, which arises from the orientation of the methyl and $CONH_2$ groups, can be directly seen in higher-resolution STM images. The principle used maybe relevant for the development of a simple method for discriminating chiral molecules in aqueous solution.

A new luminescent material 1,6-bis(2-hydroxy-5-methylphenyl)pyridine ((mdppy)BF) was synthesized and its electroluminescent properties were investigated in Jilin University[68]. Highly efficient white-light electroluminescent devices can be fabricated from this material. This offers the opportunity to construct high-performance white EL devices based on only one emitting material and with a simple device structure.

4. POLYMER CHEMISTRY AND PHYSICS

Before 1990's, many important results were obtained as oriented polymerization catalyzed by rare-earth compounds, synthesis and application of big hole ionic exchange resin, polymer degradation and graft copolymerization under ultrasonic irradiation, statistic theory of polymer condensation polymerization, addition polymerization and graft, research on conductive polypyrrole.

Synthesis of novel macromolecular materials is one of the hot topics in chemistry. Recently, an unusual narrowly distributed rod-coil diblock copolymer, poly(styrene-*block*-(2,5-bis[4-methoxyphenyl] oxycarbonyl)styrene) (PS-*b*-PMPCS) was synthesized by 2,2,6,6-tetramethyl-l-piperidinyloxy (TEMPO) mediated living free radical polymerization, synthesized in USTC[69]. The rodlike PMPCS block is a mesogen-jacked liquid crystalline polymer and is soluble in *p*-xylene at temperatures higher than 100 ℃. When a PS-*b*-PMPCS dilute solution is cooled, the copolymer chains can self-assemble into a core-shell nanostructure.

Nanostructured materials with mesoscopic dimensions have been intensely studied in recent years because they may exhibit novel electronic, optical, magnetic, and chemical properties. Scientists in Beijing University[70] reported the synthesis and structure of a novel one-dimensional nanopolymer, $\{[H_3O]^+{}_{12}\{(H_2O)MoO_{2.5}[Mo_{36}O_{108}(NO)_4(H_2O)_{16}]O_{2.5}Mo\text{-}(H_2O)\}_{12^-}\}_n \cdot 44nH_2O$, obtained from the discrete nanomolecule $[H_3O]^+{}_{12}[Mo_{36}O_{108}(NO)_4(H_2O)_{16}]^{12-}$. The synthesis and properties of this kind of polyoxometalates may lead to new quantum lines based on molecular quantum dots.

Self-assembly of macromolecules is an efficient and rapid pathway to supramolecular nanometer- or micrometer-sized objects, which are almost impossible tobe synthesized through conventional chemical reactions. In Fudan University[71], efforts have been devoted to produce micelles without the chemical bonds between the core and shell. For example, it was found that mixing polystyrene (PS) oligomers with a carboxyl end-group and poly(4-vinyl pyridine) (PVPy) in their common solvent chloroform, "graft copolymer" with the former as branches and the latter as backbone was formed due to the hydrogen bonding between the carboxyl and pyridine groups. The "graft copolymer" kept soluble in chloroform and formed micelles when a precipitant for one of the blocks was added. These noncovalently connected micelles behaved just like the conventional ones. In these investigations, both the backbone and grafts are flexible polymer coils.

A novel soluble aryl amide-bridged ladderlike polymethylsiloxane (A-LPMS) was synthesized in CAS Institute of Chemistry by stepwise coupling polymerization on the basis of amido H-bonding self-assembling template from monomer N, N'-bis(3-methyldiethoxylsilylpropyl)-[4,4'-oxybis(benzyl amide)][72]. The monomer was prepared in a high yield by the hydrosilylation reaction of template agent N, N' -diallyl-[4,4' -oxybis(benzyl amide)] with methyldiethoxysilane in the presence of dicyclopentadienylplatinum dichloride (Cp_2PtCl_2) as a catalyst.

The functionality of polymer films by the incorporation of semiconductor nanoparticles is of great importance in current chemistry, physics, and materials science because of their potential applications in fabricating devices with optical, electric, and magnetic properties. Researchers in Jilin University[73] reported a controlled fabrication of cross-linked nanoparticles/polymer compositing thin films through the combined use of surface-initiated atom transfer radical polymerization and gas/solid reaction.

Designing and constructing three-dimensional (3D) coordination polymers with well-defined pores is of current interest for the creation of new zeolite-type materials with potential applications in catalysis, gas storage, chemical separation, and ion exchange. A synthesis and crystal structures of three 3D lanthanide transition metal coordination polymers of the form $[\{M_2(Cu_3(\text{iminodiacetate})_6)\}\cdot 8H_2O]_n$, where M = La, Nd, and Eu were carried out in Xiamen University[74]. Investigation of the thermal stability of the coordination polymers shows that their frameworks retained their stability when the guest molecules were removed.

The construction of metal organic framework (MOF) coordination polymers is currently receiving considerable attention owing to their potential properties as functional solid materials, as well as their fascinating framework structures. Two novel chiral open MOFs with rare α and β quartz structures have been successfully synthesized: QMOF-1, prepared by using a simple diffusion method at room temperature has a neutral open framework with the low symmetry of a-quartz and a large left-handed channel; QMOF-2, prepared by solvothermal method at 160℃ in Fudan University[75]. A wide variety of coordination polymers can be produced by hydrothermal generation of bridging ligands in the presence of appropriate metal ions. Recently, a new method for synthesis of coordination polymers was developed in Nanjing University[76].

Researchers in Fudan University[77] have succeeded in obtaining micelles and hollow nanospheres by a block-copolymer-free strategy. The combination of poly(e-caprolactone) (PCL) and MAF (formed from methylacrylic acid (MAA), macro-monomer FA $[CH_2\cdot CHCOOCH_2\text{-}CH_2(OCOCH_2CH_2CH_2CH_2CH_2)_3OH]$ and methyl methacrylate) led to the formation of micelles possessing a hydrophobic

PCL core surrounded by a hydrophilic MAF shell. The central layer, composed of PCL branches in MAF, glues the outer shell to the core. Crosslinking of the micelles gives rise to stable shell-cross-linked nanoparticles. As this route is based on common and nonexpensive polymer products and the processing is simple, it may have great potential for further applications in fabricating supramolecuar materials for various fields such as medicine, catalysis, and photoelectronics.

Aligned polyacrylonitrile nanofibers were synthesized in CAS Institute of Chemistry[78] by a simple extrusion process with an anodic aluminum oxide membrane as the template. The surface of the synthesized nanofibers, without any modification by materials of low surface energy, proved to be super-hydrophobic. It is believed that not only the nanostructure of the nanofibers but also their lower density contribute to the very large fraction of air in the surface, which is essential to the super-hydrophobicity. This may provide a new method of controlling the wettability of nanostructured surfaces by changing the density of the aligned fibers. They also directly prepared C_{60} nanotubes from C_{60} powder[79].

In summary, the chemists in China have made plenty of important achievements in chemistry research during recent years, especially last decades. It is believed that there will be much more systematic and high-level research work in the near future.

REFERENCES

1 Han W, Fan S S, Synthesis of Gallium Nitride Nanorods through a Carbon Nanotube-Confined Reaction. Science, 1997(277): 1287-1289.

2 Li W Z, Xie S S, Qian L X, et al. Large-Scale Synthesis of Aligned Carbon Nanotubes. Science, 1996 (274): 1701-1703.

3 Sun L F, Xie S S, Liu W, et al, Creating the Narrowest Carbon Nanotubes. Nature, 2000 (403): 384.

4 Pan Z W, Xie S S, Chang B H, et al., Very Long Carbon Nanotubes. Nature, 1998 (394): 631-632.

5 Li Y D, Qian Y T, Liao H, et al. A Reduction-Pyrolysis-Catalysis Synthesis of Diamond. Science, 1998, 281: 246-247.

6 Xie Y, Qian Y, Wang W, et al. A Benzene-Thermal Synthetic Route to Nanocrystalline GaN. Science, 1996, 272: 1926-1927.

7 Zhu H W, Xu C L, Wu D H, et al. Direct Synthesis of Long Single-Walled Carbon Nanotube Strands. Science, 2002, 296:884-886.

8 Wu C, Lu C, Zhuang H, et al. Hydrothermal Assembly of a Novel Three-Dimensional Framework Formed by $[GdMo_{12}O_{42}]^{9-}$ Anions and Nine Coordinated GdIII Cations. J. Am. Chem. Soc., 2002 (124): 3836-3837.

9 Li Y, Wang J, Deng Z, et al. Bismuth Nanotubes: A Rational Low-Temperature Synthetic Route. *J. Am.* Chem. Soc., 2001(123): 9904-9905.

10 Lu Q, Gao F, Zhao D, et al. Creation of a Unique Self-Supported Pattern of Radially Aligned Semiconductor

Ag2S Nanorods. Angew. Chem. Int. Ed., 2002(41):1932-1934.

11 Yang J, Xue C, Yu S, et al. General Synthesis of Semiconductor Chalcogenide Nanorods by Using the Monodentate Ligand n-Butylamine as a Shape Controller. Angew. Chem. Int. Ed., 2002 (41): 4697-4701.

12 Wang X, Li Y, Synthesis and Characterization of Lanthanide Hydroxide Single-Crystal Nanowires. Angew. Chem. Int. Ed., 2002 (41): 4790-4793.

13 Liu X, Tian B, Yu C, et al. Room-Temperature Synthesis in Acidic Media of Large-Pore Three-Dimensional Bicontinuous Mesoporous Silica with Ia3d Symmetry. Angew. Chem. Int. Ed., 2002(41):3876-3879.

14 Wang X, Li Y, Selected-Control Hydrothermal Synthesis of α- and β -MnO_2 Single Crystal Nanowires. J. Am. Chem. Soc., 2002(124): 2880-2881.

15 Li Y, Duan X, Qian Y, et al. Solvothermal Co-reduction Route to the Nanocrystalline III-V Semiconductor InAs. J. Am. Chem, Soc., 1997(119): 7869-7870.

16 Liu L, Fan S, Isotope Labeling of Carbon Nanotubes and Formation of ^{12}C-^{13}C Nanotube Junctions . J. Am. Chem. Soc., 2001, 123(46): 11502-11503.

17 Li J, Ke Y, Zhang Y, et al. ((DABCO)·ZnGe(HPO4)3: The First Zinco-Germanophosphate with a Unique Asymmetric Cage. J. Am. Chem. Soc., 2000 (122): 6110-6111.

18 Fan J,Yu C, Wang L,et al. Mesotunnels on the Silica Wall of Ordered SBA-15 to Generate Three-Dimensional Large-Pore Mesoporous Networks. J. Am. Chem. Soc., 2001(123): 12113-12114.

19 Wang W, Geng Y, Yan P, et al. A Novel Mild Route to Nanocrystalline Selenides at Room Temperature. J. Am. Chem. Soc., 1999 (121): 4062-4063.

20 Huang L, Wang Z, Sun J, et al. Fabrication of Ordered Porous Structures by Self-Assembly of Zeolite Nanocrystals. J. Am. Chem. Soc., 2000, 122(14): 3530-3531.

21 Zhang Z, Han Y, Zhu L, et al. Strongly Acidic and High-Temperature Hydrothermally Stable Mesoporous Aluminosilicates with Ordered Hexagonal Structure. Angew. Chem. Int. Ed., 2001 (40):1258-1263.

22 Cheng W, Dong S, Wang E, et al. Iodine-Induced Gold-Nanoparticle Fusion /Fragmentation/Aggregation and Iodine-Linked Nanostructured Assemblies on a Glass Substrate. Angew. Chem. Int. Ed., 2003(42): 449-452.

23 Li Y, Li X, Deng Z, et al. From Surfactant ± Inorganic Mesostructures to Tungsten Nanowires. Angew. Chem. Int. Ed. 2002 (41):333-337.

24 Jin J, Iyoda T, Cao C, et al.Self-Assembly of Uniform Spherical Aggregates of Magnetic Nanoparticles through p ± p Interactions, Angew. Chem. Int. Ed. 2001 (40): 2135-2138.

25 Li Q, Wang X, Zhu D, et al. Super-™Amphiphobico Aligned Carbon Nanotube Films. Angew. Chem. Int. Ed. 2001(40):1743-1747.

26 Jiang X, Establishment and Successful Application of the (σ_{JJ}^* Scale of Spin-Delocalization Substituent Constants. Acc. Chem. Res. 1997(30):283-296.

27 Chu Q, Wang Z, Huang Q, et al. Fluorine-Containing Donor-Acceptor Complex: Infinite Chain Formed by Oxygen Iodine Interaction. J. Am. Chem. Soc. 2001(123):11069-11070.

28 You S, Zhu X, Luo Y, et al. Highly Regio- and Enantioselective Pd-Catalyzed Allylic Alkylation and Amination of Monosubstituted Allylic Acetates with Novel Ferrocene *P,N*-Ligands. J. Am. Chem. Soc. 2001(123): 7471-7472.

29 Wang Z, Huang M, Evidence for Higher Reactivity of B2. Theoretical Studies of Insertions of B2 and B into Methane and Water. J. Am. Chem. Soc. 1998 (120): 6758-6760.

30 Zhou J, Tang Y, Sidearm Effect: Improvement of the Enantiomeric Excess in the Asymmetric Michael Additionof Indoles to Alkylidene Malonates. J. Am. Chem. Soc. 2002(124):9030-9031.

31 Song L, Fan H, Hu Q, The First Example of Macrocycles Containing Butterfly Transition Metal Cluster

Cores via Novel Tandem Reactions. J. Am. Chem. Soc. 2002 (124): 4566-4567.

32 Hong M, Zhao Y, Su W, et al. A Nanometer-Sized Metallosupramolecular Cube with *Oh* Symmetry. J. Am. Chem. Soc. 2000 (122): 4819-4820.

33 Zhang Q, Lu X, Highly Enantioselective Palladium(II)-Catalyzed Cyclization of (Z)-4′ -Acetoxy-2′ -butenyl 2-Alkynoates: An Efficient Synthesis of Optically Active *A*-Butyrolactones. J. Am. Chem. Soc. 2000 (122): 7604-7605.

34 Dai J, Bian G, Wang X, et al. A New Method to Synthesize Unsymmetrical Dithiolene Metal Complexes of 1,3-Dithiole-2-thione-4,5-dithiolate for Third-Order Nonlinear Optical Applications. J. Am. Chem. Soc. 2000 (122): 11007-11008.

35 Fu H, Yao J, Size Effects on the Optical Properties of Organic Nanoparticles. J. Am. Chem. Soc. 2001 (123): 1434-1439.

36 Long J, Hu J, Shen X, et al. Discovery of Exceptionally Efficient Catalysts for Solvent-Free Enantioselective Hetero-Diels-Alder Reaction. J. Am. Chem. Soc. 2002 (124): 10-11.

37 Ye S, Huang Z, Xia C, et al. A Novel Chiral Sulfonium Yilde: Highly Enantioselective Synthesis of Vinylcyclopropanes. J. Am. Chem. Soc. 2002 (124): 2432-2433.

38 Chen J, Song Q, Wang C, et al. Novel Cycloaddition of Nitriles with Monolithio- and Dilithiobutadienes. J. Am. Chem. Soc. 2002(124):6238-6239.

39 Kang F, Yin C. First Completely Stable Axial Conformers of Monosubstituted Cyclohexanes at Room Temperature. J. Am. Chem. Soc. 1997, 119:8562-8563.

40 Xie S, Huang R, Deng S, Synthesis, Separation, and Characterization of Fullerenes and Their Chlorinated Fragments in the Glow Discharge Reaction of Chloroform, J. Phys. Chem. B, 105 2001 (9):1734 -1738.

41 Ma S, Jiao N, Pd^0-Catalyzed Three-Component Tandem Double-Addition ± Cyclization Reaction: Stereoselective Synthesis of cis-Pyrrolidine Derivatives. Angew. Chem. Int. Ed. 2002 (41): 4737-4740.

42 Ma S, Yu Z, Oxidative Cyclization ± Dimerization Reaction of 2,3-Allenoic Acids and 1,2-Allenyl Ketones: An Efficient Synthesis of 4-(3'- Furanyl)butenolide Derivatives. Angew. Chem. Int. Ed. 2002 (41):1775-1778.

43 Luo Z, Liu Q, Gong L, et al. Novel Achiral Biphenol-Derived Diastereomeric Oxovanadium(iv)Complexes for Highly Enantioselective Oxidative Coupling of 2-Naphthols.Angew. Chem. Int. Ed. 2002 (41): 4532-4535.

44 Zhao L, Lu X, PdII-Catalyzed Cyclization of Alkynes Containing Aldehyde, Ketone, or Nitrile Groups Initiated by the Acetoxypalladation of Alkynes.Angew. Chem. Int. Ed. 2002 (41): 4343-4346.

45 Gao C, Donath E, Hwald H, et al.Spontaneous Deposition of Water-Soluble Substances into Microcapsules: Phenomenon, Mechanism, and Application. Angew. Chem. Int. Ed. 2002 (41): 3789-3793.

46 Hu J, Zhao G, Ding Z, Enantioselective Reduction of Ketones Catalyzed by Polymer-Supported Sulfonamide Using $NaBH_4/Me_3SiCl$ (or $BF_3 \cdot OEt_2$) as Reducing Agent. Angew. Chem. Int. Ed. 2001 (40):1109-1111.

47 Song Q, Chen J, Jin X, et al. Highly Regio- and Stereoselective 1,1-Cycloaddition of Carbon Monoxide with 1,4-Dilithio-1,3-dienes. Novel Synthetic Methods for 3-Cyclopenten-1-one Derivatives. J. Am. Chem. Soc. 2001(123): 10419-10420.

48 Cheng J, Xian M, Wang K, et al. Heterolytic and Homolytic Y-NO Bond Energy Scales of Nitroso Containing Compounds: Chemical Origin of NO Release and NO Capture. J. Am. Chem. Soc. 1998 (120): 10266-10267; Cheng J, Liu B, Zhao Y. et al. Homolytic Cleavage Energies of R-H Bonds Centered on Carbon Atoms of High Electronegativity: First General Observations of O-type Variation on C-H BDEs and the Implication for the Governing Factors Leading to the Distinct O/S Patterns of Radical Substituent

Effects. *J. Am. Chem. Soc.*, 2000, 122(41): 9987-9992; Zhu X, Li Q, Hao W, et al. Dissociation Energies and Charge Distribution of the Co-NO Bond for Nitrosyl-,-tetraphenylporphinatocobalt (II) and Nitrosyl-,,.-tetraphenylporphinatocobalt (III) in Benzonitrile Solution. J. Am. Chem. Soc., 2002, 124 (33): 9887-9893; Lu J, Wang K, Wittbrodt J M, et al. NO Affinities of S-Nitrosothiols: A Direct Experimental and Computational Investigation of RS-NO Bond Dissociation Energies. *J. Am. Chem. Soc.* 2001, 123, 2903-2904 .

49 Li X, Zhang L R, Lu J F, et al, Signal peptide mimics conjugated to peptide nucleic acid: A promising solution for improving cell membrane permeability, Bioconjugate Chem. 2003, 14 (1): 153-157 ; Wang X F, Zhang L H, Ye X S, Recent development in the design of sialyltransferase inhibitors, Med. Res. Rev., 2003, 23 (1): 32-47.

50 Ma D, Yang J. Total Synthesis of Kaitocephalin, the First Naturally Occurring AMPA/KA Receptor Antagonist. J. Am. Chem. Soc. 2001 (123): 9706-9707

51 Mo Y, Zhang Y, Gao J, A Simple Electrostatic Model for Trisilylamine: Theoretical Examinations of the Negative Hyperconjugation, Bonding, and Stereoelectronic Interaction. J. Am. Chem. Soc. 1999 (121): 5737-5742.

52 Cheng Q, Peng T, Stevens R, Signaling of *Escherichia Coli* Enterotoxin on Supramolecular Redox Bilayer Vesicles. J. Am. Chem. Soc. 1999 (*121*): 6767-6768.

53 Wang J, Sun Z, Zhu X, et al.The CH_3N Diradical: Experimental and Theoretical Determinations of the Ionization Energies. Angew. Chem. Int. Ed. 2001 (40): 3055-3057.

54 Li H, Wu L, Tung C H, Reactions of Singlet Oxygen with Olefins and Sterically Hindered Amine in Mixed Surfactant Vesicles. J. Am. Chem. Soc. 2000 (122): 2446-2451; Tung, C H, Wu L Z, Yuan Z Y, Zeolites as Templates for Preparation of Large-Ring Compounds: Intramolecular Photocycloaddition of Diaryl Compounds. J. Am. Chem. Soc. 1998, 120: 11594-11602; Tung, C H, Wang H W, Ying Y M. Photosensitized Oxidation of Alkenes Adsorbed on Pentasil Zeolites. J. Am. Chem. Soc.1998, 120: 5179-5186; Tung C H, Guan J Q, Remarkable Product Selectivity in Photosensitized Oxidation of Alkenes within Nafion Membranes. J. Am. Chem. Soc.1998, 120: 11874-11879;Tung C H, Wu L Z, Zhang L P, et al. Supramolecular Systems as Microreactors: Control of Product Selectivity in Organic Phototransformation, Acc. Chem. Res. 2003, 36: 39-47.

55 Chen X, Chi D, Sun Z, et al. Synthesis, Structure, and Transport Properties of Novel Fullerides A_3C_{70} (A) Ba and Sm. J. Am. Chem. Soc. 2000 (122): 5729-5732.

56 Zhang C H, Sha G H, Double resonance spectroscopy and molecular dynamics, Science, 1993 (262): 374-377; Sha G H, He J B, Zhang C H, Evidence for Quantum Interference In Collision – Induced Intramolecular Energy Transfer Within Co Singlet – Triplet Mixed States, J. Chem, Phys., 1995 (102): 2772-2776; Chen X L, Sha G H, Jiang B, et al. Futher Study On Collisional Quantum Interference Effect In Energy Transfer Within Co Singlet – Triplet Mixed States. J. Chem. Phys. 1996 (105):8661-8665.

57 Zhu Q S, Effective rotational Hamiltonian of the local mode vibrational states, Spectrochimica Acta, 1992, 48A, 193;Zhu Q S.,Li H F and Wang X G, Anharmonicity: a Route to Vibrational Localization,Chem. Phys. Letts. 1993, 212, 403; Child M S and Zhu Q S, Local mode Coriolis quenching in tetrahedral hydrides,Chem. Phys. Lett., 1993, 207, 116;Wang X G and Zhu Q S,Vibration-rotational spectroscopy of XH_2 and XH_3 type molecules near local mode limit, J. Chem. Phys., 1996, 105 (17):8011-8019.

58 Hou J G, Yang J L, Wang H Q, et al. Topological of two-demensional C_{60} domains. Nature, 2001 (409): 304-305.

59 Xi Z, Zhou N, Sun Y, et al. Reaction-Controlled Phase-Transfer Catalysis for Propylene Epoxidation to Propylene Oxide. Science, 2001 (292):1139-1141.

60 Liu C, Fan Y, Liu M, et al. Hydrogen Storage in Single-Walled Carbon Nanotubes at Room Temperature. Science, 1999 (286):1127-1129.

61 Zhao X. J. Fabricating Homochiral Facets on Cu(001) with L-lysine. J. Am. Chem. Soc. 2000 (122): 12584-12585.

62 Zhang Y, Guo Z, You X. Hydrolysis Theory for Cisplatin and Its Analogues Based on Density Functional Studies. J. Am. Chem. Soc. 2001 (123): 9378-9387.

63 Zhan C, Zheng F, First Computational Evidence for a Catalytic Bridging Hydroxide Ion in a Phosphodi esterase Active Site. J. Am. Chem. Soc. 2001 (123): 2835-2838.

64 He J, Hu Y, Jiang L. Photochemistry of Phycobiliproteins: First Observation of Reactive Oxygen Species Generated from Phycobiliproteins on Photosensitization. J. Am. Chem. Soc. 1996 (118): 8957-8958.

65 Wang D, Jiang P. He I Photoelectron Spectroscopy Study on the Electronic Structure of Bromine Nitrate, $BrONO_2$. *J. Phys. Chem.* 1996 (100): 4382-4384.

66 Yuan L, Yang J, Wang H, et al. Low-temperature Orientationally Ordered Structures of Two-Dimensional C_{60} . *J. Am. Chem. Soc.*2003 (125): 169-172.

67 Xu Q, Wang D, Wan L, et al. Discriminating Chiral Molecules of (R)-PPA and (S)-PPA in Aqueous Solution by ECSTM. Angew. Chem. Int. Ed. 2002 (41):3408-3411.

68 Liu Y, Guo J, Zhang H, et al. Highly Efficient White Organic Electroluminescence from a Double-Layer Device Based on a Boron Hydroxyphenylpyridine Complex. Angew. Chem. Int. Ed. 2002 (41):182-185.

69 Tu Y, Wan X, Zhang D, et al. Self-Assembled Nanostructure of a Novel Coil-Rod Diblock Copolymer in Dilute Solution. J. Am. Chem. Soc. 2000 (122): 10201-10205.

70 Zhang S, Wei Y, Yu Q, et al. Mesoporous Aluminosilicates with Ordered Hexagonal Structure, Strong Acidity, and Extraordinary Hydrothermal Stability at High Temperatures. J. Am. Chem. Soc. 1997 (119): 6440-6441.

71 Duan H, Chen D, Jiang M, et al. Self-Assembly of Unlike Homopolymers into Hollow Spheres in Nonselective Solvent. J. Am. Chem. Soc. 2001 (123): 12097-12098.

72 Tang H, Sun J, Jiang J, et al. A Novel Aryl Amide-Bridged Ladderlike Polymethylsiloxane Synthesized by an Amido H-Bonding Self-Assembled Template. J. Am. Chem. Soc., 2002 (124):10482-10488.

73 Wang J, Chen W, Liu A, et al. Controlled Fabrication of Cross-Linked Nanoparticles/Polymer Composite Thin Films through the Combined Use of Surface-Initiated Atom Transfer Radical Polymerization and Gas/Solid Reaction. J. Am. Chem. Soc. 2002 (124): 13358-13359.

74 Ren Y, Long L, Mao B, Nanoporous Lanthanide ± Copper(ii) Coordination Polymers: Syntheses and Crystal Structures of [{M2(Cu3(iminodiacetate)6)}·8H2O]n (M = La, Nd,Eu). Angew. Chem. Int. Ed. 2003 (42):532-536.

75 Sun J, Weng L, Zhou Y, et al. QMOF-1 and QMOF-2: Three-Dimensional Metal ± Organic Open Frameworks with a Quartzlike Topology.Angew. Chem. Int. Ed. 2002 (41): 4471-4473.

76 Xiong R, Xue X, Zhao H, et al. Novel, Acentric Metal ± Organic Coordination Polymers from Hydrothermal Reactions Involving In Situ Ligand Synthesis. Angew. Chem. Int. Ed. 2002 (41):3800-3803.

77 Liu X, Jiang M, Yang S, Micelles and Hollow Nanospheres Based on ε- Caprolactone-Containing Polymers in Aqueous Media. Angew. Chem. Int. Ed. 2002 (41):2950-2954.

78 Feng L, Li S, Li H, et al. Super-Hydrophobic Surface of Aligned Polyacrylonitrile Nanofibers. Angew. Chem. Int. Ed. 2002 (41):1221-1224.

79 Liu H, Li Y, Jiang L, et al. Imaging As-Grown [60]Fullerene Nanotubes by Template Technique , J. Am. Chem. Soc. 2002 (124): 13370 - 13371.

Life Sciences and Biotechnology in China

Chen Zhu

Chinese Academy of Sciences

Chen Zhu, *is Vice President of Chinese Academy of Sciences; Life Tenure Professor of Rui Jin Hospital, an affiliate to Shanghai Second Medical University; Director of Chinese National Human Genome Center at Shanghai; and Director of Shanghai Institute of Hematology. He is also Member of Chinese Academy of Sciences; Fellow of the Third World Academy of Sciences; and Foreign Associate of National Academy of Sciences of US. In his academic career, Chen has made important contributions to the development of scientific research in Hematology, Genetics and Human Genome. With years' action and endeavor, what Prof. Chen has accomplished in his scientific research has had a greater impact on the scientific community both at home and abroad, and was reported frequently by such internationally well-known journals like Nature and Science etc. Up to now, he has published a total number of 60 papers in the high level international journals such as PNAS USA, Blood, and Oncogene, etc. with cited rate up to 4000 times. Chen obtained a series of awards for science and technology, including Chang Jiang Scholar Award; the French cancer research prize Prix de l'Oise de la League Nationale contre le Cancer; Chevalier de l'Ordre National de la Legion de la Legion d' Honneur from the French Government; the Second Class Award of Science and Technology Progress Prize; and the Third Class Award of National Natural Science Prize, etc.*

Abstract: This article describes the rationale and basic characteristics of developing life science and bio-technology in China, and the principle research systems and funding sources for life science and bio-technology in China. It also analysizes the major achievements, existing problems and challenges faced in this regards. With that, it defines the priority areas of basic research in China's life sciences as genomics, bioinformatics, identification of genes related to major diseases, molecular biology & biochemistry, cell and developmental biology, neuro-biology, bio-diversity, the systematics and co-evolution of animals and plants, etc. Finally, it foresees the grand perspectives of developing life science and bio-technology in China.

1. RATIONALES TO DEVELOP LIFE SCIENCES & BIOTECHNOLOGY IN CHINA AND THE BASIC FEATURES

1.1 Rationales

Over the past 20 years, life sciences and biotechnology have developed by leaps and bounds, bringing a broad prospect for the growth in medicine, pharmaceutical industry, agriculture and environmental protection, etc. As a field most important to the whole society, which could change the pattern of future industry and economy, bioscience and biotechnology have been drawing increasing attention of every country of the world. China, as a country with a population of nearly 1.3 billion, faces formidable challenges in terms of population growth, health care and agricultural development. The prevention and treatment of various health-threatening diseases needs to be improved, while family-planning policy will have to be carried out for a long time, aging is looming large, and despite of much progress made, China's agricultural development is still rather low in efficiency and poor in quality thereby exerting huge pressure on natural resources and eco-environment. On the other hand, as a big country with rich natural resources, China possesses about 10% of the world's genetic resources including microbials, plants, animals and human genetic resources. While facing the severe tasks in bio-diversity conservation and sustainable utilization, China's rich natural resources provide material conditions for the development of bioscience and biotechnology. China should rely on the contributions of life sciences and biotechnology to achieve a sustainable development.

1.2 Basic Features

The fact that China is a developing country determines that its investment priority in science and technology should only target major social problems. Therefore, emphasis must be laid on an organic integration of national goals and science frontiers, an integration of basic research and application-oriented investigation, and the accomplishment of limited but achievable goals and of breakthroughs in some key areas.

2. PRIMARY RESEARCH SYSTEM & SOURCES OF FUNDING FOR LIFE SCIENCES & BIOTECHNOLOGY IN CHINA

2.1 Research System

China' s modern life science research has maintained a very close relationship with medicine, agronomy and the utilization of biological resources. China's biotechnology research did not get started until the mid-1970s. Modern research systems on life sciences and biotechnology in China includes different influences such as institutes and facilities affiliated with the Chinese Academy of Sciences (CAS), research universities, institutes and units affiliated with different line departments and local scientific organizations. Owing to the efforts by Chinese Government for many years, a state key laboratory system in some of the key areas of life sciences has been in place as a national engineering research center system in some of the critical areas of biotechnology. Within CAS, for example, there are 21 institutes, involved in R&D of life sciences and biotechnology, which can be grouped into four. Beijing Base concentrating on agricultural and environmental issues; Shanghai Base focusing on R&D related to population and human health; Southwestern Base concerning biodiversity conservation; and Mid-south Base on biosafety and eco-system restoration. In addition, there are schools with life sciences or bioengineering related majors in many comprehensive universities.

2.2 Funding Sources

There are a number of funding sources for research in life sciences and the industrialization of biotechnology in China. Major funding agencies are the Ministry of Science & Technology (MOST) of China, the National Natural Science Foundation of China

(NSFC), CAS, other ministries, local governments and concerned enterprises. In 1986, life sciences and biotechnology were listed into the National High Technology Research & Development Program (also known as " 863" Program) as key areas to be funded, which led to a golden period of development for life sciences and biotechnology in China. The National Key Basic Research Development Program ("973" Program in brief), initiated in 1998, also included agriculture, population/health and resources/environment as key areas of funding. The support of the Department of Life Sciences of NSFC accounts for 1/3 of the Foundation's total funding both in terms of number of funded projects and amount of funding. The funding in life sciences and biotechnology from CAS accounts for 15% of its total budget directly from the Chinese Government. Also worthy of note are some national special funds related to life sciences and biotechnology (e.g. "Transgenic Special Fund" , etc.), the "Biotechnology Industry Project" funded by the State Development Planning Commission. Other funding agencies include the Ministry of Health, Ministry of Agriculture, Ministry of Education, State Administration of Forestry, State Administration of Environmental Protection, local governments and enterprises.

3. Key Development Areas & Achievements of Life Sciences in China

Priority basic research areas in China's life sciences have been defined as: genomics, bioinformatics, identification of genes related to major diseases, molecular biology & biochemistry, cell and developmental biology, neuro-biology, the systematics and co-evolution of animals and plants, etc.

3.1 Genomics

In genomics, one of the frontiers of life sciences, China has already made some internationally recognized achievements.

Chinese scientists undertook and completed 1% of international Human Genome Sequencing Project, after the United States, UK, Japan, Germany and France. It only took two years for the Chinese teams to completely finish the sequencing of 30,000,000 base pairs in the short arm of Chromosome 3 assigned by International Human Genome Sequencing Consortium, signifying that China's genomics research has reached the world advanced level. In addition, Chinese scientists have published about 1,000 full-length cDNAs of the novel genes in the public database, as a major contribution to the gene discovery effort of Human Genome Project.

The completion of a draft sequence of *Oryza sativa* L. ssp. *indica* genome is a milestone achievement of China's genomics research. The coverage of the assembled sequences was 92%. Initial analysis showed that the genome has about 466 megabases in size, with an estimated 46,000-56,000 genes. As a result, a paper entitled "A Draft Sequence of the Rice Genome (*Oryza sativa* L. ssp. *indica*)", was published in the April 5, 2002 issue of *Science* together with the Draft of *O. sativa* L. ssp. *japonica* genome completed by Swiss scientists, making a major scientific breakthrough of that time. Dr. Donald Kennedy, Chief-Editor of *Science*, made the comment that the level of China's genomics research has reached that of world class. In November 2002, another group in China finished the high accuracy sequence of *O. sativa* L. ssp. *japonica* chromosome 4, as an integrated part of the International Rice Genome Sequencing Project. The full length of the rice chromosome 4 is 35 million base pairs. The Chinese team reported a precision rate of 99.99% (leaving only seven small gaps), covering 98% of the chromosome's full sequence, which reached the internationally recognized criteria for the complete genomic sequence. The team identified 4,658 genes and worked out the full sequence of the chromosome's centromere, the first centromere sequence even obtained of higher organisms. Meanwhile, the Chinese scientists also sequenced Chromosome 4 of *O. sativa* L. ssp. *indica cv Guangluai,* another subspecies of rice. By comparing the homology of corresponding sequences of continual 2.3 million base pairs of the two subspecies of *indica* and *japonica*, the evolutional relationship of the two rice species has been discovered. The result, a paper entitled "Sequence and Analysis of Rice Chromosome 4", was published in the November 21, 2002 issue of *Nature* and other journals[2, 3], and praised by reviewing editors as a milestone event in rice genome sequencing after the draft sequence of rice genome. In December 2002, Chinese scientists again took the lead in completing the precise map of rice indica full genome. More recently, an important gene involved in rice tillering was characterized in China[4].

In microbial genome sequencing, China has also made breakthroughs. So far, Chinese scientists have completed the full genome sequencing of six microbes, including *Helicoverpa amigera* single nucleocapsid nuclopolyhedrovirus (HaSNPV)[5] and the *Leptospira Interrogans* [6]. China has launched the Swine Genome Sequencing Project with Denmark and the United States. The project is well under way.

China set up its first national gene research center in 1992, mainly focusing on rice genome research. Since then, three national human genome centers have been established in Beijing and Shanghai.

3.2 Biomedicine

With the improvement of people's living standard, increasing attention has been paid to health and life quality. Biomedical research, closest to population and health, has risen from a ramification of biology up to the front stage of life sciences and biotechnology.

Research institutions having activities in biomedicine or related areas include Chinese Academy of Medical Sciences, CAS, Military Academy of Medical Sciences, various medical schools of research universities such as Peking University, Tsinghua University, Fudan University, and major medical centers.

China started research on protein denaturalization in the 1920s. In 1965, China synthesized bovine insulin, the first protein molecule synthesized with biological activities. In the 1970s, Chinese scientists obtained the high resolution 3-D structure of insulin. These achievements were of milestone significance in the course of development in life sciences.

Since the 1990s, China's biomedical research has made remarkable progress and has witnessed significant improvement in overall research standards. An integrated research system from basic biological research to application-oriented research has been established with the initiation of the "National Knowledge Innovation Program" in 1998. Over the past five years, a lot of work has joined the ranks of international standards.

3.2.1 Research on Medical Genomics & Proteomics

The completion of full genome sequencing of human, a number of animals, plants and microbes has sent life sciences into the "Post-Genome Era". Based on the understanding of full sequence of DNA, with the genetic information, scientists will be able to study and explain functions of genomes and thus unravel the mystery of life. Research on functional genomics will provide the theoretical basis for the prevention, diagnosis and treatment of diseases and the designing and development of drugs. In this regard, China has been keeping pace with the world leaders.

In recent years, Chinese scientists have made a series of breakthroughs in the identification and cloning of major disease-conducive genes by bringing the advantage of human genetic resources into full play. The sequential breakthroughs include: gene cloning and functional research in acute promyelocytic leakemia[7]; identification of a batch of genes responsible for monogenic diseases, such as hereditary high-frequency hearing impairment[8], dentinogenesis imperfecta type II with or without progressive hearing loss[9,10], brachydactyly type A-1[11], hereditary cataract[12], heredity atrial fibrillation[13], etc. The loci of susceptibility to Type II (non-insulin-dependent) diabetes, essential hypertension, nasopharyngeal carcinoma[14] and agenesis of permanent teeth (He-Zhao deficiency)[15] have been precisely located. Recently, a batch of genes and gene markers related to the development and progression of primary hepatocellular carcinoma (HCC) have been found by using gene expression profiling and biochips[16]. A complex of a specific peptide from hepatitis B virus (HBV) and the glycoprotein 96 (gp96), a member of the heat-shock protein family, derived from liver tissues of patients with HBV-induced HCC, was isolated and purified. This discovery has potential for engineering tumor vaccines against HCC and chronic HBV infection[17]. The deployment of differential display technique found a defensin gene Bin1b in the epididymis of the rat and confirmed the natural anti-virus function of its product[18]. This work showed, for the first time in the world, that the epididymis has its own immune defensive system. Recently, China initiated, as a world leader in, the liver proteomics and structural genomics project, in an attempt to identify the structure/function of all proteins produced by this organ and to get a better understanding of various liver diseases. In the field of research on the protein structure and function, it was revealed that the protein disulfide isomerase played not only as an enzyme but also as a chaperone in protein folding[19] and that the DsbC was the first enzyme with chaperone activity in the periplasm of Gram-negative bacteria[20].

3.2.2 Research on Stem Cells & Cell Biology

Research on stem cells is one of the most exciting frontier areas in life sciences. It not only touches upon some basic issues in life sciences, but also has significant relevance to human health and disease prevention. So far, Chinese scientists have successfully conducted human somatic cell nuclear transfer, have established human embryo stem cell system spanning over 30 generations, and have mastered techniques allowing *in vitro* human embryomic stem (ES) cell differentiation towards neural and muscle cells, and techniques of gene “knock-in” and “knock-out”. All the above show that Chinese scientists have established the technological platform for further development of regenerative medicine.

In basic research on cell biology, major advancements have also been made in cell cycle control and signal transduction. So far, scientists acquired ES cell system deprived of the calcium channel function and successfully established knock-out mouse model for cardiac muscular ryanodine receptor subtype 2 genes. In addition, scientists obtained single cardiac muscular cells induced from ES cells in a stable manner, examined the dynamic changes of calcium ions in cells and their response to calcium channel modulator, observed the dynamic expression of protein regulating the contractory activities, and discovered the function of calcium signaling regulation protein. It has been proved that ES cell-turned cardiac muscular cells have similar calcium signal mechanism *in vitro* as the corresponding cells *in vivo*.

Recent years have witnessed remarkable progress in neuroscience and cognitive science research in China. After the proposition of topological visual form perception hypothesis in 1982, a recent work on visual system by using functional magnetic resonance provided biological support for the concept [21]. Chinese scientists proved that the *Drosophila* has simple decision making ability while facing contradictory visual clues, in the course of which the mushroom bodies in its brain play a major role [22]. Scientists found that activated G-protein coupled-receptors, by using two signal transduction pathways, may have opposite effects on the growth of axons. This discovery will help further explore new means to accelerate the regeneration and rehabilitation of nerves after impairment[23]. A new mechanism of neuro-signal transduction, Ca^{2+}-independent but voltage-dependent secretion in mammalian dorsal root ganglion neurons, has been found [24]. Breakthroughs have been made in clustered organization of neurons with similar extra-receptive field properties in the primary visual cortex[25] and in gene expression profile of dorsal root ganglions after neuropathic pain[26]. China has already become front runner on the international stage in this regard.

3.2.3 Biochip

One of the most important methods in gene activity research is biochip technology. By highly integrating tens of thousands of DNAs or proteins onto matrix carrier, scientists are able to do large scale parallel analysis on all gene expression products in cells, rapidly and accurately obtain a wealth of bio-information about investigated targets, thus providing the basis for disease diagnosis and treatment. In order to promote the development of biochip technology, two biochip R&D centers have been established in Beijing and Shanghai, respectively.

3.3 Ecology & Biodiversity

China has one of the longest civilizations of the world. A lot of knowledge about ecology in past millennia have been recorded in ancient Chinese literature. However, this knowledge did not form a systematic discipline until recently.

Since the 1920s, Chinese botanists have started a more systematic research on plant ecology and published papers, such as "Preliminary notes on the vegetation & flora of Huangshan, Anwei" (1927)," The woods of Chung-shan, Nanjing "(1932),etc. After that, botanists conducted investigations and characterizations of vegetation conditions in different regions of China and published a numerous papers, which are still of instructive significance to date, e.g. "A synopsis of phytogeography of northern and western China" (1934) ,"A preliminary study of the vegetation of Yunnan" (1939), "A synopsis of phytogeography of southern and western China" (1936) , "Floristic regionalization of China" (1944) , "Phytogeography of Qinling" Mountains" (1947), and "The provisional sketch of the forest geography of China" (1948), among others.

From the early 1950s to late 1980s, China has organized a dozen large-scale vegetation and ecology expeditions, covering most of the country. The scientists have basically identified the vegetation resources in China and produced a series of regional vegetation maps[27].

With the further development of research, Chinese scientists have established in the late 1980s China Eco-system Research Network (CERN) and China Biosphere Reserve Network and thus started nationwide eco-system network research. Since the 1950s, CAS has set up 64 field research stations in major ecological zones across the country, laying down a solid foundation for the establishment of CERN[28].

Setting up nature reserves is a critical way for the *in situ* conservation of biological resources. So far, China has set up nearly 1,000 nature reserves, with a total area of over 7 million hectares, covering 7.5% of China's land area and projected to cover more than 8% by 2010. In 1992, China joined the "United Nations Treaty on the Convention on Biological Diversity" and decided to build up China Biosphere Reserve Network. Currently there are 100 member zones in the China Biosphere Reserve Network and 27 in the International Biosphere Reserve Network. These nature reserves are important bases to protect biodiversity and conduct typical eco-system research. The government has also been very supportive to *ex situ* conservation of biodiversity (e.g. botanical gardens, gene banks, etc.)[29].

In order to solve the increasingly severe eco-environmental degradation problem, greater attention has been paid to the research on ecosystem restoration and re-buildup. The accumulation of data and experience over the past years has laid down the foundation for the investigation on the restoration and re-buildup of degraded ecosystems. Successful research in restoring forest vegetation in southern sub-tropical regions by means of a forestation in desolate areas pushes the restoration work of degraded ecosystem into the material stage. The integrated management and demonstration zone of sandlots in Inner Mongolia has provided another example of the new type of restoration of ecosystem. The impact of alien invasive species on local ecosystem is also beginning to attract concern[30].

Generally speaking, China's ecological research is in transition from the stage of conventional ecology characterized by structural description to that of systematic and experimental ecology whose purpose is to unveil the process and mechanism of ecosystems. The research level of China's ecology is upgrading from following international trends up to an innovative development stage of connecting with world-class level and striding toward international disciplinary frontiers. In practice, the restoration and re-buildup of China's ecosystem are running in closer accordance with national demands.

3.4 Systematic Evolution & Paleontology

Recent years have witnessed many world-class achievements from China in systematic evolution and paleontology. In the area of biological systematics, China has compiled a 255-volume corpus of Flora Sinica, Fauna Sinica and Cryptogamica Sinica, which have been highly praised by overseas peers. Paleontology has also seen numerous achievements, such as the discovery of the fossils of *Sinornithosaurus millenii*, four-winged dinosaurs and a lot of other rare dinosaurs, providing more conclusive proof for the origin of birds. Chinese scientists have also proposed a new hypothesis of bird flying function evolution – arboreal origin hypothesis[31]. Recently, the research on the fossils of head and backbone of the Early Cambrian vertebrate *Haikouichthys* and brain structure has brought about breakthrough progress in the difficult problem of the origin of vertebrates long faced by the academic community[32, 33]. In regard to the origin of humans, through analysis on special markers of 12,127 male Y chromosomes, it has been proved that modern

East Asians originated from Africa, just like other peoples[34]. The mystery of the origin of dogs has also been unveiled on the basis of in-depth investigation on the mitochondria DNA diversity of canine, proving that dogs all over the world have the same genetic base and all originated from East Asia and that they spread to other parts of the world later on[35]. In Hemudu, Jiahu, Pengshantou and elsewhere of China, a great deal of carbonized ancient rice seeds traced back to over 7,000 years ago have been found and provided direct evidence to the origin of rice cultivation.

In the field of co-evolution between insects and plants, Chinese scientists systematically described and named a new reproductive system in ginger plants, i.e. flexible style. Plants with this kind of breeding system fall into two mechanisms: hyperflexistyle and cataflexistyle. Plant individuals of these two types successfully perform allogamy by means of insect medium, thus overcoming the degeneration caused by inbreeding[36].

4. CHINA'S MAJOR ACCOMPLISHMENTS IN BIOTECHNOLOGY AND DRUG DEVELOPMENT

Priority areas for China's R&D in biotechnology and biopharmaeutical research are genetic breeding of high-yield and high-quality crops, transgenic technology and animal cloning, bio-reactor, gene and protein engineered vaccines and drugs, gene therapy and drug discovery/development.

4.1 Agricultural Biotechnology

China is a world leader with respect to hybrid rice research and application. Hybrid rice research started in 1964. China realized 3-lines setting in 1973 and successfully developed the technique of seed production for commercialization in 1975. In 1976, China began large scale commercial production of hybrid rice. Since the 1990s, China has maintained an annual cultivation area of 15 million hectares for hybrid rice, over 50% of the total cultivated area for rice, and hybrid rice yield accounts for nearly 60% of gross output of rice. The cumulative cultivated area for hybrid rice from 1976 to 2001 stood at 267 million hectares, with an average unit yield increase of more than 20% and a cumulative paddy output margin of over 380 million tons. The marginal paddy output due to the cultivation of hybrid rice nation wide could feed another 60 million people. 2-lines hybrid rice was successfully developed in 1995. The yield of 2-lines hybrid rice is 5%-10% more than that of 3-lines hybrid rice. By 2002, 2-lines hybrid rice has been applied to a total area of over 7 million hectares nationwide.

China started a super hybrid rice research program in 1997, aiming at an average unit output of 10.5 tons/hectare in 2000 and 12 tons/hectare in 2005. So far, the research has made remarkable progress, generating a number of new super hybrid rice combinations with 10.5 tons/hectare or above. Among those, "II youming 86" demonstrated a unit output as high as 17.9 tons/hectare at Yongsheng County in Yunnan Province in 2001, setting a world record of unit yield for rice. Two super hybrid combinations, "Liangyoupei 9" and "Liangyou 932", have passed new species evaluation. From 2000 to 2001, super hybrid rice was spread to a total area of 200,000 hectares, producing over 300,000 tons of high-quality paddies.

China is one of the earliest countries to upgrade crop species by adopting cell and chromosome engineering technology and to apply the technology to large scale application. Chromosome engineered high-quality wheat species have all been widely spread to farm land, such as "Xiaoyan 6", "Xiaoyan 54", "Gaoyou 503", "Zhongyou 9507", "Xiaobing 33", among others. "Xiaoyan 6" has been largely cultivated for nearly 20 years in a total area of over 6.7 million hectares. For a long time, China has been taking the lead in the world in the research on crop species improvement by means of anther culture and clone variation techniques. Based on incomplete statistics since the 1980s, altogether 18 crop species have been examined and approved and widely cultivated in a total area of 1.2 million hectares, e.g. wheat, rice, rape, sugarcane, etc[31]. China has the largest chromosome engineered crop variety community in the world.

In plant gene engineering R&D, five self-developed transgenic plants, storage property tomato, transgenic morning glory with chalcone synthetase gene, anti-virus pimiento, anti-virus tomato, pest-resistant cotton, have passed the national level approval for commercial production. Over 20 transgenic plants have reached the stage of environmental release. In 2001, the cultivation area for transgenic crops (mainly transgenic cotton) in China totaled 700,000 hectares, ranked 4th in the world. The research system and the security evaluation for transgenic plants have also been set up in China, including the upper-reach gene research, the middle-reach plant genetic transformation and the lower-reach biotechnological breeding. Also in 2001, China constituted "The Statutes for Security Administration of Agro-transgenic Organisms", laying down the legal foundation for the healthy development of transgenic plant R&D in China.

In the research of agricultural microbial gene engineering, smooth progress has been made in the genetic alteration and application of microbes that can kill pests, resist diseases, co-exist with others and coordinately fix nitrogen. Currently, China is the country with the largest area, greatest variety and widest scope of research in agricultural recombination microbe environmental release in the world. Achievements made by Chinese scientists have been drawing broad interest from scientists in various countries. More than 40 agricultural recombination microbes have applied to and been approved by the Ago-biology Gene Engineering Security Committee in China.

Encouraging achievements have also been made in animal biotechnology R&D. China is taking the lead in the research on transgenic fish and the technology has reached maturity for a potential large-scale commercial production[38]. China is capable of producing medicinal proteins from transgenic animals.

Chinese scientists have made world-renowned accomplishments in animal cloning research, successfully cloned goats and cattle, among other animals. In 1990, G1-G5 embryo-cloned Angolan goats (45 in number) and trophoblastic layer cell cloned goats were born by means of embryo cell cloning procedure. In 2000, somatic cell cloned goats were born. In 2002, scientists cloned 14 calves from cattle somatic cells, among which five are alive[39, 40]. The above achievements symbolize that China's animal cloning technique has paralleled with world advanced level. Besides the aforementioned, progress has been made in cross-species cloning research of giant panda. Primary results showed that cloned embryos could develop into blastula stage, opening a new way to the research aiming at protecting and salvaging endangered rare animal species.

4.2 Medical & Pharmaceutical Biotechnology

Through years of efforts, China's gene engineering pharmaceutical industry has developed its initial scale of economy, with 18 products approved into the marketplace, 21 into clinical phase Ⅰ and Ⅱ , and another 35 in their pre-clinical development. Market share of these drugs has been rising continually, e.g. α1b interferon has a 60% domestic market penetration. The therapeutic HBV vaccines are taking shape: blood-origined HBV antigen-antibody complex has been approved for special clinical trial; gene engineered HBV antigen-antibody complex is going into clinical trial soon and has received domestic and international patents. Currently, Chinese scientists are conducting the research and production of new triplex compound vaccines. The technology transfer of artificial blood substitutes has been successfully completed, with pilot scale experiment bases set up and numerous batches of products up to quality control criteria. Now, there are over 60 gene engineering-based pharmaceutical companies with significant production capacity in China.

A shift is occurring for biotechnology medicines in China from imitation to innovation. China now can produce eight of the top ten best selling drugs in the world. Besides, research is going on for monoclonal antibody or antibody derivatives for diagnosis and targeting therapy. China's gene therapy technology for inherited diseases has reached international advanced level. A series of breakthroughs and major progress have been made in fields such as cancer immune therapy, anti-vascular therapy, gene therapy and tissue engineering research, etc. Five kinds of therapeutic approaches have entered clinical trial such as for haemophilia B, malignant tumor, occlusive peripheral blood vessel diseases, among others. Revenues from Chinese biotechnology products increased from US$ 30 million in 1986 to US$ 2.5 billion in 2000.

Advances have also been made in the research and development of stem cell products. China established the first umbilical cord blood hematopoietic stem cell bank in 2002. Currently in storage are 6,000 samples, among which 10 were already provided for clinical transplantation. It is projected that 0.5-1 million cord blood samples will be put in stock by 2010, basically suffice the demand for Chinese patients at home and from overseas.

4.3 Drug Discovery/Development

Through focused research programs, several novel drugs have been developed. For example, artmether, a novel antimalarial drug developed in China and approved by WHO, is now widely used in the developing countries. Sodium dimercaptosuccinate, an antidote used in China against heavy metal poisonings, has also been approved by FDA and used in the United States for lead poisoning. Huperzine A, a novel alkaloid isolated from a Chinese herb medicine, was demonstrated to be effective in improving memory deficiencies in aged people and Alzheimer's disease patients. Recently, with the support of MOST and CAS, several innovation drug research projects and platforms have been established. For example, in 1997, the National Center for Drug Screening was set up in Shanghai with the main research areas in drug target identification and validation, combinatorial chemistry, high-throughput screening and computer-aided drug design. Since its establishment, more than 200 screening assays including enzyme assays and receptor binding assays have been build up and several novel lead compounds have been found in the center. Traditional Chinese Medicine (TCM) has been used in China for many years. To increase recognition of TCM on the international market, China has recently made a great effort to speed up the modernization of TCM. State-of-art techniques and equipments have been introduced to develop and standardize the TCM production. And several TCM drugs have entered international market recently.

5. AREAS OF CONCERN FOR THE FURTHER DEVELOPMENT OF LIFE SCIENCES & BIOTECHNOLOGY IN CHINA

5.1 Existing Problems and Countermeasures

China's life sciences and biotechnology is still faced by a number of problems, including: insufficient talents and lack of original innovation; backward infrastructure and lack of combined collective working; poor coordination mechanism and unnecessarily duplicated project layout; loose association between scientific research and industrial application; legal system build-up lagging behind the development of science and technology, among others. Recently, concerned governmental agencies have proposed some new thoughts and programs to address these problems. For example, MOST proposed the "Three Strategies", i.e. talent, patent and standard, whereas CAS also advocated "Triple Strategies", namely, leap development, talent and sustainability. Meanwhile, China has been gearing up the construction of national hi-tech parks and hi-tech industrialization zones, on the basis of which trying to nurture and attract leadership talent who are familiar with the whole process of biotechnology R&D. This aims at building up high-level research teams, lifting the overall capability in research and product development and, step by step, forming up the biotech innovation system of Chinese characteristics. The coordination mechanism between different research programs needs to be streamlined, funding channels to be concentrated, alliance between research institutions and enterprises to be strengthened, thus to speed up the process of R&D and marketization of biotechnologies and products with self-owned intellectual property rights. Also in need is to strengthen the legislation of relevant laws, such as enacting administrative statutes for GMOs, establishing national and institutional bioethics committee, etc. The above actions aim at promoting a rapid development of China's life sciences & biotechnology R&D.

5.2 Further Strengthening Medical & Pharmaceutical Innovation System of Chinese Characteristics

Population and health is one of the areas served by life sciences and biotechnology. Medical care market is one of the most important demand and supporting spots in the national economic growth. But current Medicare market in China is relatively small, only accounting for about 4% of GDP, or over US$ 40 billion, among which consumption on drugs stood at merely US$ 9 billion, only spending 1/60 of Japanese per capita. There is a huge potential in this market. After China's entry into the WTO, Chinese pharmaceutical companies are facing serious challenges due to their lack of innovative capability. At the same time, however, these companies are also showered with tremendous opportunities, thanks to the potential global market that has been opened to China. Not long ago, CAS pioneer-proposed to build up a national pharmaceutical innovation system with Chinese characteristics and also put forward the concept of phytochemics, i.e. to characterize in a global way, the active ingredients from the commonly used medicinal materials. Moreover, the program will study and develop new drugs by means of the concept and technological platform of functional genomics. On the basis of mega-bioscience platform research, modernization of TCM will come true, the national pharmaceutical system of Chinese characteristics will bc taking shape and China's unique contribution to the world life sciences and biotechnology will materialize.

5.3 On Bioethics Issue

Compared with other scientific areas, life sciences & biotechnology bears the closest relationship to mankind and its future. Therefore, the bioethics issue derived from the development of life sciences and biotechnology is drawing wide attention from governments, scientists and even general public in various countries. And a world wide big debate has been well underway. So far, primary areas of concern include human cloning technology, stem cell technology, gene therapy, tissue and organic engineering technology, gene modified organism, gene discrimination, etc. It is no wonder that peoples of different countries, nationalities, religions and intellectual levels may hold different views on the bioethics issue. There should be some common ground in some basic points, such as the protection of human and national dignity, the protection of gene resources and ecological environment, ensuring food safety, among others. We maintain that, while investigating and applying achievements in life sciences and biotechnology, we should observe the principles of "good-will, harmless, informed consent, informed self-choice and fair". Particular attention should be paid to safeguard the interests of developing countries in handling the relationship between the sharing of genomics research results and the protection of the intellectual property rights for biotechnology and products. In 2001, Chinese government passed the legislation of "Statutes for Agricultural Transgenic Bio-safety Administration" and is working on laws concerning human ES cell research, which will explicitly outlaw the research on human reproductive cloning.

In regard to bioethics, legislation is well behind the development of science and technology. Governments of all countries shall gear up the construction of legal system. And scientists also have the obligation and duty to make sure that biotechnology research is going on a proper track to the benefit of humanity.

6. PROSPECTS

Along with the continued rapid economic growth over the past 20 years, life sciences and biotechnology research in China has made remarkable progress. The emergence of genomics especially brought unprecedented opportunity for China to keep up with international advanced level and drive the biotechnology R&D along forward. Seizing the opportunity, China's life science community has made a historical stride forward, closing the gap between the overall research level and world advanced level, with some areas up running in pace with world leaders. What's encouraging is that, a biotechnology research team has been organized, with upper-, middle- and lower-reach research systems taking initial shape, a batch of R&D bases of critical mass have been established and funding from various sources has been increasingly pouring in. Prediction has it that China is well able to joint the ranks of world leaders in life sciences and biotechnology over the coming 15 to 20 years. It is our hope that, through the joint effort of the scientific communities all over the world, those in the Third World in particular, in dealing with the common challenges in area of life sciences and biotechnology, the whole mankind can enjoy a better life and go ahead along with a road of common development.

REFERENCES

1 Yu J, Hu S, Wang J, *et al.* A draft sequence of the rice genome (*Oryza sativa* L. spp. *indica*). Science, 2002, 296:79-92.

2 Feng Q, Zhang Y J, Hao P, *et al.* Sequence and analysis of rice chromosome 4. Nature, 2002, 420: 336-340.

3 Zhao Q, Zhang Y, Cheng Z K, *et al.* A fine physical map of the rice chromosome 4. Genome Res., 2002, 12: 817-823.

4 Li X, Qian Q, Fu ZM, *et al.* Contral of tillering in rice. Nature, 2003, 422: 618-621.

5 Chen X W, Ijkel W F J, Tarchini R, et al. The sequence of the Helicoverpa armigera sigle nucleocapsid nucleopolyhedrovirus genome. J. Gener. Virol., 2001, 82:241-257.

6 Ren SX, Fu G, Jiang XG, *et al.* Unique physiological and pathogenic features of Leptospira Interrogans revealed by whole-genome sequencing. Nature, 2003,422

7 Chen Z, Guidez F, Rousselot P, *et al.* PLZF-RAR α fusion proteins generated from the variant t (11:17)(q23; q21) translocation in acute promyelocytic leukemia inhibit ligand-dependent transactivation of wild-type retinoic acid receptors. PNAS USA, 1994, 91:1178-1182.

8 Xia J H, Liu CY, Tang B S, *et al.* Mutations in the gene encoding gap junction protein β -3 associated with autosomal dominant hearing impairment. Nature Genetics, 1998, 20:370-373.

9 Zhang X H, Zhao J, Li C F, *et al.* DSPP mutation in dentinogenesis imperfecta shields type II. Nature Genetics, 2001,27:151-152.

10 Xiao S X, Yu C, Chou X M, *et al.* Dentinogenesis imperfecta 1 with or without progressive hearing loss is associated with distinct mutations in DSPP. Nature Genetics, 2001, 27:201-204.

11 Gao B, Guo J Z, She C W, *et al.* Mutation in IHH, encoding Indian hedgehog, cause brachydactyly type A-1. Nature Genetics, 2001, 28:386-388.

12 Bu L, Jin Y P, Shi Y F, *et al.* Mutant DNA-binding domain of HSF4 is associated with autosomal dominant lamellar and Marner cataract. Nature Genetics, 2002, 31:276-278.

13 Chen Y H, Xu S J, Bandahhou S, *et al.* KCNQ1 gain-of-function mutation in familial atrial fibrillation. Science, 2003, 299:251-254.

14 Feng B J, Huang W, Shugart Y Y, *et al.* Genome-wide scan for familial nasopharyngeal carcinoma reveals evidence of linkage to chromosome 4. Nature Genetics, 2002, 31:395-399.

15 Liu W, Wang H, Zhao S, *et al.* The novel gene locus for agenesis of permanent teeth (He-Zhao deficiency). J. Dent. Res., 2001, 80:1716-1720.

16 Xu X R, Huang J, Xu Z G, *et al.* Insight into hepatocellular carcinogenesis at transcriptome level by comparing gene expression profiles of hepatocellular carcinoma with those of corresponding noncancerous liver. PNAS USA, 2001, 98:15089-15094.

17 Meng S D, Gao T, Gao G F, *et al.* HBV-specific peptide associated with heat shock protein gp96. Lancet, 2001, 357: 528-529.

18 Li P, Chan H C, He B, et al. An antimicrobial peptide gene found in the male reproductive system of rats. Science, 2001, 291:1783-1785.

19 Wang C C. Protein disulfide isomerase as an enzyme and a chaperone in protein folding. Methods In Enzymology, 2002, 348:66-75.

20 Li J, Zhang S and Wang Z Z. Effects of macromolecular crowding on the refolding of glucose-6-phosphate dehydrogenase and protein disulfide isomerase. J. Bio. Chem., 2001, 276: 34396-34401.

21 Zhuo Y, Zhou T G, Rao H Y, *et al.* Contributions of the visual ventral pathway to long-range apparent motion. Science, 2003, 299: 417-420.

22 Tang S and Guo A. Choice behavior of Drosophila facing contradictory visual cues. Science, 2001, 294: 1543-1547.

23 Xiang Y, Li Y, Zhang Z, *et al.* Nerve growth cone guidance mediated by G protein-coupled recepters. Nature Neuroscience, 2002, 5:843-847.

24 Zhang C and Zhou Z. Ca^{2+}-independent but voltage-dependent secretion in mammalian dorsal root ganglion neurons. Nature Neuroscience, 2002, 5:425-430.

25 Yao H S and Li C Y. Clustered organization of neurons with similar extra-receptive field property in the primary visual cortex. Neuron, 2002, 35: 547-553.

26 Xiao H S, Huang Q H, Zhang F X, *et al.* Identification of gene expression profile of dorsal root ganglion in the rat peripheral axotomy model of neuropathic pain. PNAS USA, 2002, 99: 8360-8365.

27 Wu Z Y. Vegetation of China. Beijing: Science Publishing House, 1980.

28 Cao Y H and Zhao S D. Global environment and ecosystem monitoring network. Beijing: Science Publishing House, 1997.

29 Huang H, Han X, Kang L, *et al.* Conserving native plants in China. Science, 2002, 297: 935-936.

30 Xie P and Chen Y Y. Invasive carp in China□s plateau lakes. Science, 2001, 294: 999-1000.

31 Xu X, Zhou Z H, Wang X L, *et al.* Four-winged dinosaurs from China. Nature, 2003, 421:335-340.

32 Shu D G, Morris S C, Han J, *et al.* Head and backbone of the Early Cambrian vertebrate Haikouichthys. Nature, 2003, 421:526-529.

33 Shu D G, Morris S C, Zhang Z F, *et al.* A new species of Yunnanozoan with implications for deuterostome evolution. Science, 2003, 299:1380-1384.

34 Ke Y H, Su B, Song X F, *et al.* African origin of modern humans in East Asia: A tale of 12000 Y chromosomes. Science, 2001, 292: 1151-1153.

35 Savolainen P, Zhang Y P, Luo J, *et al.* Genetic evidence for an East Asian origin of domestic dogs. Science, 2002, 298: 1610-1613.

36 Li Q J, Xu Z F, Kress W J, *et al.* Pollination: Flexible style that encourages outcrossing. Nature, 2001,410:432.

37 Mang K Q. Prospects of agriculture biotechnology. Annual report of high-tech development (2001, Chinese Academy of Sciences). Beijing: Science Publishing House, 2001.

38 Zhu Z Y, Hu W and Wang Y P. Bio-safety and application strategies of genetically-modified. Annual report of high-tech development (2001, Chinese Academy of Sciences). Beijing: Science Publishing House, 2001.

39 Liu J L, Wang M, Lian L, *et al.* Nuclear transfer using nonquiescent adult fibroblasts from a bovine ear. Chinese Science Bulletin, 1999, 44:1971-1974.

40 Liu J L, Wang M K, Sun Q Y, *et al.* Refrigeration of donor cells in preparation for bovine somatic nuclear transfer. Reproduction, 2001, 122: 801-808.

Recent Progress in Breeding Super Hybrid Rice in China

Yuan Longping

China National Hybrid Rice R & D Center

Yuan Longping, *was born in Beijing on September 7, 1930. He now is General Director of China National Hybrid Rice Research & Development Center, Professor of Hunan Agriculture University and Senior Plant Breeder of Hunan Academy of Agricultural Sciences. Yuan's impact to world in general, to Asia in particular, cannot be measured by how many people he saved from starvation, but also on how much he contributed to society stability through adequate food supply, as well as resource conservation through increased productivity. Only those who witnessed the high emotion of love and respect that Chinese rice farmers paid to Yuan in the paddy fields can understand the true value of a scientist to the millions of peasants he served.*

Abstract: Facing the grim situation of population growth pressure and reduction of arable land, a super rice breeding program was set up by China Ministry of Agriculture in 1996 to ensure the food security of China in the new century. By way of morphological improvement plus utilization of inter-subspecific (indica/japonica) heterosis, several pioneer super hybrid rice varieties were developed, which had attained the Phase I yield standard of the single season super rice (10.5 t/ha on a large scale) by 2000. Now, efforts are focused on breeding of the Phase II super hybrid rice and good progresses have been made. In experimental fields, some newly developed hybrid rice varieties showed a yield level of 13 t/ha and had a yield advantage of 6%-18% over the existing pioneer super hybrid rice varieties. Among them, the best one yielded as high as 12.3 t/ha averagely on a large area of about 7 ha at Longshan, Hunan in 2002. According to the above progresses, the yield standard of the Phase II super rice (12 t/ha on a large scale) can be achieved by 2005. With the application of molecular technology in rice breeding, the yield potential of rice can be further lifted. Based on the latest research results such as the discovery and utilization of yield enhancing QTLs from wild rice, successful introduction of barnyard grass DNA and C4 genes with a higher photosynthetic efficiency into rice, the Phase III super hybrid rice breeding program is proposed, in which the yield target is 13.5 t/ha on a large scale by 2010.

1. ACHIEVEMENTS AND PROGRESS

The present population in China is near 1.3 billion with less than 0.1 ha of arable land for each person. It is expected that the population will reach 1.6 billion and the crop land will be reduced to about 0.07 ha per capita. Facing such population growth pressure and reduction of arable land situation, in order to feed all Chinese people in the new century a super rice breeding program was set up by China Ministry of Agriculture in 1996 with the yield targets for hybrid rice listed below:

Yield Standard of the super rice in China

Phase	Hybrid rice			% increase
	First cropping	Second cropping	Single season	
1996's level	7.50	7.50	8.25	0
Phase I 1996-2000	9.75	9.75	10.50	Over 20%
Phase II 2001-2005	11.25	11.25	12.00	Over 40%

* Tons/ha at 2 locations with 6.7 ha each in 2 consecutive years.

By way of morphological improvement plus utilization of inter-subspecific (indica/japonica) heterosis, several pioneer two-line super hybrid rice varieties had been developed by 2000, which attained the phase I yield standard of single season rice. There were more than 20 demonstration locations with 6.7 ha (100 mu) or 67 ha (1,000 mu) each where their average yield was over 10.5 t/ha in 2000. The average yield per ha was 9.6 t in commercial production (235,000 ha) in 2000 and 9.2 t (1.2 million ha) in 2001. (The average yield of rice is 6.3 t/ha nationwide recently.) A combination, P64S/E32, created a record yield of 17.1 t/ha in an experimental plot (720 m^2) in 1999.

Now efforts are focused on breeding of the Phase Ⅱ super hybrid rice and good progress is being made.

There were three two-line indica/japonica hybrid combinations outyielding the CK (the pioneer super hybrid rice variety) by 6%-18% in replicated trials at our center in 2001. A three-line indica/japonica hybrid, Ⅱ-32A/Ming 86, yielded 12.76 t/ha at a demonstration location with 7 ha in Fujian Province in 2001, and it also created a new record yield (17.95 t/ha) in an experimental plot (800 m^2) in Yunnan Province in 2001.

Based on 2001's progress, some promising combinations are arranged for demonstration at multiple locations with 7-8 ha each in 2002. Among them the best one is P88S/0293, which yielded 12.3 t/ha on average in Longshan County, Hunan. This combination yielded 12.4 t/ha in Hainan Province again in 2003 and created a record yield in the province. Besides, there are three conbinations performing very well in our experimental plots (plot size 700-800 m^2). Their estimated yield is around 13 t/ha.

Especially, a breakthrough has been achieved in breeding of first cropping super hybrid rice. A newly developed short growth duration two-line indica/japonica hybrid (HY-S/F49) was demonstrated near Changsha in 2002. The area under demonstration was 7 ha, its average yield was 9.1 t/ha which outyielded the CK_1 (three-line inter-varietal hybrid) and CK_2 (inbred variety) by 20% and 40%, respectively. Again, there are two demonstration locations with 7-8 ha each in 2003, their estimated yield is around 10 t/ha.

2. TECHNICAL APPROACHES

Crop improvement practices have indicated, up to now, there are only two effective ways to increase the yield potential of crops through plant breeding, i.e. morphological improvement and heterosis utilization. However, the potential is very limited by using morphological improvement alone and heterosis breeding will produce undesirable results if it does not combine with morphological improvement. Any other breeding approaches and methods including high technology such as genetic engineering must be incorporated into good morphological characters and strong heterosis, otherwise there will be no actual contributions to yield increase. On the other hand, the further development of plant breeding for high target must rely on the progress of biotechnology.

2. 1 Morphological Improvement

Good plant type is the foundation for super high yield. Since Dr. Donald proposed the concept of ideotype in 1968, many rice breeders have paid great attention to this important topic and proposed several models for super high-yielding rice. Among them the famous one is the "New plant type" proposed by Dr. Khush at IRRI. Its main features are: ① big panicles, 250 spikelets per panicle; ② less tillers, 3-4 productive tillers per plant; ③ short and sturdy culm. Whether this model can realize super high yield or not, it has yet to be proved by practices.

Based on our studies, especially inspired by the striking characteristics of a high-yielding combination, P64S/E32, which has created a record yield of 17.1 t/ha, we have found that the super high-yielding rice variety has the following morphological features:

2.1.1 Tall erect-leaved canopy

The upper three leaf blades should be long, erect, narrow, V-shaped and thick. Long and erect leaves not only have larger leaf area but also can accept light on both sides and wile not shade each other. Therefore, light is used more efficiently. Narrow leaves occupy relatively small space thus allow to have a higher effective leaf area index. V-shape makes leaf blade more stiff so that not prone to droopy. Thick leaves have higher photosynthetic function and are not easily senescent. These morphological features mean a huge source of assimilates essential to super high yield.

2.1.2 Lower panicle position

The tip of panicle is only 60-70 cm above the ground during ripening stage. Because the center of gravity is quite low, such architecture enables the plant to be highly resistant to lodging. Lodging resistance is also one of the essential characters required for breeding super high-yielding rice variety.

2.1.3 Bigger panicle size

The grain weight per panicle is around 5 grams and the number of panicles is about 300 panicles/m^2. Theoretically, the yield potential is 15 t/ha if they are attained.

Grain yield = Biomass × HI. Nowadays the harvest index (HI) is very high (above 0.5). Further lifting of rice yield ceiling should rely on increasing biomass because further improvement of HI is quite limited. From view point of morphology, to raise plant height is the effective and feasible way to increase biomass. However, this approach will cause lodging. To solve this problem, many breeders are trying to make the stem thicker and sturdier, but this approach usually results in HI decrease. Therefore, it is difficult to obtain super high yield by this way. The plant model of taller canopy which consists of leaves can combine the advantages of higher biomass, higher HI and higher resistance to lodging together.

2.2 Raising Heterosis Level

The heterosis level in rice has the following general trend: indica/japonica > indica/javanica > japonica/javanica > indica/indica > japonoca/japonica, according to our studies. Indica/japonica hybrids possess very large sink and rich source, the yield potential of which is 30% higher than inter-varietal indica hybrids being used commercially. Therefore, efforts have been focused on using indica/japonica heterosis to develop super hybrid rice. However, there exist a lot of problems in indica/japonica hybrids, especially very low seed set, which must be solved in order to use their heterosis in practice. By means of wide compatibility (WC) genes and using intermediate type lines as parents instead of typical indica or japonica lines, a number of inter-subspecific hybrid varieties with stronger heterosis and normal seed set have been successfully developed as mentioned above.

2.3 Biotechnology

This is another important approach to develop super hybrid rice. So far two very promising results have beed obtained in this research field.

2.3.1 Utilization of favorable genes from wild rice

Based on molecular analysis and field experiments, two yield enhancing QTLs from wild rice (*O. Rufipogon* L.) were identified. Each of the QTLs contributed to a yield advantage of 18% over the high yielding CK hybrid Weiyou 64 (one of the most elite hybrids). By means of molecular marker-assisted backcross and field selection, an excellent R line (Q611) carrying one of these QTLs is developed. Its hybrid, J23A/Q611, outyielded CK hybrid by 35% in a replicated trial for the second cropping rice in 2001. Its yield potential on a large scale is being evaluated for the time being. Preliminary data show that its estimated yield is 13 t/ha in experimental plot and 11 t/ha on farmers' field planted as second cropping rice.

2.3.2 Using genomic DNA from barnyard grass (Echinochioa crusgalli) to createnew source of rice

Total DNA of barnyard grass was introduced into a restoring line (R207) by Spike-stalk Injection Method, variants occurred in D_1. From these variants, new elite stable R lines have been developed. The most outstanding one is RB207-1, its agronomic characters including number of spikelets per panicle and grain weight are much better than those of the original R207. Particularly, its hybrid, GD S/RB207-1, has good plant type and very strong heterosis, the estimated yield was over 15 t/ha in our experimental plot in 2002.

3. PROSPECTS

The yield standard of the Phase Ⅱ super rice (12 t/ha) can be achieved by 2005. By reaching this target, 2.25 t/ha more rice can be produced, which will increase 30 mt of grains yearly and can feed 75 million more people when it is commercialized up to 13 million ha.

The development of science and technology will never stop. Rice still has a great yield potential, it can be further tapped by advanced biotechnology. Excitingly, C_4 genes from maize have been successfully cloned and transferred into rice plant by HK Chinese University. Using this transgenic plant as donor to introduce C_4 genes into super hybrid rice parents is under way. If this approach is achieved the yield potential of rice could be further increased by a big margin. Relying on this progress, the Phase Ⅲ super hybrid rice breeding program is proposed, in which the yield target is 13.5 t/ha on a large scale by 2010.

Progress of Crop Genetics and Breeding in China

Li Zhensheng   Zhang Aimin

State Key Laboratory of Plant Cell and Chromosome Engineering, Institute of Genetics and Developmental Biology, Chinese Academy of Sciences

Li Zhensheng, *Academician of the Chinese academy of Sciences, Fellow of the Third World Academy Sciences, Chairman of the Academic Committee of State Key Laboratory of Plant Cell and Chromosome Engineering.*

Abstract: Grain production that humans live on was initiated from domestication of wild plants. In the course of collecting wild plants or fruits as food our ancestors accumulated their knowledge of plant growing and fruiting and domesticated the wild plants to cultivated plants and created the primitive agriculture and crop genetic improvement. Modern plant genetics and breeding is the artificial remake of the plant to fit the needs of humans by genetic manipulation. Plant cultivars are the most important means of production for agriculture and also the carrier for continued developing of agricultural technology. Therefore plant genetics and breeding occupy the core position in agricultural sciences. Chinese government and scientists pay appropriate attention to genetics research of plant yields and agronomic characters and the development of technology for plant breeding.

1. INTRODUCTION

China is an age-old agricultural country. The earliest historical record about cultivar selection could trace back to Zhou Dynasty, at least 2,500 years ago. But the modern plant genetics and breeding was learned from western countries since the beginning of the last century. After the founding of New China in 1949, the basic and applied research on plant genetics and breeding was highly regarded. With the efforts of Chinese scientists, plant breeding has made great achievements in the recent 50 years. Chinese scientists invented hybrid rice, achieved a green revolution independently by using the dwarfing gene in both rice and wheat, finished the first draft of the genetic map of rice, created a large number of plant genetic germplasm resources, developed a great quantity of new plant cultivars, and made the total output of crop production seven fold increase together with the other agricultural techniques compared with that in 1949. China feeds 22% of the world population with only 7% of the arable land and has made great contributions to world food safety and development of economics. In this paper we review the progress of plant genetics and breeding in China, and look forward to the future of plant genetics and breeding and exchange our ideas with other scientists in order to enhance our plant genetics and breeding in China.

2. ATTACHING IMPORTANCE TO COLLECTION, CONSERVATION, EXPLOITATION, AND UTILIZATION OF PLANT GENETIC RESOURCES, CHINA IS ONE OF THE COUNTRIES IN THE WORLD POSSESSING A LARGE NUMBER OF PLANT GENETIC RESOURCES

Genetic resources are the basis of plant breeding and an important factor of BIODIVERSITY as well as the material basis of human life and development. Plant genetic resources consist of all cultivated plants anywhere and anytime and their relatives, including half domesticated plants, and wild plants. The more plentiful are the genetic resources with high levels of diversity, the more successful will be achievements in genetic research and plant genetic improvement.

China is one of the origin centers of plant germplasm and has an abundance of plant genetic resources with their special characters. Since the founding of the P.R. China, the government set the policy and key scientific projects for plant genetic resources collection and conservation as well as utilization, established the special research institute for genetic resources, and Chinese scientists conducted successful work in plant genetic resources. Up to now, the total number of plant genetic germplasm preserved has reached 370,000 accessions for 161 different plants. There are 318,000 accessions preserved in national long-term conservational gene banks equipped with the newest instruments and duplicate collections are conserved in Qinghai. These banks made China the second largest germplasm-preserving country and the first in number of long-term conservation in the world.

Since the 1950s, local varieties collection was conducted twice to avoid loss of local varieties from release of improved cultivars. 200,000 accessions of 43 crops were collected and 12,000 accessions were introduced abroad. The third time local varieties collection was started in1979, and after five years hard work, 110,000 new accessions of more than 60 different crops were collected. Three collection efforts made the number of germplasms reached up to 300,000 and among them there were 130,000 accessions of wheat, rice and maize. Chinese scientists have also completed precision work on identifying and putting in order the germplasms collected. 54 volumes of catalogs with 18 millions characters were published. 200,000 accessions of different crops were identified for their quality, biotic and abiotic resistance, 300,000 accessions were recorded for their agronomic characters and 1.15 millions data records were collected. A larger number of germplasms with some important target characters were provided to plant breeders for their genetic research and breeding. A National Crop Germplasm Information System consisting of 4 subsystems has been set up: characters subsystem, gene bank management subsystem, international germplasm exchange subsystem and comprehensive eva luation subsystem. The system holds the information of 20 million data records from 350,000 accessions of 160 different crops and serviced to the agricultural scientists and plant breeders in 42 service stations.

The exploitation of Chinese plant genetic resources has made a great contribution to global crop production. The rice dwarfing gene from Chinese cultivar "Di Jiao Wu Jian", the most important gene resource for the green revolution, has been used widely in rice breeding program and derived a quantity of new dwarf rice cultivars. Chinese local wheat variety "Chinese Spring", which was identified to have 3 Kr genes for distant hybridization, was introduced to the west countries early in the last century and from which derived most of the genetic materials including monosomic and other aneuploid. Most of the wheat genomics research program also used Chinese Spring as material. Another famous Chinese wheat variety "Sumai 3", the most popular and the best resistant source to scab, was used in almost all of the scab resistant breeding programs. Just as "Sumai 3", it is well-known that the Chinese soybean germplasm plays a key role in the American soybean disease resistant breeding programs.

Recently, Chinese scientists employed genomic approach for genetic germplasm research. They do fingerprinting of important germplasm of main crops and mapping the important genes with molecular markers. A standard method for fingerprinting of main crops was established and a map bank of gliadin for Chinese wheat varieties was created. Molecular markers for quality and disease resistance genes were obtained, for example, molecular marker for rice blight resistant gene and wheat powdery mildew resistant gene. Sampling methods system for "core collection" was established and primary core collections of wheat, soybean, rice, rapeseed, and so on were constructed.

Based on the analysis of trends of agricultural development, we proposed the exploitation of the "green gene" which promote yield increase and reduce the pollution in the germplasm and developed the new variety with high efficiency use of nature resources and high yield by using the green genes. So we set the purpose of genetic germplasm research to prepare green gene resources for new green revolution. What is green gene? The green genes in a narrow sense are the genes concerning high efficiency use of resources, abiotic stress durable genes and defense genes, but high yield genes should be included in a broad sense. If we pyramid the genes for heterosis and genes for high efficiency, we can solve the problem of food safety and at the same time protect the ecological environment and make the agriculture development sustainable. We believe that more and more genes will be discovered along with the germplasm research that facilitates the plant breeding and plant production.

Table 1 Green genes and their functions

Genes	Name of genes	Function
Genes for high efficiency	P high-efficiency gene, N high-efficiency gene, K high efficiency gene	Increase the efficiency and reduce the dosage of fertilizers, increase the yield potential and protect the environment
Abiotic stress durable gene	Salt stress durable, drought stress durable and cold, heat stress durable genes.	Make the plant adapt to stress environment and inferior soils
High yield potential genes	Genes concerning heterosis genes, High yield QTLs, High light-efficiency genes	Increase the yield potential of hybrid, increase the efficiency of light utilization.
Defense gene	Disease resistant and insects resistant genes	Reduce the dosage of pesticide and produce the green products

3. GREAT ACHIEVEMENTS IN PLANT BREEDING, CONDUCT THE GREEN REVOLUTION INDEPENDENTLY AND MADE THE CROP CULTIVARS 5-6 TIMES REPLACEMENT

Plant breeding is artificial evolution using the genetic information and improving the yield, quality, resistance to disease and insects and adaptation to environment to fit the need of humans. It was estimated that the new cultivars account for 30%-40% of the contribution to yield increases in China. Nowadays, in China the population is increasing, but the arable land is decreasing, genetics is the only way to increase the yield potential per unit area for ensuring the total output increase and food safety.

Since the founding of P.R. China, plant genetics and breeding was taken as a key project of agricultural research by the government and scientists. With the efforts of plant geneticists and plant breeders as well as scientists in related areas, more than 6000 new cultivars belonging to more than 40 different crops were developed which made the crop cultivars 5-6 times replacement. The new cultivars of every replacement raised the yield per unit area by at least 10%. Let us take the last decade as an example. From 1991 to 1995, 207 new cultivars of main crops, 184 new cultivars of economic crops, 82 new cultivars of vegetable crops were developed, those new cultivars were not only better in yield potential but also in quality compared with the old generation of cultivars. Based on the statistic, the new cultivars of six main crops including rice, wheat, maize, millet, barley, and sorghum cover areas of 91 millions ha in total and made the grains increase 31,9644 million tons in total. From 1996 to 2000, 411 new cultivars were developed and 719 new elite lines of genetic materials were created, the new cultivars cover 22.6 millions ha and increased the social and economic benefit by 135,800 millions Yuan.

When the green revolution was carried out in the world by using dwarfing gene to develop new dwarf, high yield cultivars, China had conducted the green revolution independently. Since 1950, the Chinese scientists began the utilization of the dwarfing gene from Aizizhan, Dijiaowujian, Aijiaonante and so on to develop plant height reduced cultivars. Along with the release of new dwarf cultivars, the yield of rice reached 3380 kg/ha from 2466 kg/ha and later the yield of dwarf rice varieties reached 4539 kg/ha. The successful of dwarf rice not only opened a new page of rice breeding in China but also led the world rice breeding turn to a new direction.

Wheat is the second important crop in China and first crop in northern China. Achievements of wheat genetics and breeding have been obvious. By planting new wheat cultivars, the yield of wheat reached 3945 kg/ha in 2000 from only 645 kg/ha in 1949. Taking Beijing as example, the yield of wheat in 1949 was around 465 kg per ha. Up to 2000, the main varieties have six times of replacement, and the yield reached 5925 kg/ha. That means in 50 years the yield increased by 12 times. Breeding for reducing plant height in wheat has been a remarkable success. The plant height of Chinese wheat cultivars were dramatic, reduced from 107.9 cm in 1950 to 88.8 cm in 1990, but the 1000 kernel weight was increased from 31.4 g in 1950 to 40.5 g in 1980.

Research on maize breeding was developed very quickly after the P.R. China was founded. From 1949 to 1965, Chinese maize breeding went through a typical path that goes from local cultivars, hybrid between cultivars, double cross hybrid, three-way hybrid, top cross hybrid to single hybrids. China is one of the pioneer countries in the world to cultivate single cross hybrids in large scale. In roughly 40 years, from 1965 to the end of the last century, single cross hybrid cultivars have taken five times of replacement, and make an all-important contribution to maize production in China. There are some top class hybrids that cover huge areas and have been, popular decades. For example Danyu 13, covered more than 3 millions ha, 14.84% of total production areas in 1991. Zhangdan 2, released to farmers in 1977 was popular for more than 20 years and the areas covered reached 2.3 millions ha. Recently, CAU108, the most popular single cross hybrid, reached an area of 3 millions ha in 2002.

High oil content corn is one of the high value-added crops and one of the important developmental orientations of corn breeding. Maize oil is high quality plant oil, and high oil maize is the maize with high oil content and high energy as well as high protein content. Right now, some new high oil hybrid maize cultivars have been developed in China, for example High-oil 115, with its oil content is higher than 8% with high yield. The oil content of BHO, one of the high oil populations developed in China, reached 13.9% from 4.71% originally after 15 generations of selection. The oil content of another population AIHO reached 18% after 10 generations of selection with some single plants achieving 25% of oil content. We could say that the research on high oil corn germplasm development and breeding in China has kept ahead in the world.

The genetics and breeding of resource crops and vegetable crops in China made great progress. The Chinese scientists developed a great number of cotton cultivars and hybrids and made the cotton production reach 4 millions tons from only 0.444 millions tons in 1949. The successful development of hybrid cotton and transgenic cotton made the cotton genetics and breeding in China move ahead the world cotton breeding. Development of rapeseed is also a great success. Since 1980 Chinese scientists initiated quality rapeseed improvement and now more than 20 high quality rapeseed cultivars have been released to farmers.

After the founding of P.R. China, the approach of research and methods of breeding for plant genetics and improvements have taken a revolutionary change. In the 1950s the main method used in plant breeding was pure line selection from cultivars and hybridization taking place first in 1960s. Beginning from the 1980s, the different methods including hybridization breeding, heterosis, induced mutation breeding, cell and chromosome engineering and genetics engineering were used to bring a new breakthrough in plant genetic improvements. The mutation breeding, even thought was used in China later than western countries, developed very quickly and fruitfully. Based on the statistics, more than 500 cultivars were developed by mutation breeding in China and both the number and area covered by those cultivars ranked No. 1 in the world. Chinese scientists developed a large number of new cultivars like space green pepper, space cucumber, space wheat and so on by using space induced mutation. The outstanding achievement of breeding by chromosome engineering is octploid triticale in China. Biotechnological breeding including anther culture, haploid breeding, screening of somatic cell mutation and cell hybridization has became an effective approach for genetic improvement in more than 50 kinds of different crops with the recognized cultivars "Zhonghua 8" rice and "Jinghua 1" wheat and so on. In the last five years from 1996 to 2000, there were 45 new cultivars developed by biotechnological approaches in China.

The progress of plant genetics and breeding strongly facilitated the development of agriculture in China and led to remarkable economic, social and ecological benefits as well as fitted the need of increased population and raised living standard for the Chinese peoples. Now, the developing biotechnology is giving the plant genetics and breeding new connotation, we believe that more and more and greater fruits will be achieved in the new century with great efforts of Chinese scientists to ensure the food safety of a 1.6 billion population in the 21st century in China.

4. THE UTILIZATION OF HETEROSIS IN CROPS IS IN THE TOP POSITION OF THE WORLD

Utilization of heterosis could improve the yield, biotic and abiotic resistance and quality of crops. China is one of the countries in the world that use the heterosis widely and effectively in different crops. Here we only take rice, rapeseed, soybean, and wheat as examples to introduce the advances of hybrid breeding in China.

4. 1 Invention and Development of Hybrid Rice

Rice is the first crop in China with the area around 33 millions ha, one third of grain crops area and 40% of total output of grain crops. China is the origin and diversity center of rice and the cultivation of rice in China has more than 8000 years of history. With the invention and development of hybrid rice, rice investigation in China has kept ahead of the world.

Rice is a self-pollinated crop. Hybrid rice had not been used in rice production until the 1960s. Even the heterosis of rice was reported in the beginning of the 20th century. Chinese scientists found the wild type of cytoplasm male sterility in rice 1970 and three years later they completed the "three lines" and initialed the hybrid rice research in China. In 1975, the first hybrid rice cultivar was released to farmers and hybrid rice quickly popularized in rice production. In the 1990s, the areas of hybrid rice reached 17 millions ha, 50% of rice production area and the popularization of hybrid rice increased the yield of rice to 6500-7500 kg/ha。

The invention of photoperiod-temperature sensitive genic male sterility (PGMS or TGMS) and two-line hybrid rice is another great contribution of Chinese scientists to the world rice investigations and production. Since the discovery of PGMS in 1973, great progress has been made in two-line hybrid rice and a new field of rice research was created. First, a number of PGMS or TGMS lines like Peiai 64, Xiang125S, 7001S, 5088S, CD2s and Zhuguang 612S were developed and based on that more than 20 two-line hybrids like Peiai64/Teqing, XiangLY68, Liang-You-Pei- Jiu were developed and released to farmers. Those hybrids not only achieve high yield, 10%-15% yield advantage over three-line hybrid or pure line rice cultivars, but also achieve better quality. In recent years two-line hybrids cover an area of about 2.5 millions ha with the yield level of 7.5 tons/ha and one of the hybrids made a new record of rice yield in China, 17.1 tons/ha. In the past 2-3 years, super-rice with yield potential of 10.5 tons/ha started commercialization and it is predicted that the yield of super-rice could reach 12.0 tons/ha in 2005.

The commercialization of hybrid rice in China achieved huge economic and social benefits and made a far-reaching influence on the theory and practice of rice breeding in the world. From 1975 to 2002, the hybrid rice was extended by 284 millions ha in total and increased rice grain production by 400 millions tons. The commercialization of hybrid rice led to the reform of the rice cropping system and a revolution of rice production.

4.2 Advances of Hybrid Rapeseed Breeding

China is one of the big producers of rapeseed in the world with cultivation areas of about 6 millions ha and 1/3 of output world total output. Rapeseed is one of the important oil crops and oil of rapeseed take around 55% of total oil production, so development of hybrid rapeseed is veryimportant to oil crop production in China.

China initiated his hybrid rapeseed research long ago. “San-Tian A” and “87A” were discovered in 1965 and Polima male sterility was found in 1972. The three lines of Polima male sterility was first completed in China in 1976 while USA, Canada and Australia completed these 10 years later.

China is the pioneer country for commercialized hybrid rapeseed with the largest area in the world. In 1980 Chinese scientists found “Shan 2 A” CMS, and five years later the first hybrid rapeseed “Qin You No.2” was released to farmers. Qin You No.2 covered only 53,000 ha in 1988, but it reached 1 million ha in 1992. The commercialization of Qin You No.2 made a great influence on the rapeseed production in the world.

Along with the trends of rapeseed breeding to double zero, China released its “double lower” rapeseed hybrid. From 1988 to 1994, 37 rapeseed hybrids were developed and commercialized, 25 of them were single low or double low hybrids. Right now, more than 50 high quality rapeseed hybrids have been developed and they covered 2.5 millions ha, taking 38% of rapeseed production area. The Chinese rapeseed breeders now combine the heterosis and biotechnological approaches together to develop new hybrid rapeseed, to exhibit a more fruitful foreground of rapeseed production.

4.3 Invention of Soybean CMS and Release First Hybrid Soybean

China is one of the biggest producers in the world and the original center of soybeans. But the yield and quality of Chinese soybean cultivars lie behind the cultivars of USA and Brazil. Soybean is not hybridized. Utilization of heterosis in soybean to increase yield potential is the dream of Chinese soybean breeders. The Chinese soybean breeding program set the goal of creating a three-line system in the late 1980s, and in 1993, the first soybean CMS line was developed by crossing wild soybean with cultivated soybean. Two years later, through cross test and backcross, the first soybean CMS with cytoplasm of cultivated soybean was developed and its maintainer was obtained simultaneously. The restored line was found by test cross at the same time.

In 2002, the first soybean hybrid " Za Jiao Dou No.1" was developed. This hybrid showed high yield potential, 4000 kg/ha and 21.9% advantage over CK cultivar in regional registration tests.

At the same time, a seed production system by insect pollination was also developed in China and established the basis for hybrid soybean popularization. The commercialization of first hybrid soybean was a new breakthrough in soybean production made by Chinese scientists.

4.4 Advances of Hybrid Wheat

Promoted by the utilization of heterosis in rice and corn, Chinese scientists kept interest on hybrid wheat. Hybrid wheat breeding, one of the core breeding programs in China, was initialed in 1965 and great progress has been achieved.

Because limited resources of restoration genes and poor agronomic characters of R. lines in T. timopheevi cytoplasm, scientists were still interested in looking for new sources of cytoplasm that can induce male sterility by nuclear-cytoplasm interaction and more easy to restore. Chinese scientists developed more than ten kinds of new CMS including cytoplasm of Ae. Crassa, Avena fatua, T. aestivum and T.durum and the restoration. The research enriched the sources of cytoplasm for hybrid wheat breeding.

Aiming to use the genetic male sterility in hybrid wheat breeding, a blue-kernel marker system was developed by chromosome engineering. The system is similar to XYZ system but the male fertile and male sterile were distinguished by the color of the seed. Later the scientists transfer the dominant dwarfing gene Rht10 to additional chromosome 4E and created a dwarf-blue marker system in wheat.

Chemical hybridization agents (CHAs) induced male sterility is another way to achieve hybrid wheat seed production. In the 1970s, Ethrel and DPX3778 had been investigated as potential CHA in China. Later two CHAs, WL84811 and SC2053 were introduced and investigated for hybrid seed production. Based on the chemical hybridization agents tested, a new homemade compound BAU9403 was synthesized and demonstrated with perfect effects to introduce male sterility in wheat. The invention of BAU9403 ensured the source of CHA for hybrid seed production. Some hybrids produced by BAU9403 are now in regional yield trails.

Following the creation of two-line system in rice, both photoperiod-sensitive male sterility and a temperature-sensitive male sterility were found in wheat by distant hybridization in China, and so there pioneered a new way for hybrid wheat investigation.

With the efforts of hybrid wheat breeders, six hybrid wheat, three produced by CHA (Jin Hua No.1, Xi Za No.1, Xi Za No.5), one produced by CMS (Xi Nong 901) and two produced by TGMS (MS1 and Yun Za No.3) were released to farmers with the yield advantage of 12.5%-20%.

5. FRUITFUL DISTANT HYBRIDIZATION

Distant hybridization is an important pathway to germplasm enhancement and development of new cultivars, so Chinese scientists always attach importance to distant hybridization. In 1926, Chinese rice breeders developed Zhong Shan No.1 by crossing cultivated rice with wild rice, and then developed Zhanshanzhan, Zhongshanhong, Zhongshanbai, Baoxuan No.2 and Baotaiai *et. al.* using Zhong Shan No.1 as the parent. After that, Chinese scientists created a great number of new germplasms and new cultivars by distant hybridization in rice, wheat, cotton and other crops.

In the distant hybridization of rice, Chinese scientists developed a number of cultivars by crossing O.rufipogon with O.officinalis. They created some important germplasm by crossing cultivated rice with O.rufipogonand and then developed a series of Yue Ye Zhan, Gui Ye Zhan, Ye Zhan Qing. They developed Jian 8 by crossing cultivated rice with O.officinalis. Jian 8 performed High yield and resistant to brown-back rice plant hopper. The resistant gene comes from O.officinalis.

The most research work of distant hybridization has been done in wheat. In the distant hybridization of wheat, scientists used not only the close relatives in the genus of Triticum, but also relative species in the Triticeae including the genus of Secale L., Elytrigia Desv., Agropyron Gaertn., Aegilops L., Leymus Hochst, Roegneria C. Koch., Hanynaldia Schur., Psathyrostachys Nevski. and Eremopyrun (Ledeb. Jaub. Et Spach).

Crossed wheat with El.intermedia, the Chinese scientists developed Long Mai 1, Long Mai 2, Xin Shu Guang 6 and octoploid Triticum-Elytrigia and from this octoploid Xiaobing 33, Longmai 8, Longmai 9, Longmai 10, Shanmai150, Shanmai 879, Shanmai 611 and Zaoyou 504 were developed. We pioneered distant hybridization between wheat and Elytrigia elongata and developed a series of Xiaoyan new cultivars, for example Xiaoyan 6 and Xiaoyan 54. Xiaoyan 6, proved high yield potential, disease resistance, good quality and wide adaptation,was one of the most popularized cultivars in Huang Huai region in the 1980s and the area reached more than 700,000 ha. Xiaoyan 54 shows good adaptation and good quality, as well as high efficiency in N, P and high pothosynthesis rate. Using the germplasm developed from this distant hybridization, another high quality variety Gaoyou 503 with high yield record （11430 kg/ha） variety Gaoyuan 506 were developed.

The Chinese scientists have successfully crossed wheat with more than 32 species of 11 genuses and Allo-amphidiploid, Allo-additional lines, Allo-substitution lines or translocation lines have been derived from those crosses. A typical example is 6AL/6VS translocation line with powdery mildew resistant gene Pm 21 developed from amphiploid of T. turgidum/Haynaldia. This is the first event of transferring resistant genes from Haynaldia to wheat. Chinese scientists also developed a multiple disease resistant (yellow dwarf, powdery mildew and rusts) translocation line 7DL/7XL YW243 from an Allo-additional line of El.intermedia.

Cotton is the all-important economic crop in China. The distant hybridization of cotton was also fruitful. Chinese scientists established the technique system for cotton distant hybridization, and successfully crossed 26 wild cotton species with cultivated cotton and developed a lot of high quality, high yield cotton cultivars and germplasms. Changrong 3, Jiangsu 1, Yuan 2, Yuan 3, Yuan 394, Shiyuan 345, Shiyuan 159, Luyuan 5617 were developed by crossing G. hirsutum with G.

arboreum. Shiyuan 312 was developed by a distant cross of G.barbadense/ G.hirsutum/G.thurberi. Shiyuan 321 showed 19.7% yield increase in national registration test and 26.5% of yield increase in pre-production test. Qinyuan 4 was developed by crossing of G.hirsutum with G.sturtianu. Yuan 820 was developed from cross of G.hirsutum/ G.thurberi.// G.hirsutum/G. amomalum. By crossing white color cotton with half wild species G. hirsutum richmondii brown-colorcultivars Zongxu 1, Zong 1-61 were developed. Distant hybridization in cotton improved the quality and disease resistance of new cultivars, enriched the germplasm of cotton and created a successful pathway for genetic improvement of cotton.

6. GENETIC IMPROVEMENTS BY BIOTECHNOLOGY AND CULTIVAR' S MOLECULAR DESIGN

The 21st century is the century of life sciences. The development of biotechnology will play an important role in improving the human health level and ensuring food safety. The Chinese government and scientists pay high attention to biotechnological research and its application in plant genetics and breeding. The government strongly supports the biotechnological research by setting up high technology projects (863) and special programs of transgenic plants and animals. With the efforts of both government and scientists, significant advances have been achieved in China.

6.1 Achievements and Contributions of Tissue Culture

China is one of the pioneer countries to improve varieties using cell and tissue culture techniques. The research on anther culture and clone variation is in the front line of the world. Chinese scientists invented the culture medium M8 for rice anther culture, and developed the first rice cultivar by anther culture in 1975. In the past 30 years, more than 40 varieties were developed by anther culture in rice, among them 19 were developed between 1991 and 1998. The cultivation area of those varieties reached around 1 million ha. In wheat, the Chinese scientists developed the first cultivar "Jing Hua No.1" by anther culture in the world in 1984. This work was recognized as the new breakthrough in techniques of wheat breeding. Later, Jing Hua 3, Yu Hua 1, Ji Hua 555, Kui Hua 1, Hua Pei 28, Shan Nong 757, Beijing 8686, Jimai 24, Yumai 60, Yumai 66 and so on were developed by anther culture. Yumai

66, a recently developed cultivar, performed high quality and high yield, created a new record of wheat yield with 12,325kg/ha. Dan 2, a barley cultivar developed by anther culture, shown excellent quality for brewing and became the most popularized barley cutivar in China. From 1996 to 2000, 44 new cultivars were developed by cell engineering technique. Those cultivars covered 6.6 millions ha and contributed actively to the sustainable development of agriculture.

6.2 Development of transgenic crops

In research on plant genetic engineering, China has set up a genetic transformation system for different crops, invented transgenic techniques of pollen-tube pathway, and developed transgenic plants in more than 100 kinds of plants. Six kinds of transgenic plants including transgenic tomato (anti-ACC gene), transgenic petunia (Anti-CHS gene), transgenic pimiento(CMV CP), transgenic tomato(CMV CP), and transgenic cotton (Bt) were commercialized and more than 60 kinds of transgenic crops in field trails. The transgenic crop cultivation in China covered 70,000 ha in 2001, ranked No.4 in the world.

Transgenic cotton is the only GMO commercialized in large acreage in China. Chinese scientists synthesized BtCryIA gene in 1991, constructed the plant expressed vector in 1992 and then transferred the gene to commercialized cotton cultivars mediated by agro bacterium tumefactions or by pollen-tube pathway. China became the second country, after USA, to hold the property rights of Bt gene and successfully developed transgenic Bt cotton independently. Up to 2001, more than 20 transgenic Bt cotton cultivars (lines) and one transgenic Bt and Cpti cotton cultivar, SGK 321, were developed. Hybrid Bt cotton was also developed by using the Bt cotton as parents. Now, eight transgenic cotton cultivars have been authorized, 11 have been commercialized, 41 received permission for environmental release. The acreage of transgenic cotton increased from 10,000 ha in 1998 to 500,000 ha in 2001.

6.3 Molecular Breeding and Cultivar molecular design are on the upgrade

Development of biotechnology leads to a new revolution of plant genetics and breeding. The isolation and cloning of target genes meant the genetic improvement could use the direct gene transformation. Establishment of molecular markers of important traits made the selection be based on genotype instead of on phenotype.

Understanding the function of the gene meant the scientists could up-regulate the expression of some genes and down-regulate the expression of the other genes. Plant breeding came into a new era of molecular breeding, where artificial plant genetic improvements are. Using the technology of plant molecular biology, the genetic resources for molecular breeding expanded from intra-species or intra-genus to whole living things. The technology of molecular breeding including identifying gene function, gene cloning, molecular marker of the gene, gene targeting, gene transfer and site-specific gene insertion as well as regulation of gene expression. The breeding goal of molecular breeding is the gene buildup of the genotype instead of morphological characters that means cultivar molecular design. Based on the design of a variety at the molecular level, scientists could build the gene needed to a variety and regulate the gene expression to produce expected products to fit the need of the humans.

Chinese government pays much attention to plant molecular breeding, and they give strong support to molecular breeding and related techniques by setting national foundation and special projects like 863, 973 and transgenic plant program. The projects of 973, "Functional genomic research of important traits in rice", "Establishment of core collection of main crops, discovery and utilization of important genes" ,"Physiological and molecular basis of abiotic resistance and water, fertilizer high efficiency in crops" , "Functional genomics of quality traits in main crops" all focused on the basis and methods of molecular breeding and their results will be directly or indirectly used in plant molecular breeding.

Great progress has been achieved in molecular breeding in China. More than 100 permanent segregation populations including DH, RIL were constructed; the molecular makers for important traits like disease resistance, high quality, induced plant height were established. A great number of genes concerning important traits, for example, yield potential, end-products quality, disease and insects resistance, abiotic durable were isolated and cloned. Transgenic crops including wheat (quality, aphid resistance, herbicide resistance, male sterility), rice (Bt rice, quality, and disease resistance), corn (Bt corn, high lysine content), cotton (Bt cotton) were obtained. Chinese scientists conducted MAS (molecular assistant selection) for quality traits, disease resistance and yield potential to pyramiding those important genes to one genotype. We believe that we could design the new variety on molecular levels and develop the new cultivars with ideal genes along with the advances of structure genomics, functional genomics, proteomics, and metabolomics.

7. CONCLUDING REMARKS

Research on plant genetics and breeding in China has made big progress in the 50 years since 1949, especially in the past 20 years, but Chinese scientists still face austere challenges. China is an agricultural country, and it is always the first great intent on how to keep the food safety of the 1.3 billion population. The Chinese government set the developmental objectives for 2020 to realize the well being of society. The safety of food and sustainable development of agriculture are the guarantee to this objective. The population is increasing, arable land is decreasing, the society is developing and the need of humans is increasing, so we must do our best, working harder to develop more high yields, high quality, high resistance, and high efficiency varieties to fit the needs of the human and contribute to the new objectives of social development.

REFERENCES

1 Liu Xu. Crop germplasm resources and agro-science revolution Reviews of China, Agricultural Science and Technology, 1999, 1(2):31-35.

2 Dong Yuchen .Today and tomorrow of crop germplasm resources in China Reviews of China, Agricultural Science and Technology, 1999, 1(2):36-40.

3 Jia Jizeng. Applications of theories and approaches of plant genomics to discover crop genetic resources in China,Reviews of China Agricultural Science and Technology, 1999, 1(2):41-45.

4 Fang Jiahe. Great advances in the field of crop germplasm resources in China during 1996-2000, Journal of plant genetic resources, 2002, 3 (3): 37-40.

5 Liang Zhenlan. Genetics and breeding of Cotton distant hybridization, Science Press, Beijing 1999

6 Dong Yuchen. Wheat breeding through distant crossing, Prospects of wheat genetics and breeding for the 21st century, China Agricultural Scientech Press, Beijing, 2001, p17-22.

7 Fu Tinfdong. Breeding and utilization of rapeseed hybrid,Hubei Science and technology Press, Wuhan, 2000.

8 He Zhonghu ,Zhang Aimin. Advances of wheat breeding in China,China Science and technology Press, Beijing, 2002.

9 Dai Jingrui .Reviews and prospects of maize genetics and breeding in China, Prospects of maize genetics and breeding for the 21st century, China Agricultural Scientech Press, Beijing, 2000,p1-7.

10 Zhang Aimin. Strategy for hybrid wheat development in China Prospects of wheat genetics and breeding for the 21st century, China Agricultural Scientech Press, Beijing, 2001, p96-102.

11 Yuan Longping. Hybrid Rice, China Agricultural Press, Beijing, 2002.

12 Jia Shirong, Guo Sandui, An Daochang. Transgenic cotton, Science Press, Beijing, 2001.

13 Zhu Zhiqing. Contribution of Chinese botanists to plant tissue culture in the 20th century, Acta Botanica Sinica,2002, 44(9):1075-1084.

14 Liu Dajun Genome analysis in wheat breeding for disease resistance, Acta Botanica Sinica 2002,44(9):1096-1104.

15 Li Zhensheng.Wheat distant hybridization.Science Press, Beijing, 1985.

Recent Advances in Medical Sciences in China

Gu Fangzhou

Chinese Academy of Medical Sciences

Gu Fangzhou, *virologist, was born in Ninbo, Zhejiang Province, China in 1926. He graduated from medical college of Peking University in 1950. In 1951, he was graduated from the Institute of Virology, Academy of Medical Sciences, USSR with a candidate-doctor of medical sciences degree in 1955. After returning to his motherland in 1955, he was appointed to be a deputy-chief of laboratory of encephalitis of Institute of Microbiology and Epidemiology, Ministry of Health, China. In 1958 he moved to the Institute of Virology, Chinese Academy of Medical Sciences (CAMS), and was responsible for research work of poliomyelitis. From 1957 to 1958 he and his colleagues firstly in China isolated 116 strains of poliovirus by using monkey kidney cell cultivation technique and identified the predominant serotype of poliovirus causing epidemic in 1957 in Shanghai. Since 1959, he led a group for a pilot production of 20 million doses of oral polio vaccine (Sabin type). Through clinical trials in 11 cities among 4.5 million children under 7 years old. The immunological and epidemiological effectiveness of oral polio vaccine (OPV) has been proved to be very satisfactory. From 1960 to 1962, Dr. Gu asked by presidency of CAMS to set up a new institution in Kunming,Yunnan Province, the Institute of Medical Biology. One of its tasks is on the research and production of OPV. In 1964, he was appointed the deputy director of this institute. Now this institute produces annually about 150 million doses of trivalent OPV. During the period of 1962-1964 , he and his colleagues successfully developed a new type*

of formulation of OPV; OPV in candy dragee. This type of OPV prolonged the storage time at room temperature as compared with diluted liquid OPV.

Based on the experiences of pilot production of 20 million doses of OPV, Dr. Gu and his colleagues worked out the first issue of the minimal requirement of production and control of OPV in China. Particularly, they formulated the standard of safety evaluation of pathological changes in the brain tissue of monkey after intracerebral inoculation with OPV. Dr. Gu was honored as the founder of the Institute of Medical Biology, CAMS and of China's poliovaccine. He is now a member of the national commission for the certification of poliomyelitis eradication in China. Since April 1995, there has been no polio case caused by indigenous wild poliovirus. That means the circulation of wild poliovirus in Chinese community was interrupted. Dr. Gu has made a great contribution to the eradication of poliomyelitis in China.

His scientific reputation and administration skill have led to his appointment to a series important posts within and outside of the Academy and the college. He is now the honorary member of the Chinese Association of Sciences and Technology; member of National Foundation of China; fellow of Royal College of Physicians (London); member of European Academy of Art and Science; fellow of the Third World Academy; honorary member of Chinese Medical Association; honorary chairman of Beijing Association of Science and Technology; and honorary president of Chinese Society of Biomedical Engineering and of Immunology.

From 1976 to 1984 he served as vice-president of CAMS and PUMC. In 1985, he became the president of these two institutions. From 1993, he left the position as president of CAMS and PUMC. He won the award of National Science Congress in 1978 and HoLeungHoLee (HLHL) prize in 2001. He published 50 scientific papers and 6 books.

Abstract: In over 50 years since the founding of the country, China has had significant development in health care. Particularly, in the past 20 years, with the adoption of reform and opening policy, great changes have taken place in people' s health condition and health status.

During the period of 1991-2000, new-borne mortality rate decreased from 33.1‰ to 22.8‰; infant mortality rate dropped from 50.2‰ to 32.2‰ and maternal mortality rate declined from 88.9 per 100,000 to 58.7 per 100,000. In 2001, the mean life expectancy was 71.4 years old. Compared with the results of the national census of 1990, there is 2.8 years increase. According to the data presented by the Center of Diseases Prevention and Control, China, in the past 10 years, the prevalent rate and morbidity of infectious diseases has been declined dramatically. There is no doubt that the progress in medical sciences has been played an important pole in the development of health care.

1. CONTROL OF MAJOR INFECTIOUS DISEASES

1.1 Eradication of Poliomyelitis

In the 1950s, paralytic poliomyelitis was a serious problem in China. In 1959, F. Z. Gu and his colleagues successively developed liquid oral polio vaccine (OPV, Sabin type). In 1962, a sucrose-lactose based dragee was developed by F. Z. Gu and his assistants and replaced the liquid form. The safety and epidemiological effectiveness of OPV has been proved by large scale clinical trials in 12 cities with 4.5 million children. In 1962, when China-manufactured OPV was licensed for use, China possessed the capability to produce 200 million doses of OPV yearly. In 1988, the Chinese government joined the WHO's worldwide program of eradication of poliomyelitis. In 1993, the first of three coordinated national immunization days was conducted with 80 million children immunized. By 1995, a total of 350 million children had received trivalent OPV. In 1994, the last case of poliomyelitis due to indigenous wild poliovirus occurred in Hubei province. In July 2000, the National Committee of Certification of Eradication of Poliomyelitis concluded that since September 1994, there has been no indigenous wild poliovirus in China for at least 3 years and transmission of indigenous wild poliovirus has been interrupted in China. In October 2000, in Manila, the West Pacific Regional Commission on Poliomyelitis Eradication announced that China as well as other countries of West Pacific Region were certified polio-free.

1.2 Eradication of Leprosy

In China, leprosy has a history of two thousand years. In the 1950s last century, there were 25 provinces with prevalence >0.1 per 100 thousand. Since then, H. Y. Li and others have started to study the methods for prevention and treatment of leprosy and they made significant achievements.

In the 1950s, a county-level nationwide vertical administrative system for leprosy control was established, including active case finding,treatment of multibacillary leprosy with Dapsone in isolated wards and treatment of paucibacillary leprosy outside the hospital.

In the 1980s, combined chemotherapy (WHO MDT) was introduced in to China in order to solve the problem of resistance to long term use of Dapsone. Fixed duration MDT was conducted in Yunnan, Guizhou and Sichuan provinces among 10 thousand patients. The FD-MDT has shown very significant therapeutic effect. After 10 years surveillance, the relapse rate decreased to only 0.3/1000 person-years. Thereafter, FD-MDT was applied to all high prevalence counties in China throughout the 1990s.

At this time, techniques of immune-pathology, Dot-ELISA/ECL and PCR are used for early case finding and early diagnosis. In 1991, the National Surveillance System of Leprosy Control was established.

The targets of prevention and treatment as well as the standard and method of evaluation of leprosy control measures were formulated. H. Y. Li has proposed that operation of eradication of leprosy should be done at the county level.

After 20 years efforts, prevalence of leprosy for the whole country in 1999 was only 0.051 per 10 000. Compared with the high prevalence year of 1966 (1.149 per 10,000), the prevalence of leprosy was dropped by 95.6%. The endemic areas were greatly reduced. 98.6% of the counties have reached the WHO standard of eradication of leprosy (prevalence <1/100); 90% of counties realized the national standard (prevalence <0.1/10,000 ; case finding rate <0.5/10,000).

Fig. 1 Distribution of prevalence of registered leprosy cases in China,1950-1960

Fig. 2 Distribution of prevalence of registered leprosy cases in China, 1997

1.3 Basic Elimination of Lymphatic Filariasis

China is one of the countries with significant cases of lymphatic filariasis (LF). It is estimated that there were 30,994,000 cases before implementing a control program. In 1956, LF was picked as one of the priorities of 9 diseases controlled by the central government. A series of achievements have been recognized internationally, through several generations' efforts, with over 40 years of field studies by the scientists and health workers, on the strategy and technical measures for interrupting the transmission of LF.

Through field comparative studies and enlarged control practice, it was demonstrated that the effect of single elimination of the source of infection with diethylcarbamazine citrate (DEC) was similar to that of elimination of the source of infection integrated with vector control. Therefore, in the 1970s the strategy for elimination of the infection source as the major focus in LF control was established.

A six-month mass administration regimen was formulated using 0.3% DEC-fortified salt. It had less therapeutic reaction and is effective, safe and easy to use, which provided a strong technical guarantee for implementing the source elimination strategy.

Based on the findings of the residual microfilaremia cases with low density (5Mf ±/60μl) which has provided no practical significance for keeping LF transmission and the results of studies on the threshold of LF transmission, the criteria of basic elimination of LF was first raised. The microfilaria rate decline to below 1% after control practice, and could be regarded as an indicator of interruption of transmission.

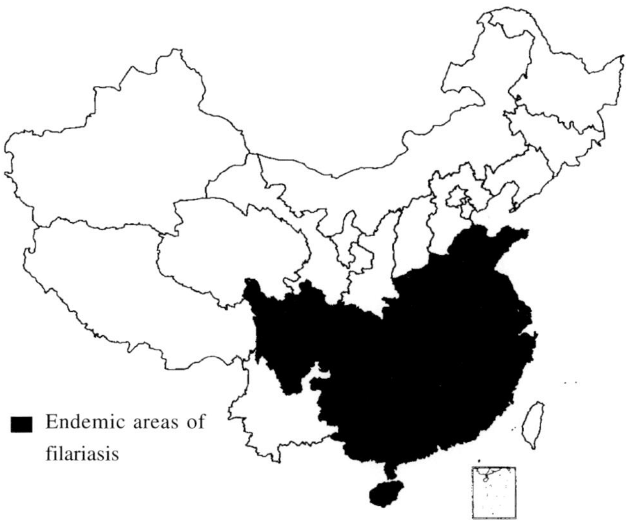

Fig. 3 Distribution of lymphatic filariasis in China (Pre-control)

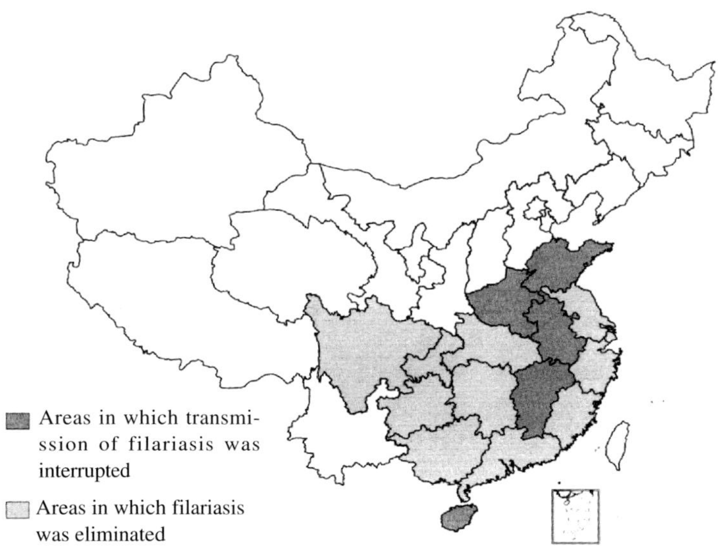

Fig. 4 Distribution of lymphatic filariasis in China (2002)

The longitudinal and cross-sectional surveillance was established system after transmission interrupting, which provided further evidence of LF elimination.

The criteria of basic elimination and elimination of LF and correspondent technical indicators were formulated for guiding the realization of the goal of basic elimination and elimination of LF in the whole country.

After the application and dissemination of the above research results across the country, all the 16 endemic provinces/autonomous regions/municipalities had reached the criteria of basic elimination of LF in 1994 and 11 provinces/ autonomous Regions/municipalities had reached the criteria of elimination of LF in 2002.

1.4 Control of Viral Diseases

1.4.1 Attenuated Live Japanese B Encephalic Vaccine (JE vaccine)

Since 1960, Y. X. Yu and others started to study attenuated live JE vaccine. After 100 serial passages of a JE virulent virus (SA14) in primary hamster kidney tissue culture for attenuation, 17 times for plague-cloning of attenuated virus selection and several times for suckling mice peripheral passages for increasing immunogenicity, one JE virus clone was obtained which appeared with no intracerebral virulence for mice and monkeys, and kept stable after *in vitro or in vivo* passages and showed good immunogenicity. Several human trials showed that the live vaccine was not only safe but also highly immunogenic for human. The seroconversion rate and protection rates were estimated to be 85%-100% and over 95%, respectively. The vaccine was licensed for production by Ministry of Health in 1989. This vaccine is the first attenuated JE vaccine developed in the world.

1.4.2 New Human Rotavirus

In the 1980s, a large scale outbreak of adult diarrhea occurred in China. In 1982-1983, the number of patient reached to more than one million. Endemics of this disease were occurring in mining areas with poor drinking water supplies. T. Hong discovered a new group of rotavirus, named adult diarrhea rotavirus, from feces of patients. It is a unique human group B rotavirus world discovery.

2. CONSTRUCTION OF ECOLOGICAL SANITATION IN RURAL AREAS

Since 1997, a group of health engineers, J. G. Wang, F. Y. Sun and S. C. Pan started to study the feasibility to develop and apply urine diversion for non-flushing ecological toilet systems in China. Ecological sanitation applies the fundamentals of ecology and takes the non-hazardous, decrement as well as resource recovery of human excreta as the guiding principles for the construction of ecological facilities and the rational utilization of treated human excreta as a useful and abundant manure. Based on an extended period investigation and practical application, various types of rural household toilets that meet the requirement of eco-sanitation were popularized. This included the three-compartment septic tank toilet, double urn funnel-shaped toilet, triplet biogas toilet, ecological toilet with urine diversion, double pit toilet and so on.

The characteristics of ecological toilet with urine diversionary is as follows:

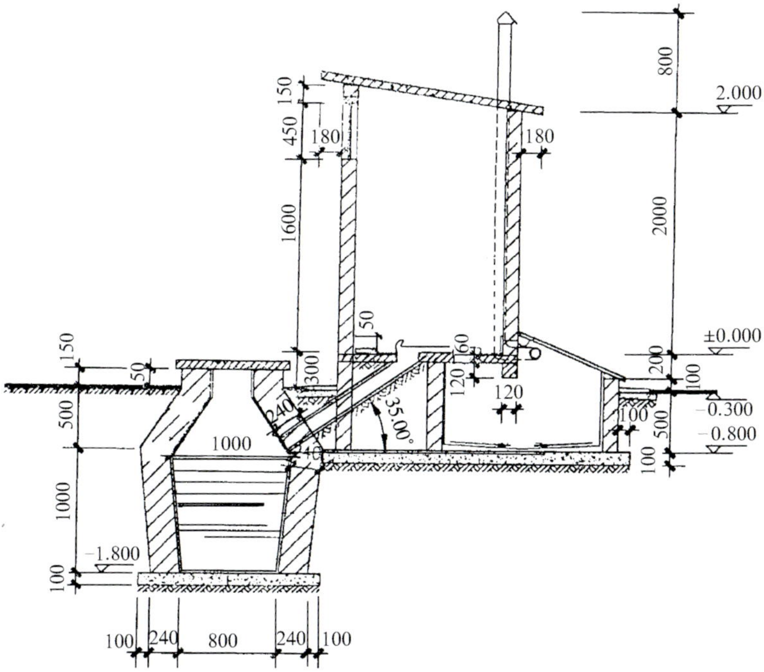

Fig. 5 Sectional drawing of ecological toilet with urine diversion

based on the distinct qualities of urine and feces, excreta are sanitized on site yielding odourless, no airborne or ground pollution, while conserving water, urine and feces are fully recovered and recycled, which benefits agriculture. Sucha system is freeze-resistant, promising to fill the gap where there is no suitable toilet system in cold rural areas. Up to 2000, the rate of popularization of various types of rural household toilet reached 44.8%.

3. NEW APPROACHES TO TREAT MALIGNANT TUMOR

In 1986, Z. Y. Wang successfully treated acute premyelocytic leukemia by means of inducing differentiation with Chinese manufactured all-trans-tretinoin, which was the first success in the world. As the first case in the history of cancer research that cancer cells in the human body were reformed into normal cells by natural products instead of toxic chemicals, this has since been widely proved and recognized in clinical treatment around the world. The remission rate reached up to 85% – 90%. In addition, it also led to the preliminary understanding of the mechanism of the reaction of all-trans-tretinoin in the body of leukemia patients. This new approach has been acclaimed as a revolutionary achievement.

In 1978, Z. Y. Tang raised a hypothesis of the sub-clinical small liver cancer, which deduced that the natural history of liver cancer is 2 years, instead of 3-6 months. Based on this theory, the early detection and treatment of liver cancer was conducted and proved to be very beneficial to liver cancer patients. The 5-year survival rate after radical resection of small liver cancer (<5 cm) increased to 67%. Cytoreduction and sequential resection is another idea proclaimed by Z .Y. Tang, which makes the 5-year survival rate increase to 62% for some big liver cancer. Among various treatment choices to make the tumor shrink, radio-immunotherapy with radiolabelled antibody to liver cancer cells will obtain greater possibility to make tumors shrunk in selected patients.

4. GENETIC ENGINEERING AND EMBRYONIC ENGINEERING

First in the global, Y. T. Zeng elucidated the DNA sequence of the sex determining gene SRY of cows and goats as well as controlling the sex of livestock by identification of the embryonic SRY gene and embryo transfer. The rates of transfer pregnancy for cows and goats reached 56.8% and 31% respectively. In the study of the transgenic animals Zeng and his colleagues have successfully obtained transgenic goats with expression of human coagulation factor 1X in the milk and a transgenic cow integrated with human serum albumin gene, which established a solid base for building the "animal pharmaceutical factory".

5. NEW ANTI-MALARIA DRUGS

Malaria remains a major hazard to the health of people in the developing world .Meanwhile, new strains such as P. falciparum that are resistant to existing antimalarials emerged emphasizing the importance of the search and development of new antimalarial drugs. Several new anti-malarial drugs have been developed recently augmenting the well known Chinese new antimalarial drug, Qinghaosu. A new compound, designed and synthesized by R. X. Deng from1972 to1985, was proven an effective anti-malarial drug after pharmaceutical and biological assessments and clinical trials. Its registered name was Benflumetol, which was suitable for all forms of malaria, including chloroquine resistant falciparum malaria. The main characteristics of Benflumetal are: novel structure, strong plasmodicidal action, high cure rate and low toxicity.Progress has also been made in the research in the field of radical curative drugs for vivax malaria.

In 1999, another group of Chinese pharmaceutical chemists, led by Y. C. Ni, found that the Chinese herb product, Daphnetin (DPNT), a known iron chelator, has anti-hemolytic and anti-erythrocytic membranous lipid peroxidation effect. Moreover, Ni and his colleagues studied the schizontoocidal activity of DPNT against malaria parasites. Their study indicates that the DPNT exhibit potent schzontocidal activity against malaria parasites both *in vitro* and *in vivo*. The iron chelator DPNT has shown a dose-dependent schizontocidal activity to malaria parasites. All of the studies have shown that DPNT is a promising candidate for antimalarial drugs.

6. NEW FINDINGS, NEW THEORY, NEW FIELD AND NEW ACHIEVEMENTS

6.1 New Achievements in Neurosurgery

Z. C. Wang, director of Beijing Institute of Neurosurgery, is credited with pioneering neurosurgical development in China. He is the first man in the world to complete more than 1,000 operations of cranial aneurysms with lowest operative mortality rate (<2%). He has performed the largest number of operations in arteriovenous malformation cases in the world and is the first man with the lowest operative mortality rate (<1%) for brain stem and intramedallary spinal cord tumor surgery. During the past 52 years, the total number of neurosurgical operations he did exceeded 10,000 cases.

6.2 New Findings in Study of Adrenoceptors

Q. D. Han, director of the Institute of Cardiovascular Research, former Beijing Medical University, took the lead in verification of the hypothesis that α1-adrenoceptor (α1-AR) contained two subtypes, α1A and α1B, by their different pharmacological characteristics; established the approach to distinct these two subtypes, and revealed different signal transduction mechanisms between the two subtypes in smooth muscles. In addition to the further verification of existence of the third subtype (α ID) and studies on basic characteristics of different subtypes, he mainly focused his efforts on the distribution, expression, mediated biological function, regulation, cross-talks with B-adrenoceptors and pathophysiological alterations of different α1-AR subtypes in heart and blood vessels. The results are of important value in exploring the physiological and pathophysiological significance of the existence of the multiple subtypes of α1-AR in heart and blood vessels.

6.3 New Theory of Brain Mechanisms of Vision

C. Y. Li, research professor of the Institute of Neuroscience of Chinese Academy of Sciences, and his colleagues have demonstrated an extensive disinhibitory region outside inhibitory surround, which may contribute to the transmit area contrast and luminance gradients of visual images. Based on this observation, Li proposed a new theoretical model, suggestion triple concentric organization for the receptive field. The new model has been used to explain the origin of visual evoked potentials induced by visual stimulus, and applied in the field of neuronal network and artificial intelligence. He also found an extensive integration field beyond the classical receptive field of cortical neurons and demonstrated different structures of the integration field that might be adapted to the analysis of different visual texture patterns. In addition, Li and his colleagues found, for the first time, that the structure of receptive field of lateral geniculate neurons was not fixed, but could be dynamic, adapting to the contextual spatial (shape) and temporal (speed) characteristics. This self-adapting ability of visual neurons is crucial for the processing of various natural images.

6.4 New Field of Chemistry – Cellular Inorganic Chemistry

K. Wang, professor of Department of Chemical Biology, former Beijing Medical University, and his colleagues endeavored to bring the bioinorganic chemistry from molecular lerel to cell level. In order to follow the events in the course of interaction between metal ions with the cells, they established several experimental methods. Based on the result of experiments, they generalized several concepts and basic principles and explored the field of cellular inorganic chemistry. They tried to understand the chemical basis of patho-logic pharmacological and toxicological processes on the basis of the studies on metal-cell interactions. They proposed chemical mechanisms for the development of Kaschin-Beck's disease, caries, pigment gallstone and for the across membrane transport of metal-based drugs. Several basic principles underlying the bioresponses of various metal species were summarized on cellular chemical basis.

In recent years, in the fields of medical biotechnology, human genomics, tissue engineering, pharmaco-genomics, proteomics, bio-chips, stem cells, etc., Chinese medical scientists have also achieved significant progress.

Development of Population and Reproductive Medicine in China

Liu Yixun[1] Xiao Shaobo[2]

[1]Institute of Zoology,Chinese Academy of Sciences

[2] Family Planning Committee of China

Liu Yixun, *born on 5, October, 1936. Research professor, Member of the Chinese Academy of Sciences; Chairman of Academic Committee of State Key Laboratory of Reproductive Biology; Director of WHO/Rockefeller Foundation Implantation Network, China Center; member of Editorial Board of "Human Reproduction (Cambridge)", "Archives of Andrology (USA)" , "Biology of Reproduction & Endocrinology (USA)", "Science in China", Vice Editor-in-Chief of "Developmental Reproductive Biology"; Professor Liu has published more than 180 papers, of which 80 paperswere in SCI journals and more than 500 SCI papers well quoted. He has been awarded with the CAS first class of Natural Science Prize once, and the Second Class of Natural Science Prize for 5 times.*

Abstract: In the past thirty years, the Chinese government has achieved great success in implementing the Family Planning (FP) program, which relied on the independent FP service network and contraceptive measures. The total fertility rate of women of child-bearing age dropped from 5.81 in 1970 to 1.22 in 2000, resulting in a change for the Chinese population from a condition of high birth and low death to that of low birth and low death matching that of the developed countries. Because of this rapid drop in the fertility rate, the day which marked the world population milestone for 6 billion was delayed by four years. Most of existing contraceptive methods in China were developed during the 1970s to

1980s with female sterilization and male vasectomy utilized as the main contraceptive means. The security problems of validity and the side-effects of these methods, however, have to be considered and solved. In the new century, with highly developing science and improvements in human society, we should pay more attention to development of safe and effective new multiform contraceptive methods to meet the range of needs for reproductive health and personal choice. The basic research in the reproductive medicine should be reinforced for the goal to discover which genes or molecules can provide the key step or process of reproduction, in order to exploit new non toxic, innocuous, safe, effective and convenient prophylactics. Chinese scientists have made great achievements in search for such gene(s) or molecule(s) related to implantation and spermatogenesis, and have begun to successfully develop anti-implantation contraceptive methods for women and anti-sperm prophylactics for men.

Environmental degradation, increasing population and overuse of resources are the serious problems that threaten the survival of human beings and the economic development in the new century. It was estimated that the environmental resources and space in China could only afford 1.5 to 1.6 billion people. In the past three decades, although the increase in China population has been effectively restricted by the powerful administrative and contraceptive means, the customary inveteracy such as "having both sons and daughters" and "valning the make child only" remain ineradicable. Most of existing contraceptive methods were developed during the 70s and 80s. The problems and shortcomings, such as security, validity and facility, have to be solved to meet the standard for reproductive health and personal choice in China's new century. The 21st century is one of advanced developments and great achievements in human society. Birthrate control in China should therefore not be done by the bureaucratic or administrative means using older contraceptive methods [1,2], as those approaches will not be appropriate for the needs of the new generation. We should pay great attention to "Chinese letters" and develop safe and effective new multiform contraceptive methods to meet the various needs for the purpose of reproductive health and personal choice in China [1].

1. ACHIEVEMENT AND CONTRIBUTION OF FAMILY PLANNING (FP) PROGRAM IN CHINA

The Chinese government has made a great effort in FP program in order to popularize and introduce the knowledge and techniques to put contraceptive measures into practice. As a result, visible outcomes have been achieved as summarized in Fig. 1, 2 and 3. The total fertility rate of women at the child-bearing age in 1970 was 5.81 and dropped rapidly to 1.22 in 2000 (the result of the fifth population census), reaching the level of the developed countries. The rate of natural increase in China's population was 2.583 in 1970 and dropped dramatically to 1.569 in 1975, 0.914 in 1998 and 0.695 in 2001. The total population of China in 1975 was 924 million, which accounted for 23.1% of the world total population. From 1975 to 1999, the increase in population of China totaled 334 million, which accounted for only 16.7% of the world population increment. In the past 30 years population in China accomplished a shift from high birth and low death, creating a high growth rate, to low birth and low death leading to a low growth rate. This change made a significant contribution to reducing the growth of world population. Because of the FP program in China, the day that marked the milestone of reaching 6000 million people for the world population was delayed by nearly four years. According to the United Nations forecast, the total population of China is expected to drop slowly after reaching the peak value of 1.491 billion during 2035 − 2040 as shown in Fig. 3. At that time, China's population will account for 18.5% of the world population.

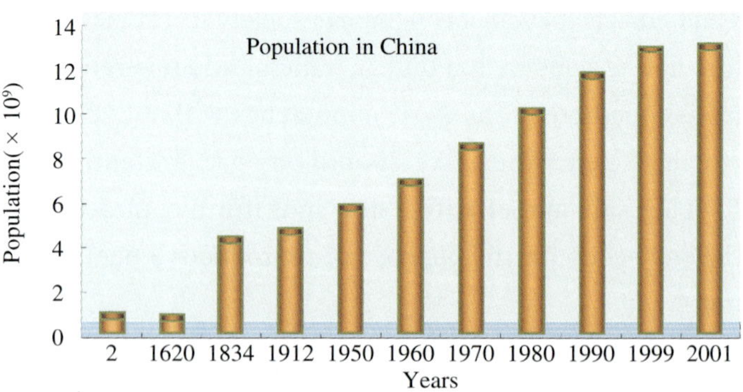

Fig. 1 Development of population in China

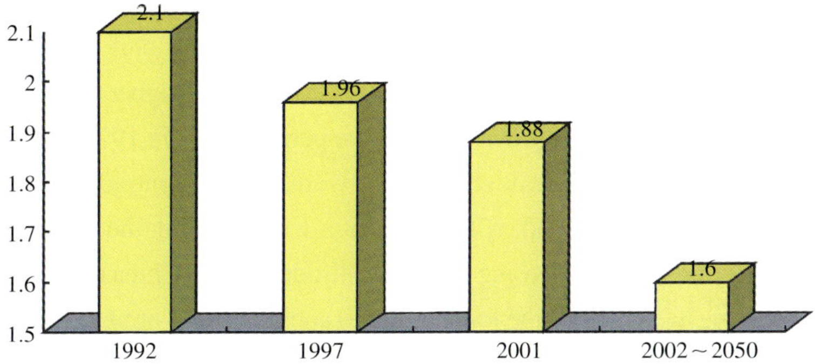

Fig. 2 The average birth rate of Chinese women for whole life

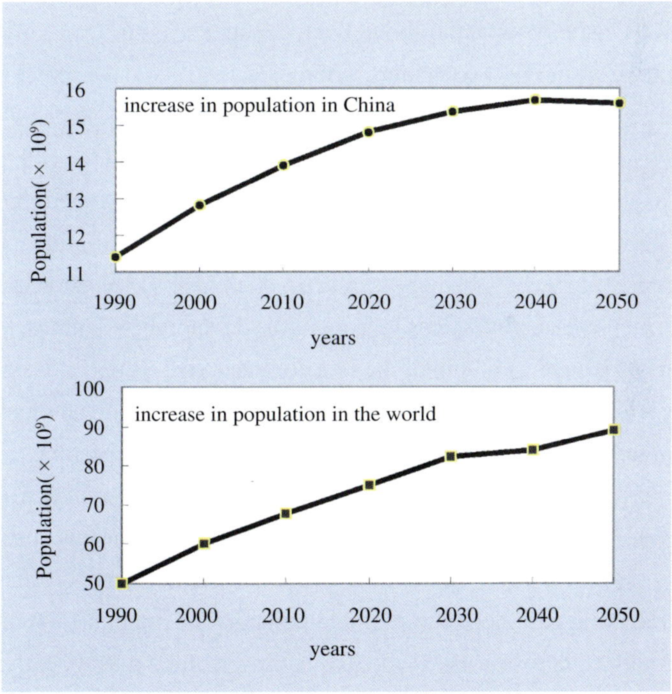

Fig. 3 Population growth in China and in the world

2. DEVELOPMENT OF CONTRACEPTIVE METHOD AND TECHNOLOGY OF FAMILY PLANNING IN CHINA

Relying on the achievement of the FP independent scientific research and the development of the FP service system for contraceptive drugs and devices to meet the domestic needs has yielded solid results. According to the sampling survey data, the general contraceptive prevalence rate of the married women of child-bearing age in China reached 83.4% in 1992 and 86.9% in 2000 [1], which were higher than the average levels of the developed countries during the same period (72% in 1995). The female sterilization and the intrauterine devices (IUD) were the two contraceptive methods commonly used in China (80%), with 80 million women using the reversible IUD method. When the T-copper releasing IUD was introduced into China in the 1980s, the scientists in China had changed the loop IUD into another shape similar to the T or the lumen of the womb and put the brass wire loops into the inside track that was enlaced by the stainless steel wire, which became the Chinese original copper ion releasing stainless IUD. On the basis of the above IUD, Chinese scientists have invented a new type of copper releasing IUD containing indomethacin to avoid the shortcomings of the IUD with large copper surface covered which caused side effects, such as bleeding and pain [3]. Moreover, efforts have been made in China to develop various other contraceptives, such as the oral, external-used, emergency contraceptives and the sub-dermal implants. Chinese scientists have also invented the vacuum uterine aspiration for terminating early pregnancy. In the research, production and evaluation of safety of the steroid contraceptives, China stands in front of the developing countries. For example, the study on reducing the amount of the oral contraceptive compounds and the design and synthesis of effective compounds including ST1435. The studies by Chinese scientists on mifepristone for terminating early pregnancy, hastening menstruation, preventing pregnancy and emergency contraception have brought worldwide attention. Achievements in sub-dermal implants (Norplant) in China have also been made. The two and six root types of sub-dermal implants have been produced as well as the single-root releasing gestodene sub-dermal implants. The pre-clinical trial research on the sub-dermal implants for use by breast feeding women was recently completed. Meanwhile, the phase 1 clinical trial has been accomplished with satisfactory results on mifepristone for the sub-dermal implants releasing L- methylnorethindrone, which can be bio-degraded and therefore not required to be removed [2].

Development of male contraceptive technology and plant medicine for the birth control, as well as the studies on the immune system contraception in China has been recognized throughout the world. The no-scalpel vasectomy invented by Chinese scientists has been popularized among the developing countries. The Chinese latex condom has begun distribution on the international market after strengthening the quality of the products and adequately addressing the needs of the domestic market [2, 3]. The composition of the major female and male contraceptive methods is shown in Fig. 4.

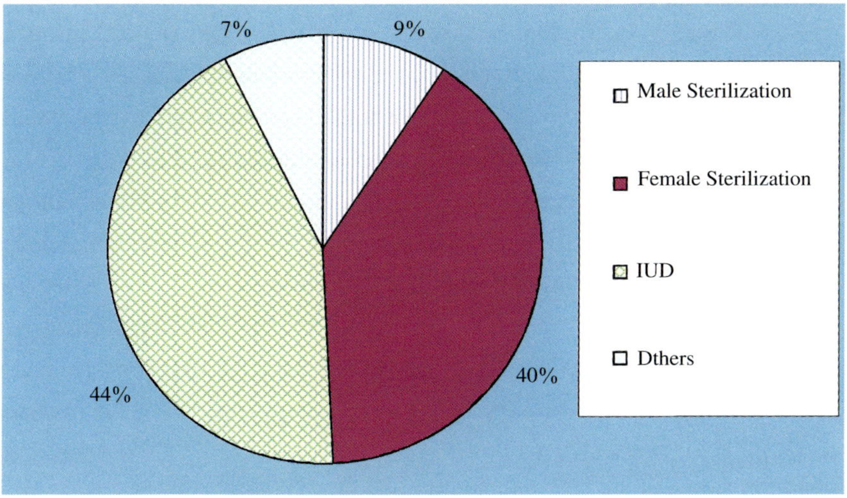

Fig. 4 Compositon of the main contraceptive methods for the married male and female in China

The Chinese government placed significant effort at promoting the reproductive health activities oriented towards informed choice for contraception, intervention on reproductive tract infection and reduction of birth defects, development of herbalist doctor and Chinese traditional medicine, the technology of genic CMOS chip and the immune contraception, and other new technology and products that prevent both sexually transmitted diseases (STD) and HIV/AIDS. Human-centered and client's need-oriented reproductive health services have been creating a great opportunity to integrate the technologies of information and communication and the service management sciences.

China was the first country in introducing the concepts and methods of the system control into the field of demography research, and in using the theory of the population control to direct the tasks of FP and population. Since 1953, China has conducted five population censuses[3]. The rapid development of the science and technology of the communication (information technology) in China provided the technical support for systemic and large-scale population census and data analysis. The network of FP and the basic data accumulated year by year has provided the conditions for accomplishing the population census task.

Different from other countries, China has established a relatively independent service network of FP. There were 40 thousand FP service delivery centers with about 150 thousand service providers among this FP service network. The network of FP centers in China is supported by the national FP system of Science & Technology Research Institutions and the system of contraceptive production. This FP system was strategically linked with the fields of population, research, reproductive health and contraceptive technology. This FP system gradually becomes an important sub-system of the health care service system in rural areas.

The development of the contraceptive techniques and the network advantage of FP for the reproductive health has realized the low fertility rate and made a great contribution to world population control[1].

3. REINFORCEMENT OF BASIC RESEARCH OF REPRODUCTIVE MEDICINE AND DEVELOPMENT OF NEW CONTRACEPTIVE METHODS

Development of new prophylactics is closely related to the breakthrough in theory obtained from basic research in reproductive medicine [4,5]. Because of the important discoveries in reproductive endocrinology in the 50s and 60s, Professor Zhang Ming-Jue, a Chinese American, and his colleagues exploited the first generation of feminine steroid oral contraceptives. With the research breakthrough in hormone receptors in the 80s, the second generation of feminine steroid prophylactics was exploited. Because inadequate recognition was paid by government to the importance in the basic research of reproductive biology for discovering new contraceptive method and knowledge of understanding human reproductive mysteries, reproductive research did not keep up with the developments of modern biosciences in the last 20 years and the contraceptive revolutionary "wave" wore off [4]. The existing contraceptive methods are still resting on the advances of the 80s.

The aim of contraception is to ensure a lifestyle for human beings that can pursue a long-term healthy body and mind . The reproductive health is the most strategic important matter on the earth. Therefore, we must think it over with strategic thought and developmental insight. Contraception is, in fact, aimed at disruption of the normal physiological rules of the body with 100% certainty without obvious side effects. We should also consider the problems, associated with attending to individual diversity of prophylactic use and the culture and economic background of different groups for different needs of the contraceptives.

It has been suggested that basic research in reproductive medicine and health should be reinforced in the new century for the goal of exploiting new non toxic, innocuous, safe, effective, and convenient prophylactics [4,5]. Comprehending various newest theories in reproductive medicine is the key step to resolve the important reproductive problems, such as sterility, abnormal pregnancy, and diseases of reproductive tract, for designing and exploiting prophylactic [5]. Scientists from the multi-disciplinary research fields, molecular biology, cell biology and molecular immunology should work together to initiate the second contraceptive revolutionary "wave" [4,5].

Exploiting new contractive methods should be based on the precise understanding of certain key points that can easily be blocked and have no side-effects on health. Contraceptive method development should use the newest molecular measures and the achievements from the human genome to find out key genes specifically expressed at a particular easily-blocked weak step during a reproductive process and then choose ones that are the most specific and functional and has no effect on other body functions. Worldwide agreement has been reached that the work should be started from two aspects: ① to look for molecule(s) which play key roles in the process of embryo implantation with which to develop an anti-implantation contraceptive method; and ② to clone key gene(s) or look for molecule(s), related to spermatogenesis and germ cell maturation for developing male prophylactics. Disturbance or blocking of these reproductive weak points will not affect human health and will be the most ideal approach to developing safe and valid contraceptives.

3.1 Development of Anti-Implantation Contraceptives in Women by Blocking Implantation

Embryo implantation is a downstream process and weak point for fertility control of female reproduction, and is a complex and multifactor-related long process [6,7]. Blocking implantation will not have any effect on the maternal body and is believed to be the weakest target for developing woman contraceptives. Supported by WHO and Rockefeller Foundation six international cooperating research centers (including our center in Beijing) have been founded since 1999. The aim of the six Centers is to exploit jointly a non-steroid, non toxic, innocuous, steady and monthly-used, convenient for women, anti-implantation prophylactic method on the basis of exploiting molecular mechanisms of implantation. It is the most ideal point to exploit and design a female prophylactic through blocking establishment of the implantation "window," which is a temporal "physiological embryo-acceptive state" occurring in the endometrium just before implantation, as shown in Fig. 6. The maternal-embryo recognition, signal exchange, and implantation take place during this limited period. The formation of the "window" is regulated mainly by estrogen and progesterone secreted by the corpus luteum and cell factors, growth factors, apoptosis-related factors and PA/MMP activator/inhibitor produced by the uterus[8, 9]. We have established a rhesus monkey model, similar to humans, to investigate and confirm the key molecules related to implantation in primates. Our results showed that apoptosis, tissue remodeling and related extra-cellular matrix break down and angiogenesis occur in the endometrial or decidual glandular epithelium, endothelium and villous epithelium in the early implantation site. It seems that apoptosis, tissue remodeling and angiogenesis occur in the same tissues or cells at the same time, and are special temporal-related and reliant on one another. Preliminary data have shown that a compound which can inhibit one or two of the three processes is capable of effectively inhibiting implantation[9]. These findings

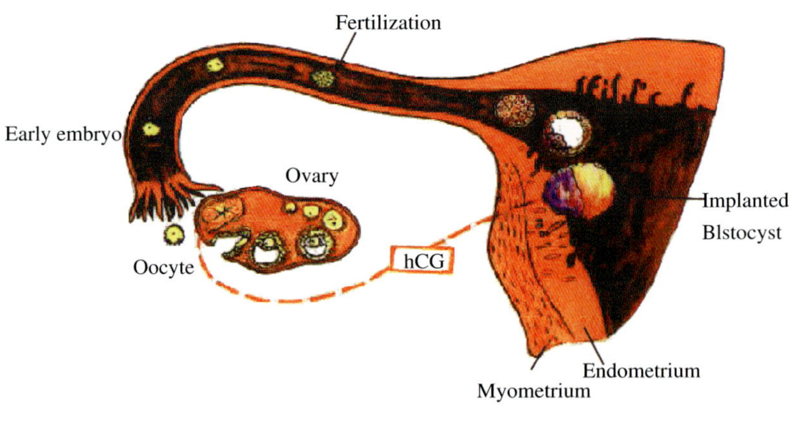

Fig. 5 A reprehensive diagram showing the establishment of implantation widow and the processes of implantation in the uterus

have created an important basis for searching for anti-implantation molecules for developing prophylactic methods.

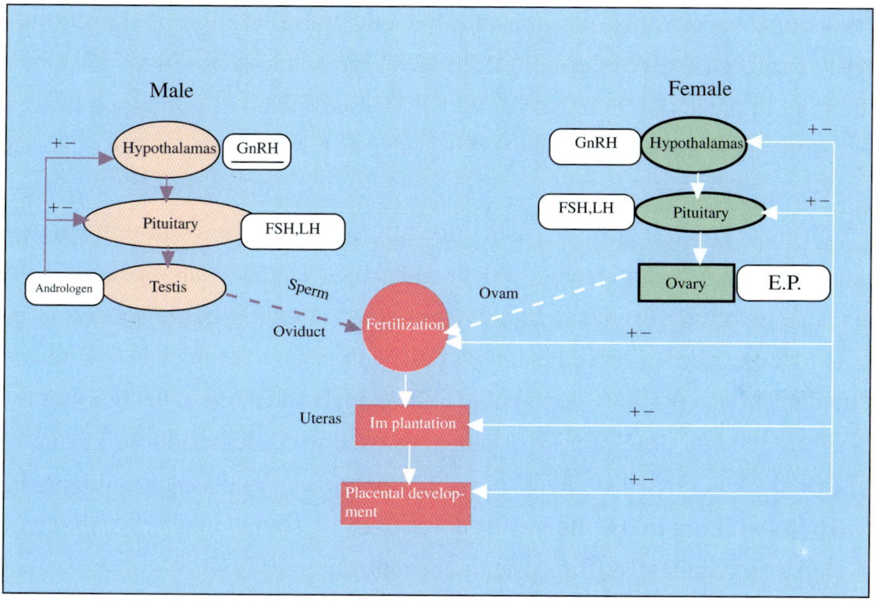

Fig. 6 Regulation of male and female reproductive processes
Hypothalamus secretes Groh to control pituitary gonadotropin (FSH and LH) production which regulate the ovarian hormone synthesis (estrogen and progesterone) and oogenesis in female and testicular testosterone production and spermatogenesis in male. After fertilization of the ovum and sperm which takes place in the oviduct, the embryo transports into the uterus where implantation and the embryo development take place. The steroid hormones secreted by both the ovary and the testis also regulate the function of both hypothalamus and pituitary through a positive or negative pathway

3.2 Anti-implantation of Mifepristone, Gossypol and Caffeic Acid

Mifepristone (RU-486) has been clinically broadly used as an abortion and anti-pregnancy drug. RU-486 at high doses, however, induces a persist uterine bleeding. A low dose of RU-486 without side effects has been tested to be capable of inhibiting implantation[9] by interfering with the endometrial apoptosis and with the regional ECM break down that normally occurs during the implantation period by stimulating the production of endometrial PA and MMPs while inhibiting their inhibitors on one hand, and inhibiting expression of the apoptosis-related genes in the uterus on the other.

Gossypol is a yellow polyphenolic compound extracted from the cotton seeds, and has been clinically used to treat endometritis and ectopic pregnancy. Clinical studies showed that gossypol functions in the uterus mainly through inhibiting the endometrial expression of estrogen and progesterone receptors to decrease locally the actions of both steroids, but did not affect ovarian secretion of the steroids [10]. Preliminary data showed that a high dosage of gossypol was capable of completely inhibiting implantation in mice, but such high doses could also induce a hypokalemic paralysis as reported in men[11]. This drug at lower dose was not effective to completely block implantation.

Caffeic acid (CA) has been also clinically used as an anti-tumor medicine by inhibiting certain growth factors and the molecules related to angiogenesis. CA has been also reported to enhance luteal regression and inhibit implantation.

The studies above showed that it was difficult to get a result of 100% inhibition of implantation by RU486, gossypol or caffeic acid used alone at the innocuous low doses, but the low dose of each compounds used in combination may effectively inhibit implantation and avoid the possible side-effect. Following this direction, we have been working to use the lowest doses of these compounds in combination to develop a new female anti-implantation prophylactic.

3.3 Human CG Vaccine and GnRH Analog

Reproductive biologists have been trying to look for an earliest message molecule produced by the embryo for developing an anti-implantation agent. Professor Zhiyi Zhang from our lab suggested in as early as the 70s that chorionic gonadotrophin (CG) may be one of the candidates from the embryo. An hCG vaccine has been reported to be capable of specifically blocking implantation[12], but for the reasons, such as needing series of injections to get a sufficient immunity in the body and local pain at the injection area caused by the injected materials, it is therefore not used until now. Using hCG DNA vaccine made by modern biotechnics can entirely avoid the above shortcoming, but whether the vaccine DNA would be integrated into the genome still needs to be further clarified. Further study would be hopeful that an hCG DNA vaccine may be one of the most effective anti-implantation contraceptives.

Studies also demonstrated that gonadotropin releasing hormone (GnRH) produced by the embryo is different from the GnRH produced by the hypothalamus and may play an important role in controlling CG release and corpus luteum function[13]. Under the support of Conrad Foundation, we have investigated a specially designed artificial small GnRH analog in collaboration with Professor Siler-Khodr in USA to look for the possibility to develop an ideal anti-implantation agent. The GnRH analog is steady in the blood by resisting hydrolyzation of the blood proteinases, is capable of specifically inhibiting corpus luteum function and embryo hCG secretion, and inhibiting uterine PGE production. Our data obtained from nearly two years' studies with more than one hundred monkeys showed that the analog can effectively block the embryo implantation and prevent pregnancy with a 100% of efficiency, and no obvious side-effect could be observed. Therefore, we believe that this peptide would be hopefully further approved and accepted by the WHO and the Conrad Foundation as an ideal oral non-steroid, monthly-used anti-implantation contraceptive.

3.4 Development of Male Contraceptives by Blocking Spermatogenesis

Research in the male reproduction has not been emphasized for many years, as no theoretical basis can be directly used for developing a male contraceptive agent. Only 15 percent of the total current contraceptive methods (85 percent in the female), such as spermaduct ligation and condom, have been used in the male. Looking at the reproductive processes in both male and female, as shown in Fig. 7, the main steps, such as oogenesis, sperm and ovum transportation, fertilization, fertilized egg transportation, embryo implantation and development, are fulfilled in the female reproductive organ or tract. The only way for developing male contraceptive methods is to try finding the genes or molecules related to controlling

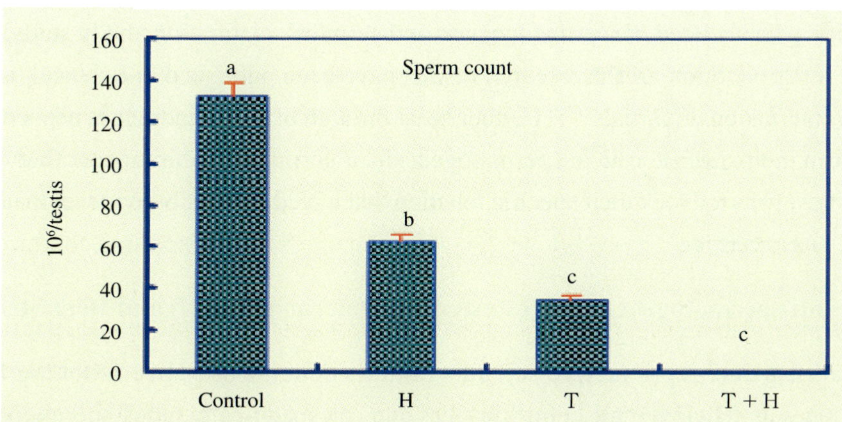

Fig. 7 Effect of testosterone (T) and heat stress (H) on semen sperm account

spermatogenesis in the testis or sperm maturation in epididymis. In 1995 and 1999, World Health Organization, Rockefeller Foundation, Mellew Foundation, Conrad Foundation, and NIH have jointly organized special meetings with the worldwide well known reproductive biologists to discuss the direction of basic research for male reproductive biology in the 21st century and to emphasize and develop new male contraception[14]. The meetings have greatly promoted progress in the basic research of male reproductive medicine .

Spermatogenesis begins with a mitotic proliferation of the stem-cells, spermatogonia, and half of these mitoses then enter a lengthy meiotic phase as preleptotene spermatocytes. The latter cells complete the meiotic phase from spermatocytes to round spermatids that subsequently differentiate into elongated spermatids. It takes about 56 days to finish these phases. During the long process the sequential expression of a series of special genes and the interaction of the expressed products are involved[15]. Heredity, variety and bad environmental factors have been demonstrated to interfere with normal expression of these genes and the interaction of their products, leading to blockage and abnormality of spermatogenesis, and male infertility or birth defect may take place. One may ask nohe ther or not any specific genes exist which control the opening of blood-testis barrier for spermatogonia enterring into the tube for meiosis, or which initiate the meiosis of spermatogonia or control the different phases of the meiosis in the tube? To answer these questions, Chinese scientists have designed various animal models to screen and clone the possible specific genes or look for specific molecules that control these processes. Great progress has been made in screening and cloning such specific genes related to spermatogenesis and germ cell maturation and in studying their function. Some of the research results have been published in Science[16] and other international journals[17-22]. Further breakthrough of these findings is needed to confirm their specific role in spermatogenesis or germ cell maturation or find out specific agents to block their specific function that would be possible to develop new male contraceptive.

3.5 Anti-spermatogenesis of Testosterone Undecanoate (TU) and Heat Stress

For a long time reproductive scientists in this country have been looking for hormones as efficient male contraceptives that can specifically block spermatogenesis, while harmless to the body. At the beginning of 1990s Chinese scientists with the international cooperative groups supported by World Health Organization succeeded in making use of high dose of TU as the male contraceptive agent to block spermatogenesis, and have applied it clinically[23,24]. High doses of exogenous testosterone are capable of suppressing release of gonadotropins from the pituitary with a resultant deprivation of the intra-testicular T and spermatogenesis finally blocked. Our experiments on rhesus monkeys showed that germ cells begin to apoptosis on day 7 after TU injection and reached a peak level on day 30[25]. Fas/FasL proteins play an important role in the germ cell apoptosis[26], but heat stress protein Hsp-7-2 and heat stress factor HSP are not[25]. Although TU is a specific and efficient male contraceptive, the side-effects in some volunteers for the long-term use appeared, such as local pain, breast enlargement, decreasing high density lipoprotein, bone metabolism abnormal, even testis atrophy and sex function decline. Because of these side-effects TU is not widely used clinically.

It is clinically well known that men with cryptorchidism are infertile and the young men who wear jeans frequently decrease fertility. These facts suggest that temperature can affect spermatogenesis. Our study on the cryptorchidism model in rat and rhesus monkey[27] indicated that all germ cells in cryptorchidism testis (spermatocytes and round spermatids) appear apoptosis in 5 days after operation, but the normal division of spermatogonia is not influenced. The expression of Fas/FasL and Bcl-2 regulated the processes of the cell apoptosis and Bax did not play a role. The recent cooperative research with Dr. Christina Wang in the University of California further demonstrated that the amount of sperm in the semen reduced to the degree of infertility in a two month period when the testes were incubated at 43℃ water bath for 30 min each day for two days as shown in Figure 7. If heat treatment is in combination with the implant ation of testosterone capsules, the amount of sperm reduced to zero in two months. The combined treatment induces germ cell apoptosis via different molecular and cellular pathway and an additive or synergistic action could be observed[28]. The former resulted in the cell apoptosis in the middle stage (VII-VIII) of the seminiferous epithelial cycle, whereas testicular warming affected the cell apoptosis at the other stages (I-IV, X-XII). These findings have provided an important theoretical basis for designing male contraceptive agents.

3.6 An Anti-sperm Agent, Gossypol

In the 1960s, Chinese researchers discovered that cooking with crude cottonseed oil led to infertility in men. Experiments showed that gossypol could act on testis directly and selectively damage testicular mitochondrial function[29, 30], which resulted in the blockage of spermatogenesis in testis and germ cell maturation in epidydimis. Gossypol bound to the mitochondrial with various enzymes and inhibited their activities. It is interesting that the damage to the mitochondrial was found to begin with the spermatids around the seminiferous lumen and extend to the base membrane with increased duration of the drug administration [31, 32]. These findings were landmark and gossypol had been showed to be an effective anti-fertile agent in men. Supported by grants from World Health Organization and the Rockfeller Foundation, the systematical studies by Chinese researchers from 14 provinces or cities were carried out and over 10000 men as volunteers used the drug for clinical trial [30, 31]. The statistical analysis from 8806 cases showed that the men became infertile in 75 days if the pills of 20 mg gossypol were orally taken every day and the validity approached 99.89%. After 75 days, taking 7 mg of the drug every day would maintain the contraceptive results. It is also shown that gossypol could suppress spermatogenesis and sperm mobility, but did not influence the normal circulating level of T and sex function [31]. However, with prolong use of the drug administration (up to 6 years), following the increase of gossypol a cumulation in the body, difference of the individual constitution and different reaction to the drug, two kinds of side-effects appeared[11], hypokalemic paralysis (0.75%) and irreversible infertility (9.9%). These stopped the clinical trial and research funding from the nation and abroad was cut, the study on gossypol turned down from the middle of 1980s.

Discovering problems is the beginning of resolving them and is the key of success. Chinese scientists suggest that the combination use of low dose gossypol and testosterone may overcome hypokalemic paralysis and irreversible infertility. Large-scale rat experiments indicated that the low dose of gossypol in the combination could specifically affect the sperms in the epidydymis and quickly exert its action, did not damage the testicular and other viscera function because of the presence of the testosterone. The validity of anti-fertility attained 100% [34, 35].

From the series of research achievements as shown above, a male anti-fertile contraceptive method may be hopefully developed soon in China. The combination design for the methods would be acceptable and applicable. The low dose of gossypol in combination with testosterone plus a local testicular water-bath at 43℃

in a specially designed apparatus for 30 min twice every two months, no sperm would be found in the semen after two month treatment. If you want to recover for fertility, just simply to stop these processes. Gossypol, TU and heat stress in combination act on different sites and stages of the spermatogenesis in testis and sperm activity in epididymis. Apoptosis induced by TU occurs at stages Ⅶ–Ⅷ and heat stress induces germ cell apoptosis at stages Ⅰ–Ⅳ and Ⅹ–Ⅻ. Low dose of gossypol mainly influence the sperm in epididymis, in addition to acting on the testicular spermotids, therefore they will gain the best combined efficiency. This kind of combination meets the standard requirement of contraception that is safe, effective and irreversible and no obvious side-effects because of decreases in dose, action intension and period as used alone.

Acknowledgments: The authors should thank Fu Wei, Zhang Liming, Liu Qing ,Wei Peng, Zhang Zhehong and Li Yinchuan for their help in collecting the materials for this article.

REFERENCES

1 常平主编 ,20 世纪我国重大工程技术成就，暨南大学出版社，2002，P127-133.

2 中国计划生育年鉴编委会 , 中国计划生育年鉴 ,《中国计划生育年鉴》编辑部，2002，P256-272.

3 Population Research Information Center, Theses Collection of 1997 National Population and Reproductive Health Survey, China Population Publishing House, Beijing 1999.

4 Holden C, Research on contraception still in the doldrums, Science 2002 Vol. 296 : 2172-2173.

5 Greaves S and Wang C A, Fertile field , Nature, fertility supplement 2002, In: Nature/cell Biology Vol. 4 and Nature/Medicine Vol. 8 : fertility Supplement 1 (01 Oct. 2002).

6 Liu Y X, Molecular basis of implantation, Bulletin of the Chinese Academy of Sciences, 2002,7(5):331-334.

7 Feng Q, Liu K, Liu YX (2001): Plasminogen activators and inhibitors are transcribed durirr early macaque implantation Placenta, 22 (2-3):186-199.

8 Gao F, Chen XL, Wei P, Gao H J, Liu Y X. Expression of matrix metalloproteinase-2, tissue inhibitors of metalloproteinase-1, -3 at the implantation site of rhesus monkey during the early stage of pregnancy. Endocrine, 2001, 16(1): 47-54.

9 Gao F, Zhou X C, Liu YX. Mifepristone regulates expression of apoptosis related genes Fas and FasL in mouse endometrium. Acta Pharmacol Sin. 2001, 22(6):524-529.

10 Liu YX, The effect of gossypol acetate on the content of the sex steroid hormone in endometrium. Chinese Journal of Dbstetrics and Gynecology 1981, 16(3) :129-131.

11 Qian S Z. Gossypol-hypokalemia interrelationship. Int J Androl, 1985, 8: 313-324

12 Gao Z, Pan S, Liu X. Advances in the study on human chorionic gonadotropin-based birth control vaccines Zhonhua Fu Chan Ke Za Zhi 1996, 31(9): 569-570.

13 Siler-khodr T M and Grayson M. Action of Chicken II-GnRH on the Human Placenta J. Clin Endocrin

Metab 2002 86:804-810.

14 Depaolo LV, Hinton B T, Braun R E, Male contraception: Views to the 21 century, Trend Endocrinol Metab 2000, Vol 11 (2): 66-69 2000.

15 Mu X M, Liu YX, Collins L L. The P53/Rb-mediated repression of testicular orptan receptor-2 in the Rhesus monkey with cryptorchichidism J. Biol Chem.. 2000, 275 (31): 23877-23883.

16 Li P, Chan H C, He B. An antimicrobial peptide gene found in the male reproductive system of rats. Science 2001 Mar 2;291(5509):1783-1785.

17 Baker P J, Sha J H. O' Shaughnessy PJ, Localisation and regulation of 17beta-hydroxysteroid dehydro genase type 3 mRNA during development in the mouse testis. Mol Cell Endocrinol, 1997 ,133(2):127-133.

18 Cheng L J, Li J M, Chen J. NYD-SP16, a novel gene associated with spermatogenesis of human testis Biol Reprod 2003 Jan;68(1):190-198.

19 Tian X Y, Sha Y S. Zhang SExtracellular domain of YWK-II, a human sperm transmembrane protein, interacts with rat Mullerian-inhibiting substance. Reproduction 2001, 121(6): 873-80Wang LF, Wei SG, Miao SY, Calpastatin gene in human testis. Biochem Mol Biol Int 1994 ; 33(2):245-251.

20 Guo C X,Tang T S, Mu XM ,Liu Y X . Cloning of novel temperature-related expressed sequencetag in rat testis during spermatogenesis . Biochem Biophs Res Comm, 1999, 258(2):401-406.

21 Nie G Y, Hampton A L, Fu G Q, Liu Y X.A potential molecular mechanism for regulating premRNAsplicing of implantation-related genes through unique uterine expression of splicing factor SC35 in women and rhesus monkeys. Reproduction. 2002; 124 (2): 209-217.

22 Han X B, Zhou X C, Hu Z Y, Zhang Z H, Liu Y X.Cloning and characterization of temperature-related gene TRS1. Arch Androl. 2002 , 48(4):273-280. World Health Organization, Task Force on Methods for the research on male fertility :contraceptive efficacy of testosterone-induced azoospermia in normal men Lancet ,1990, 336,955-958.

23 World Health Organization.Task Force on Methods for the research on male fertility: contraceptive efficacy of testosterone-induced azoospermia in normal men Lancet , 1999, 336, 955-958.

24 Zhang G Y,Zhang G H, Zhu X H et al. Studies on testosterone enanthate-induced oligozoospermia and azoospermia in normal fertile men J Chinese reprod Med, 1992, 1: 7-11.

25 Zhou X C, Zhang Z H, Hu Z Y et al. Expression of Hsp70-2 in rhesus monkey testis during germ cell apoptosis induced by testosterone undecanoate. Contraception. 2002: 66(5):377-382.

26 Zhou X C, Wei P, Hu Z Y et al. Role of Fas/FasL genes in azoospermia or oligozoospermia induced by testosterone undecanoate in rhesus monkey. Acta Pharmacol Sin. 2001 ;22(11):1028-1033.

27 Zhou X C, Han X B, Hu Z Y et al .Expression of Hsp70-2 in unilateral cryptorchid testis of rhesus monkey during germ cell apoptosis. Endocrine 16 (2): 89-95.

28 Lue Y, Sinha HAP and Wang C et al, Testicular heat exposure enhances the suppression of spermatoge nesis by testosterone in rat: The two hit approach to male contraceptive development Endocrinology ,2000, 141(4): 1414-1424.

29 Liang D C, Fei R R, Liu Y. Studies on distribution of 14C-gossypol in subcellular fractions of rat teastes and site of gossypol action.Zhongguo Yi Xue Ke Xue Yuan Xue Bao ,1981.3 (3):153-157.

30 Xue S P. Studies on the antifertility effect of gossypol, a new contraceptive for males. In: Chang CF, Griffill D, Woolman A eds. Recent advances in fertility regulation Geneva: ATAR SA 1980,122-146.

31 Xue S P. Gossypol contraception and mechanism of action In: Lobl T, Hafez ES, eds, Male fertility and its regulation. Boston : Mtp Press Limites 1985, 155-174.

32 Xue S P,Liang D C, Fei R R, Chen X M (1983): Subcellular site of antispermatogenic effect of gossypol are its possible molecular mechanism of action. Sci. Sin (B). 26(6):614-633.

33 Ei RR, Liang DC, Guo XY, Xue SP , Effect of gossypol on rat testicular mitochondrial (1983): Yi Xue

Ke Xue Yuan Xue Bao 1983, 5(4): 219-222.

34 Ye W S, Dan L, Guo Y.The antifertility effect of gossypol plus testosterone and estrogen, Yao Xue Xue Bao, 1996 31(4) :313-315.

35 Due S P .A Beam of dawn light of study on gossypol as a sale, effective and reversible male ant fertility contraceptive- Evaluation of the studies by using low dose gossypol combined with steroid hormone for male contraception Act Academia Medicine since 2000, 22:211-213.

Chinese Astronomy in the Century Transition

Ai Guoxiang

National Astronomical Observatory

***Ai Guoxiang**, Born in Feb. 1938 in Hunan Province, China. An academician and the director of the Division of Mathematics and Physics, Chinese Academy of Sciences. The director and a research fellow of the National Astronomical Observatories, Chinese Academy of Sciences. An advisor and a member of the Standing Committee of the Association of Science and Technology of China. A member of the Expert Committee of the State High Technology Spaceflight Area of China. The Chairman of the Tenth Committee of the International Astronomical Union (IAU) in the period of 1997-2000.*

Abstract: We present a review on the progresses of Chinese astronomy since the middle of last century, in aspects of construction of observatories and stations, layout of study fields, cultivation of young astronomers, development of equipments and achievements of astronomical scientific studies. A comprehensive introduction of the developing program of Chinese astronomy in the beginning period of the new century, is also itemized, including the particular strategy action of astronomy in the new time, the challenge we faced and the countermeasure we may take, several Chinese big astronomical projects in building, their layout, significances and progress circumstances.

Chinese astronomy has a very long history. The earliest record of astronomical phenomena in China can be traced back to about 4000 years ago. In the beginning of the 20th century, traditional Chinese astronomy began coming into contact with the modern astronomy. Since 1949, Chinese astronomy has formed a comprehensive research system. In the 21st century, astronomy is one of the important basic sciences possessing important characteristics such as frontier science, a fountainhead of high technologies, and forming an essential sector for the quality of national education programmes. Chinese astronomy now faces great challenges and development opportunity.

1. THE PRESENT CONDITIONS OF CHINESE ASTRONOMY

1.1 The Distribution of Chinese Astronomical Observatories

Since the middle of last century, the government has attached importance to the application of astronomy in social development, state security, and environmental protection, in addition to understanding nature and universe. In the Chinese Academy of Sciences (CAS), the Purple Mountain Astronomical Observatory was rebuilt in the beginning of the 1950s for expanding the studies on celestial bodies in Solar System, man-made satellite observations, calendar and calendar calculations. Subsequently, the fundamental construction of Chinese Astronomy went along four aspects. First, for developing astrophysics and cosmology study, Beijing and Yunnan Astronomical Observatories were built in 1958 and 1972, respectively. Second, for astrometry and time service work, the Shanghai Astronomical Observatory was built in 1962, upon the base of old Xujiahui Observatory. Mainly working on time service, the Shaanxi Astronomical Observatory was built in 1966 in West China. Third, for the observations and studies on man-made satellite, the founding of Urumqi Astronomical Station and Changchun Man-made Satellite Station can be traced back to 1957 and 1958. Fourth, to develop the astronomical instrumentation, the Nanjing Astronomy Instrument Factory (afterwards having been reconstructed to the Nanjing astronomy instrument research center) was built in 1958. Employed staff had reached more than 2000 during this period. Also during this particular period, for astronomy teachings and studies, the Department of Astronomy of the Beijing Normal University, the Astronomical Specialty of the

Geophysics Department of Peking University, and the Astrophysics Research Branch in Chinese Science and Technology University were also founded. These have formed the research system for Chinese astronomy.

The Chinese Academy of Sciences put into effect the Knowledge Innovation Engineering beginning at the end of the 1990s. In 1999, the Chinese National Astronomical Observation Center of the CAS was formed by the Beijing Astronomical Observatory, Purple Mountain Astronomical Observatory, Shanghai Astronomical Observatory, Yunnan Astronomical Observatory, Shaanxi Astronomical Observatory, Urumqi Astronomical Station, Changchun Man-made Satellite Station, Guangzhou Man-made Satellite Station, and Nanjing Astronomical Instrumentation Research Center. In April 2001, the National Astronomical Observatories of China (NAOC) was built on the foundation of the Chinese National Astronomical Observation Center.

In this process, the study subjects have been optimized, and the study layout rearranged. Nine Principal Fields which received enhanced support are: large scale structure of universe, dark matter and the dark energy of universe, formation and evolution of galaxies, high energy and cataclysmic processes of celestial objects, formation and evolution of stars, solar magnetic activities and heliogeospace environment, astrogeodynamics, dynamics of celestial bodies in Solar System and artificial bodies, observational facilities for space astronomy and space exploration, and development of new astronomical techniques and methods. In the meantime, to give special support to astrophysics, solar physics, astrogeodynamics and precise orbit calculation of man-made satellites, and space astronomy have been selected as Four Superiority Domains. Chinese astronomy has entered a new particular period of development with emphasis on maintaining unified planning and laying out of study subjects, unified arranging of operation and fundamental construction of large-scale astronomical instruments, unified arranging employee posts, personnel organization and attracting outstanding scientists, unified distributions of funds and other resources, and unified invitation of chief scientists and arranging of positions.

1.2 Present Observational Equipment on Large Scale

During the last 20 years, significant development of Chinese astronomy has been archeived. Some large and middle scale observational equipments for astronomy have been built and used successfully. Beijing Astronomical Observatory built an Aperture Synthesis Radiotelescope Array for metric wave that has 28 antennae in 9 meter diameters in 1984, a Solar Magnetic Field Telescope and a 1.26 meter infrared telescope were constructed in 1988, and the largest optical telescope (2.16 meters) of China was built in 1989. The Shanghai Astronomical Observatory built a 1.56-meter optical telescope in 1987 for astrometry and a 25-meter radiotelescope used for VLBI (Very Large Baseline Interferometer). The Purple Mountain Astronomical Observatory successfully installed in 1990 the 13.7-meter millimeter wave telescope in Delingha, Qinghai Province. The Urumqi Astronomical Station accomplished its 25-meter radiotelescope in 1993. These telescopes have all been equipped with advanced terminal detectors, image and data processing systems, etc. The construction of these observational equipments has made modern astronomical observations in China enhanced greatly.

1.2.1 Solar Magnetic Field Telescope

The Solar Magnetic Field Telescope (SMFT) installed at Huairou Solar Observing Station (northeast of Beijing), was invented, designed and manufactured by Chinese astronomers and engineers. In 1966, Chinese astronomers first suggested that we could use the two spectral lines formed in photosphere and chromosphere respectively to study the magnetic field and velocity field of the Sun almost at the same time. In the primary part of the terminal system of the telescope completed in 1985, a narrow band birefringent optical filter is used as a two-dimensional monochromator,which increases the total efficiency of instrument by 250,000 times as compared with the single point system developed in the 1960s, and 500 times compared with the single slit system developed in the 1970s. In 1987, the receiving system of SMFT was improved significantly. SMFT is advanced in many aspects compared with the similar instruments around the world, including the Photosphere Vector Video Magnetograph at Marshall Space Flight Center and the Photosphere Video Magnetograph at Big Bear Solar Observatory of the California Institute of Technology in USA, both of which can only observe the photospheric magnetic field. Advances also include the following:

For its eliminating of instrument polarization, scientists can obtain first-rate materials of solar photospheric vector magnetic field and velocity field,which is very significant for studying the vector magnetic field structure of solar active regions, particularly the solar flare-producing ranges.

With its high spatial resolution and sensitivity, scientists can study the weak and fine structures of magnetic field.

It is the only instrument in the world that can do the conventional observations of solar chromospheric magnetic field. The data of chromospheric magnetic field with high spatial and temporal resolutions are very important for studying the spatial configuration of solar magnetic field.

Based on the data obtained from SMFT, scientists have acquired many new results about the Sun,including the fibril structures of chromospheric magnetic field, the fact that the superpenumbral magnetic field of sunspots possesses strong horizontal components, the reversal structure of chromospheric magnetic field relative to photospheric one, the very complicated relations derived from the theory of photospheric vector magnetic field that the maximum distribution of current is in the bright kernel of a flare, the increase of magnetic shear after the solar flares in super active regions, the fact that generally solar flares alway appear in the redshift areas in the active regions, the opposite sign distribution of current helicity between solar northern and southern hemispheres and their evolution with the solar circle, the intranetwork magnetic field in solar quiet region, the evolution character of magnetic field in coronal hole regions, to infer the spatial configuration of magnetic elements in quiet region through the two layers of magnetic field in photosphere and chromosphere, in addition to other findings.

1.2.2 The 2.16-meter Optical Telescope.

The 2.16-meter Optical Telescope was installed at Xinglong Station, Hebei Province in 1989. In 1993, the Coude Echelle Spectrometer was added to the telescope. The 2.16-meter Optical Telescope is the largest astronomical telescope in Asia,which is opened to the world astronomers. Several important results have been achieved by this telescope, for example, from the spectrum of supernova 1993J an unsymmetrical structure of the explosion was observed and consequently "the finger model" was suggested to explain the blue shift of the spectrum of the supernova, the result was published in "Nature."

1.2.3 The Very Long Baseline Radio Interferometer (VLBI) system.

Two 25-meter diameter radio telescopes, installed in Sheshan, Shanghai Astronomical Observatory, and Nanshan, Urumqi Astronomical Station, form the VLBI system joined with the European Very Large Baseline Interferometer Net (EVN), and other eight international projects. In 1996, a cooperative VLBI observation by

Shanghai, Urumqi, Italy and South Africa, found five sources with apparent superluminal velocity. The two radio telescopes can be used as a single antenna, for example, to observe pulsars or interstellar molecular spectra.

1.2.4 A millimeter wave telescope.

This telescope with the diameter of 13.7 meter and working frequency of 110 GHZ – 115 GHz, is operated at the Delingha Station, Purple Mountain Astronomical Observatory. The station is located in the Wild Horse Beach, 35 kilometers east of Delingha City, Qinghai Province. Until now, more than 100 new sources of water maser, about one-seventh of which were found by this telescope. The telescope can study molecular clouds and stellar formation regions in the Galaxy, planetary nebulae, circumstellar envelopes, interstellar media, sky survey of molecular spectra, solar continuum and extra-galaxies, etc.

1.3 Astronomical Study Groups in China

The Chinese Academy of Sciences has actively attracted many outstanding Chinese astronomers. NAOC attracts and recruits first-rate persons for astronomical observation and theory research, according to the principles of "opening positions according to necessity, invting applicators with making it known to the public, going to the post by competitions, giving positions through contracts, and managing personal matters actively." As a result of these principles 49 persons have been openly, fairly and justly selected as chief research fellows of study groups and chief scientists of observational bases or laboratories. About 480 astronomers or managers entered the system of NAOC, their average age is about 45 years. Half of chief research fellows or scientists are younger than 45 years, 60% of them are of doctorate level, most of them have worked in foreign countries more than one year. There are 29 study groups, seven observational bases and seven laboratories have been created. Study work with different characteristics have been carried out in NAOC, Academic groups with international influences are purposely bred here. A loose and comfortable environment for scientific study is being purposely created ,which includes puting into practice the Chief Scientist Responsibility System and strengthening a relatively large scale funding support for scientific studies.

Many Chinese universities have also concentrated quite a few outstanding astronomers. One of important acts of NAOC has been to expand astronomical studies in cooperation with universities. Besides continuing to support the Joint Astrophysics Center of Peking University and Chinese Academy of Sciences, NAOC will cooperate with Nanjing University and other universities to arrange for the East China Astrophysics Center, in order to attract outstanding youth astronomers and to strengthen the competitive ability of China for astronomy studies.

2. THE ACHIEVEMENTS OF CHINESE ASTRONOMY STUDIES IN LAST 20 YEARS

In the last 20 years Chinese astronomy studies have obtained important achievements. Here we give two representative examples.

2.1 Studying the Distributions of Dark Matter in the Universe from Gravitational Lensing Effect

One of the fundamental questions in modern physics is the nature of dark matter in the universe. Astrophysical observations have provided the most competing evidence for the existence of dark matter. Namely, more than 90% of the matter in the universe is undetectable, while the evolution of our universe is governed by this unseen material. One of the most effective ways to detect dark matter in astronomy is the so-called gravitational lensing effect. The light-rays emitting from background sources will be deflected by foreground objects regardless of matter content. This allows one to explore not only the distribution, composition and evolution of individual dark halos but also to measure the cosmic density, the cosmological constant, the Hubble constant, and the overall matter distribution of the universe.

(1) Study of statistical properties of gravitational lensing and matter distributions of galaxy clusters: Using arc statistics in combination with observations, Chinese scientists showed that dark matter should be more peaked in the centers of clusters than baryonic matter. Such a claim was later confirmed by numerical simulations and X-ray observations.

(2) An alternative model for the explanation of the microlensing experiment results towards LMC: In 1993, two groups, EROS and MACHO, reported the exciting discovery of microlensing events toward LMC, which was attributed to the MACHOs in the halo of Milky Way, indicating that the unseen matter in the Galactic halo has been discovered. NOAC realized that part of the events could simply be the

result of self-lensing, i.e., the lenses are actually massive stars in LMC itself rather than inside the Milky Way. This has important impacts on the understanding of dark matter, lending strong support to the non-baryonic particles as the promising candidate of dark matter.

The traditional method of mass estimate in galaxy clusters relies strongly on the assumption of hydrostatic equilibrium, while many of clusters are still in the process of formation and thus deviate from the dynamic equilibrium. Gravitational lensing provides an independent way of evaluating cluster masses regardless of matter content and dynamic state. NOAC has developed a joint analysis of cluster masses from gravitational lensing, optical observations of galaxy distribution and motion, and X-ray measurements of intracluster gas. It is shown that there exists a significant mass discrepancy in the central regions of clusters between strong lensing and conventional methods. This finding is crucial for the study of matter distribution and cluster dynamical evolution. Using the largest lensing sample of massive clusters, NOAC demonstrated independently that the total mass density of the present universe is Ω=0.3.

2.2 Studies on Solar Magnetic Field and Activity

In the last 10 years, there appear two important directions in solar physics research: to understand the solar magnetism on the smallest scale and weakest limit, and to learn in detail the vector magnetic field evolution in solar active regions. By these studies we may understand the physical process in which solar magnetic energy is accumulated and released in an eruptive way in solar activity, and establish a foundation on which we may be able to predict the solar activity and its effects on human living environment. As the Solar Magnetic Field Telescope has provided us the state-of-art observations of solar magnetic and velocity fields, Chinese astronomers have made remarkable progress in the studies of solar magnetism and activity.

For the first time, international colleagues in collaboration with Chinese astronomers identified more than 2500 solar intranetwork magnetic elements, and derived their flux distribution functions. It has been demonstrated that the intranetwork magnetic field appeared to be a kind of intrinsically weak magnetic field and contributed more than 20% of total solar magnetic flux at any given time in the solar surface. The lifetime, evolution character, moving pattern and polarity law of this was field component are derived, and that the interaction between this weak field and network elements could provide enough energy needed for coronal heating was further revealed. These systematic efforts seem to have closed the long-lasting debates on whether the small-scale solar magnetic fields are intrinsically strong or weak.

Vector magnetic fields of many thousands of active regions were mapped in the last two solar cycles. A systematic method of vector magnetogram analysis was developed, including a new method to calculate the vertical current and a new concept of vector shear angle, which were widely followed by international colleagues. Observational configurations of magnetic shear generation (magnetic energy accumulations) were summarized, and theoretical equations of shear development were derived under the approximation of force-free magnetic field, and physical modes of shear development were revealed. Moreover, we clarified the physical processes that controll the magnetic helicity evolution in solar active regions, and made the first quantitative estimation of helicity changes in an observed active region. By the observations of vector magnetic fields in hundreds of solar active regions, the solar cyclic variations of solar electric current helicity have been systematically investigated. This is of essential significance for understanding solar dynamo and the global evolution of solar magnetic field.

Based on the observed structure and evolution of vector magnetic fields, it is clearly demonstrated that the well-known magnetic flux cancellation observed at the photosphere fit the physical scenario of magnetic reconnection in the lower atmosphere. A new idea in the flare energy process, a two-step magnetic reconnection, was proposed to account for the observed flare-associated magnetic changes, and this has been confirmed by independent space observations of ultraviolet jets. More recently, the magnetic reconnection in the lower atmosphere was found to be closely associated with the initiation of coronal mass ejections.

3. CHALLENGES, OPPORTUNITIES, AND GOALS ASTRONOMY FACES TODAY

Today astronomy faces both new challenges and opportunities. Elements of this situation can be shown as follows:

3.1 Space has become the key control point of strategy of a state. In the aspects of navigation and positioning, communication, remote sensing, scouting, tracing, and space antagonism, etc., astronomy possesses important applications. The development of space-travel has also provided a new historical opportunity for astronomy application. Astronomy has become an important component of national science and technology.

3.2 In the present time, one of the focal points of mankind's exploration of the natural world has shifted to understanding the rules of extreme physical conditions on the scale of time and space of cosmos. Astronomy has shown once more its strategic and pioneering characteristics.

3.3 The economy development of the nation and the increase in the national living standards promote strong demands by the public to raise themselves in social civilization and to embrace innovations. The incomparable properties of tangible and intrinsic mystery have made astronomy become the most important basic science to arouse teenagers' creative thought. To popularize the significance of astronomy will also improve the nation's innovation ability.

Because of its strategic position as a discipline, a fountainhead of high technologies and an essential aspect of the national character of education, developed countries have attached more and more importance to astronomy, and the financial support for astronomy has also achieved an unprecedented level. In China, modern astronomy is also treated as an important part of science and high technology of the state.

Emphasized items of China's astronomical studies in the next 10 years will focus on efforts to cultivate the two study groups with international influences: the group of dark matter, dark energy and large-scale structure of galaxy, and the group of solar magnetic field. Astronomers will also focus their efforts on aspects of the Galaxy magnetic field and polarization, γ-ray eruptions, large sample statistics on binary stars and the element abundance and evolution of celestial bodies. In addition to construction of the Observation Base of South China in Yunnan Province, NOAC will actively push to start the survey project for an excellent observing site in Tibet. Under the present operation conditions, NOAC will try to achieve double results: acquiring useful observational data, promoting astronomical study, and breeding astronomical high technology.

In the construction of big astronomical facilities, NOAC will accomplish LAMOST, SST, FAST and Moon Exploration Satellite, the four state level scientific projects that will promote the astronomical high-technique field of China.

In astronomy applications, China' s contributions will satisfy the requirement of the national strategy, with respect to the supervision and space exploration objectives, new systems of passive navigation and positioning, second generation of navigation system, large aperture and high resolution remote sensing system, new system for satellite supervising and control with accuracy as high as 10^{-6} degree, new standards and devices of time and frequency used in space environment (i.e. Hydrogen Clock), new system for the forecast of space environment, etc.

4. IMPORTANT ASTRONOMICAL DEVICES AND PROJECTS IN CONSTRUCTION AND PLANNING

NAOC is pushing forcefully forward on several large-scale astronomy equipments.

4.1 LAMOST (Large Sky Area Multi-Object Fiber Spectroscopy Telescope)

As a grand-science project in the Ninth Five-Year Plan period (1996-2000) of China, LAMOST is now in construction. The meridian reflecting Schmidt telescope will be laid down on the ground with its optical axis fixed on the meridian plane. Taking the active optics technique to control the reflecting Schmidt corrector, this gives the telescope both the large aperture (4 meter) and large field of view (5 degree). Its 4-meter aperture enables it to obtain the spectra of faint celestial objects down to 20 m. 5 with 1 nm spectral resolution in 1.5 hours exposure, and its 1.75 m focal plane corresponding to the 5 degree field of view can accommodate several thousands of optical fibers. This telescope can obtain 4,000 spectra of celestial objects simultaneously, becoming the telescope of the highest spectrum acquiring rate in the world. LAMOST can be used for the survey of thousands faint extra-galaxy celestial bodies and will bring significant developments on the study of galaxies, quasars and the large-scale structure of the universe. Its spectroscopic survey combined with the observations in other wavebands, such as radio, infrared, X-ray and γ-ray will make substantial contributions to the cross-identification of multi-waveband celestial objects. The telescope will be located at the Xinglong Station of NAOC, CAS. As a national facility, LAMOST will be opened to the whole Chinese astronomical community.

4.2 SST (Space Solar Telescope)– the First Chinese Built Astronomy Satellite

It will be equipped with a 1-meter aperture primary mirror with 0.1 arcsec space resolution and a Magneto-Analyzer with an accuracy of 10^{-4} in polarization measurement. The telescope can carry out a full waveband and continuous observation, operating jointly with the on board Ultra-violet Telescope with 0.5 arcsec space resolution, White Light Telescope, Wide Band Light Spectrometer and Radio Spectrometer, for obserring solar coronal active regions, magnetic field and speed field on the solar surface, coronal and solar-terrestrial, interplanetary space, etc. This system will be significant for the studies of fine structures and physical evolution of magnetic filed and speed field of solar active regions, the processes of energy accumulation and eruptive release, coronal matter throwing, the formation mechanism of solar wind, and other kinds of transient physical phenomena in the solar-terrestrial space, and will provide important observational data for the development of solar physics and the forecast of heliogeospace. SST represents the first space astronomy project in China. From the construction of SST, the techniques of large aperture space optics, 2-dimentional real time solar spectrum measurement, relevant tracing, object locking and high speed processing for great data flow, will be significantly developed, and will contribute more experience and lay a solid foundation for the space removing measuring, space remote sensing in China.

4.3 FAST (Five-hundred-meter Aperture Spherical Telescope)

In the new century, the development of radio astronomy requires much larger apertures of radiotelescope. The international radio astronomy community has proposed a very large radiotelescope, with a receiving area as big as 1 kilometer (Square Kilometer Area, SKA). For realizing the SKA project, Chinese astronomers suggested that use of the karst depressions in Guizhou Province, China would achieve a 1 kilometer receiving area for a radiotelescope. The proposed Five-hundred-meter Aperture Spherical Telescope (FAST) is the engineering forerunner of SKA. FAST will offer great contributions to the studies of cosmos origination, deep sky survey on pulsars, VLBI network and even SETI (Search for Extra-Terrestrial Intelligence). FAST plan has many essential technical innovations. Although it is designed along the model of a fixed radiotelescope, like the Arecibo radiotelescope in USA, it has taken use of active optics in the radio waveband, making its feed system much easer to be designed and cover much wider sky region. FAST has chosen the karst depression in southern Guizhou as its observing site.

4.4 The 2.4 -meter Telescope of the South China Astronomical Observation Base in Yunnan

Through many years of survey and observations, an excellent observing site, Gaomeigu, in the southwest province of Yunnan, with very good viewing has been found. With the elevation of 3200 meters above the sea level, and average viewing is 0.71 arcsec measured by DIMM method. In addition, it is located in low latitudes district, and can conveniently observe the center of Milky Way. A plan to build a 2 – 3 meter aperture optical telescope in Gaomeigu has been suggested with an estimated detecting limit of the telescope to be at 25 V magnitude,which will enhance the capability of astronomical observation in China.

4.5 The One-meter Infrared Solar Tower at the Bank of Fuxian Lake, Yunnan Province

The Fuxian Lake Observational Station is at the northeast bank of the Fuxian Lake, 60 kilometers from Kunming, the capital of Yunnan Province. The lake is the second deepest in China, and is one of the best solar physics observation site in the country. A one-meter Infrared Solar Tower will be builtthere, with working wavebands covering from 0.3 μm to 2.5 μm. The tower will provide important observational capacity for solar physics throughout the upper part of the 21st century.

4.6 The Lunar Exploration Project of China

Since the launch of the first satellite DFH-1 in 1970, China has greatly improved its space technology, and now has the ability to try to attemtp a space probe to the Moon. The lunar exploration will be the third milestone of the Chinese space exploration, after the launch of the Earth orbiter and manned spaceflight. China's first lunar exploration project will focus on the survey of potential and usable resources of the Moon. The main scientific objectives include delineating a 3-D atlas of the Moon's surface, abundances of usable elements, thickness of the lunar regolith, and space environment near the

Moon. The probe will be a three-axis stabilized lunar satellite, with a payload of seven scientific instruments, including an SMHSI spectral imager, a 3D CCD camera, a laser altimeter, a Gamma/X Ray Spectrometer, a microwave radiometer and two space environment detectors. The robotic spacecraft will fly a polar orbit at a working altitude of about 200 km above the Moon surface for at least one year. In the project, NAOC,together with other institutes subordinated to the Chinese Academy of Sciences, will be charge of the arrangement of the scientific objectives, design and manufacture of instruments, probing orbit measurement, and then processing and interpreting of down-loaded data. China's lunar exploration project is an open item, any suggestion of international cooperation will be warmly welcome and accounted in detail.

4.7 The project of surveying new site on the Qinghai-Tibet plateau

The average elevation of the Qinghai-Tibet plateau is 4000 meters or more, so the plateau is called " the World's Roof." It is generally believed that in relation to the standards for seeing night sky light and atmosphere transparency, an excellent, first-rank observing site for astronomy may be found on the plateau. In the favorable situation of China government starting the "Project of Great Exploration of Western Part" , the work in the initial stage includes organizing a special group and investigating the most proper study method and equipment for the site survey. In the summer of 2003, an international conference with the main topic of Site Survey in the Qinghai-Tibet Plateau was held. NAOC has begun the project of new site survey in the Qinghai-Tibet plateau in searching for a best observing site for the new generation ground optical, infrared and millimeter telescopes.

In the meantime, NAOC is now paying attention to arranging the construction of several middle size facilities. For example, the 50 meter aperture radiotelescope in construction will be installed in Miyun Station, Beijing, which will be the largest, entirety traceable radiotelescope in China. It can be used to study many advanced topics, such as interplanetary scintillations, supernova remnants, extra-galaxy radio sources and gravitational wave, and also can be used for the communication, measuring and controlling in deep space.

After the completions of the facilities and projec mentioned above, Chinese astronomy will be significantly pushed forward in the near future. Chinese astronomy is going by big strides to the stage of the world.

REFERENCES

1 Ai, G. and Hu, Y., 1986, Publ. Beijing Astron. Obs., 8, 1
2 Li W., Ai G. and Zhang H., 1994, Solar Physics, 151, 1
3 Wang, Jingxiu, Shi, Zhongxian; Wang Huaning; Lue Yanping, 1996, ApJ, 456, 861W
4 Wang, J ; Wang, H, ; Tang, F; Lee, J. W. ; Zirin, H.,1995, SoPh, 160, 277W
5 Wang, Jingxiu, Shi, Zhongxian, 1993, Solar Physics, 143, 119W
6 Wu, Xiang-Ping and Hammer, F, 1993, MNRAS, 262, 187W
7 Wu, Xiang-Ping,1994, ApJ,436L,115W
8 Wu, Xiang-Ping, 1994, ApJ, 435, 66W
9 Wu, Xiang-Ping,1994,A&A,286,748W
10 Wu, Xiang-Ping, Deng Zu-Gan, et al.,1995,ApJ,448L,65W
11 Wu, Xiang-Ping, and Hammer,F,1995 A&A,299,353W
12 Zhang, H., Ai, G., Sakurai, T. and Kurokawa, H., 1991, Solar Physics, 136, 269
13 Zhang, Jun ; Wang, Jingxiu; Deng, Yuanyong ; Wu, Dejin, 2001, ApJ, 548L, 99Z

The Progress and Prospect of Space Science of China

Liu Zhenxing

Center for Space Science and Applied Research, Chinese Academy of Sciences

Liu Zhenxing, *Male, Sept.14, 1929 born in Changle, Shandong Province. Academician of CAS, Space Physicist. Graduated from the Department of Meteorology, Nanjing University in 1955, Received vice-Ph.D. from Chinese Academy of Sciences in 1961. Research Professor of the Center for Space Science and Applied Research, CAS; Chief Scientist of Chinese Geospace Double Star Program; Director of Chinese Cluster Data and Research Center; Chairman of Chinese Committee of IAGA. Liu has long been engaged in the investigation of space physics, he has achieved many important results, and gained several prizes.*

Abstract: The research of Chinese space science started in the early 1960s. The exploration and research of Chinese space science has made great progress over the past 40 years. The purpose of this article is to summarize this progress. Its main contents include four parts: (1) Introduction; (2) Progress and main achievements of Chinese space scientific research; (3) Chinese manned spaceship space scientific experiment programs and main achievements; (4) Prospects of Chinese space science.

1. INTRODUCTION

The new era of exploring the universe by mankind started when the first man-made satellite was successfully launched in Oct 1957. Using varied spacecrafts to explore and conduct research on space has greatly broadened human's view on the universe and promoted the development of space science, space technologies and space applications.

Space science is a new, comprehensive and overlapping subject whose development is accompanied by the development of space technologies, and mainly includes space astronomy, space physics, space astronomy, and space life sciences.

The exploration and research of Chinese space science began to develop from the early 1960s. The research has made much progress during the past 40 years. This article will summarize the progress and prospect of Chinese space science.

2. Progress and main achievements of Chinese space scientific research

2.1 Progress of Observation and Research of Space Physics of China

China's observation and research of space physics developed under the initiation and organization of the famous scientist and academician, Prof. Zhao Jiuzhang after the International Geophysical Year (IGY) in 1957. In the past 40 years, the observation of Chinese space science has been developed to a system including ground, balloons, rockets and satellites and laid a solid foundation for the research of space physics.

2.1.1 Development of China's observation of space physics

2.1.1.1 Ground based observation

Networks of observation stations for geomagnetism and the ionosphere have been founded throughout the country, including Zhongshan Station and Great Wall Station in the South Pole. And an observation chain for geomagnetism and the ionosphere has also been built along east latitude 120° , north to Mohe and south to Sanya. In the aspect of ground-based observation of middle and high atmosphere, observation equipment for ozone, gas sunlight, airgrow VFH radar, laser radar and inclined filter photometer, etc. has been set up successively. Two cosmic ray observatories have been built in Beijing and Guangzhou. In the field of sun observation, our country has built a multi wave band sun observation system. During the period of the 20th solar activities maximum, "Research on global behaviour of Solar-Terrestrial System" , the grand program organized by the

Chinese Academy of Sciences, performed a five-year joint observation of Solar-Terrestrial space physics and acquired comprehensive and systematic observation data.

2.1.1.2 Balloon observation

The Chinese Academy of Sciences has set up balloon space exploration system, which has the observation ability of large payload, high altitude and long distance and can carry out observation of astronomy and space physics.

2.1.1.3 Development of rocket exploration

Rocket exploration of China developed under the initiation and organization of Prof. Zhao Jiuzhang in 1958. Much progress has been made with respect to manufacturing and exploring of meteorologic rockets and sounding rockets in space over the past 40 years. In the field of meteorologic rockets, China has manufactured four types of rockets: T7, “Peace II ” , “ Peace VI” and “ Vega I”. Numerous middle layer atmosphere observations have been carried through and a large amount of exploration data of middle layer atmosphere wind, temperature, pressure and density have been received. Valuable scientific data have been supplied to the research of middle layer atmosphere. In the field of sounding rockets for space, experiments of space physics (ionosphere, magnetic field of space and cosmic rays) and space biology have been done by using T7A rockets since the middle 1960s. China has been manufacturing “Vega” sounding rockets for space since middle 1980s. Two sounding rockets with the name of “Vega III” were launched at the Hainan Rocket Launching Site in 1991. The payload cones weighed 42kgs and the rockets rose to a height of 147kms and laid a solid foundation for the scientific and technological experiments by using sounding rockets.

2.1.1.4 Satellite exploration

Since the first man-made satellite was launched successfully in 1971, our country has launched “Shijian I (Practice I)”(March 1971), “Shijian II” (Sep. 1981), two balloon satellites (1990) and “Shijian IV” scientific satellite (1994) successively. Many instruments for the exploration of space particle radiation and sunlight radiation (including sun X-ray, ultraviolet ray and infrared ray) were loaded on the “Shijian I” and “Shijian II” and valuable exploration data were obtained. Two balloon satellites measured the density of the upper atmosphere through the changes of orbit.“Shijian IV” mainly explored high-energy particles and single particles reversal events caused by the former. The results of the measurement showed that reversal of single particle happened 3-4 times everyday in the districts of L<2 and L>5, in most cases.

"Shijian V" scientific satellite was launched in May 1999. Its orbit was a sun synchronous orbit and 11 exploration instruments were loaded on it. The main purpose was to detect the dosage of particle radiation in the abnormal magnetism region of South Atlantic Ocean and effects of single particle event caused by the former, and to do some relevant scientific experiments on space fluid in the micro-gravity environment. Some results of scientific significance and valuable application were acquired (see section on micro-gravity).

Furthermore, sun X-ray detectors, high-energy particle detectors and static electricity potential difference meters have been loaded on communication satellites and meteorologic satellites. Some important observation results were obtained:

The data of sun X-ray breakout and high-energy electrons in the synchronous orbit region event was gained for the first time; it was also the first time for China to measure sun proton events, Heavy Ion events and abnormal components of the Milky Way cosmic rays.

2.1.2 Development of research of space physics

Theoretical research of space physics has made much progress since the 1980s More than ten scientific units have been cultivated in China. And some important research achievements at the forefront of each branch subject have been made, some of which are in a leading world science position and have definite influence in the world.

(1) Main achievements of sun upper layer atmosphere and interplanetary physics

Theory of solar wind Alvin energy series, "without field of force-electric current flakes" model of solar flare and trigger style model of releasing energy, unsymmetrical characteristics of spreading of solar flare stirred waves, stable-like structure of disturbing of Solar-Terrestrial revolution and sun atmosphere and interplanetary numerical value simulation methods, etc. put forward by Chinese scholars take the leading position and have definite influence in the world.

(2) Main achievements of research of magnetosphere physics

Chinese scholars have taken the leading world position with regard to the theory of multi X-ray magnetic field reconnection, theory of vortex induced magnetic field reconnection, theory of K-H instability at the magnetopause, mode of Jupiter magnetosphere magnetic disk, magnetospheric sub storms equivalent electrocircuit method , the instability theory of near earth magnetotail and neutral sheet models.

(3) Main achievements in research of ionosphere physics

As early as 1947, Chinese scholars found the abnormal phenomenon of the East Asia ionosphere equator, observed low latitude whistle in Hainan and set up the spreading theory of low latitude whistle. They also discovered that inner gravity waves whose period is several hours had up phase speed components. They brought forward and validated the influence to ionosphere structure by atmosphere planet waves, discovered the district characteristic of the ionosphere structure and characteristics of the E-F concave, etc. These research programs received much attention in the world's physics community.

(4) Main achievements of research of middle and high layer atmosphere physics

Reversion method of middle layer atmosphere was put forward by Chinese scholars. According to the radar observation data, they found that the gravity wave critical layer existed in the top region of the middle layer of polar region. Discovery of the mechanism for the middle layer atmosphere planet wave that spans the equator and spread, and middle layer atmosphere turbulent flow, and the coupling mode of thermosphere and ionosphere was regarded highly internationally.

(5) Space environmental mode and forecast research

Chinese space physics investigators supplied space environmental parameters and cooperated in the experiments of ground environmental mode since the first man-made satellite was designed and manufactured. On this basis they edited a book titled "Man-made Satellite Environmental Manual." This manual played a very important role in the design of Chinese aircrafts. After the 1980s, in order to meet the needs of space undertakings, the research of space environmental mode integration and the effects on aircrafts by space environment were developed. On the basis of the research, "Low Orbit Aircrafts Space Environment Manual" was written. In the aspect of forecast research of space environment, the space environment forecast of solar activities has been developed since middle 1960s, including alarm of solar activities, short-term, medium-term and long-term forecast, geomagnetism activities forecast, sun proton event forecast and high layer atmosphere parameter forecast. These forecasts have played an active role in our country's space undertakings.

Regarding the progress on space chronometers, see the part on Chinese astronomy progress.

2.2 Progress on Research of Chinese Space Chemistry

The birth of the research field on Chinese space chemistry- aerolites in spite of China maintaining one of the most detailed records of aerolites , our country's space chemistry research was still underdeveloped before the 1960s. Beginning with the March 8, 1976, meteorite shower, which appeared in Jilin Province, Chinese space chemistry community developed a comprehensive research program. Successive discoveries of rare gases in aerolites and of abnormal composition of oxygen isotopes were a great breakthrough in the research of composition of solar system material isotopes in the 20th century. Furthermore, these discoveries promoted and accelerated the course of another research field, solar system material like silicon carbide, graphite, TiC and Si_3N_4, etc. During the course of the research of aerolites (EH3) from Qingzhen, Chinese scholars discovered and confirmed the existence of pre-solar system material, X-SiC and Si_3N_4. They also made systematic analysis and research of the mechanism on the source of the material and put forward the possible super new star source area theory, supplementing and perfecting the theoretic system. Recently the Chinese Science and Observation Team in South Pole retrieved 32 aerolites from Glover Mountain, and Chinese scholars did the research on geognosy, mineralogy and their structures. On the basis of elementary appraisal of geognosy type and chemistry type of the 26 aerolites, they discovered and appraised that GRV99027 was the rock from Mars. This new discovery is not only of academic research value in terms of aerolites, but also supplies important prospects for the research of Mars' aerolites and even Mars' chemistry.

2.2.1 Theory of striking out of earth

Along with the constant discovery of evidence of material striking out of earth and thorough research of the striking effects, a new research field theory of striking out of earth has formed since the 1980s. The new thesis proposed by Chinese scholars, that small celestial bodies struck the earth at least six times and led to the global climate and environment disasters and the extinction of most living things on different extent since Cenozoic, has greatly enriched the theoretic system Chinese research has been based on, especially the chemistry record of striking the earth of 65Ma and 2.4Ma and the research results from middle-sized and little glass aerolites in the loess of 0.73Ma.

2.2.2 The theoretic frame of the uneven characteristic of primitive chemistry of the earth and planet is now beginning to take shape.

On the basis of synthesizing the latest research data of earth chemistry, physical geography and cosmochemistry, Chinese scholars have derived the theory of the uneven cause of formation of the earth. They discussed the planetesimal theory, the model of multi stage piling up of the earth, and the uneven characteristic of chemistry of the earth and their relevant relationships, expounding from the angle of planet evolvement and forming the theoretic frame of the uneven cause of formation of the earth.

2.2.3 Samples of the Moon and the research of comparative study of planets

The new era for the development and refulgence of the modern space science and space technologies begun with the late 1950s. People have begun to explore and research the solar system celestial bodies like the Moon, Mars, and other planets in deep space. Samples of the Moon weighing 382kgs have been acquired, so people's knowledge of solar system formation and evolution has been greatly increased. Chinese scholars not only took part in the research work of the samples from the Moon, but also they made the research of the solar system and comparative study of planets systematic. On the basis of the long-term tracking and research of the international Moon, Mars exploration and research trends, Chinese scholars have conducted the research on exploration objects of science of the Moon. This not only supplies an important scientific foundation for the exploration of deep space, but also provides direction for the future development of China's space chemistry.

2.2.4 Research of cosmic dust and space fragments

Outstanding achievements have been made by Chinese scholars on chemical classification and distinguishing signs of cosmic dust. They documented the comprehensive distinguishing characteristics of the chemical ingredients and structure. Space fragments are very dangerous to the safety of aircrafts, which has become a conspicuous problem of the space environment. In recent years, Chinese scholars have made systematic research on the chemical classification, characteristics of distribution, source of material and the risk to the space environment of space fragments and they have acquired elementary results.

2.3 Development of Scientific Research of Chinese Micro-Gravity

Scientific research of Chinese micro-gravity started from the National High

Technological Development Program during the late 1980s. A very good beginning after many years' efforts resulted in research groups of micro-gravity science in different subject fields, given priority by the Chinese Academy of Sciences (CAS) with more than 30 labs set up to deal with the main fields of micro- gravity. Elementary research and facilities include the science of space material of physical research institute, experiments of growth of monocrystal of space GaAs of semiconductor research institute, research of micro-gravity fluid physics of mechanics research institute, combustion research of micro-gravity of engineering heat physical research institute, research of space growth of protein crystals of biological physics, experiments of the cultivation of cells in space of animals research institute, gravity biology research of Shanghai foliage physiology research institute, among others. These research groups switched to the field of micro- gravity on the basis of accumulation of many years' ground research. In a sense, development of Chinese micro-gravity is promoted by the development of Chinese space technologies. Chinese returned satellites have become the experimental platforms and supplied technical guarantee for the space micro-gravity experiments. Lanzhou Physics Research Institute under the Space Industry Head Office cooperated with the Institute of Semiconductors, CAS and made great contribution to the promotion of experiments of micro-gravity by using returned satellites.

On the scientific satellite "Shijian V" launched in 1999, space fluid experiments were under the condition of micro-gravity and acquired important results. These experiments validated the characteristics of multi laminar flow dynamics and the influence of micro-gravity environment. In the experiments scientists found the coupling effects of fluid-interface capillarity driving effects and gravity, buoyancy convection. They also found the new phenomenon of Marangoni driven liquid-gas and liquid-liquid multi interface in micro-gravity environment and heat capillary coupling convection. At the same time, they put forward a new model of fluid mechanics. See part 3 of this article for the manned spaceship micro- gravity experiments.

2.4 Research Development of Chinese Space Life Sciences

More attention to the effects of life systems under the space condition of micro-gravity, strong radiation and high vacuum has been associated with the development of space technologies and manned space ships undertakings. China's research of space biology started from the launch of biological rockets T7A-S1 and T7A-S2, in cooperation with the Institute of Biophysics Academia Sinica and Institute of Mechanical and Electrical Engineering. The altitude reached by the flight was 70-80 km. The Biology chamber, which returned successfully, of the first launched T7A-S1 rocket was loaded with experimental animals including small and big white mice. In 1965 and 1966, two T7A-S2 rockets were successfully launched and returned for the third time with the experimental animals including small dogs and small and big white mice. All of these activities created a precedent and laid a solid foundation for the research of space life sciences. In 1987, the Chinese Academy of Science and other relevant units launched and utilized the world's first recovery style scientific satellites equipped with 68 pieces of experimental instruments for study of the biology effects on 32 pairs of plant seeds, 6 kinds of plant cured tissue, two kinds of insects, one virus type, and 7 kinds of micro-organism. During the following 20 years, Chinese scholars have made extensive research on life sciences including space medicine, cell biology, botany, zoology, micro-biology, aquatic biology and radiation biology. Significant progress has been achieved.

Scientists of the Chinese Institute of Space Medicine and Medicine Engineering and of relevant units have conducted some research on biology effects of bones and muscle metabolism through experiments simulating space weightlessness. They found that the ectoblast formation rates of shoulder bones and thighbones of mice which were suspended for 21 days was lowered by 37% and 59%. The shankbone formation stopped and the teeth formation rate was lowered by 6%. The content of calcium, phosphorus, and magnesium of the suspended mice was obviously lowered. Enzyme relevant metabolism of calcium in bones and the formation and metabolism of hormones changed and alkalescence phosphoric enzyme activity of the thighbone was also lowered. After the white mice were suspended for 13 days, citric acid compound enzyme activities of soleus, gastrocnemius, bicipital muscle of arm, and the content of non-collagenous protein NCPs were lowered by 19%and 24% respectively. Long-term simulation of weightlessness can cause the contractibility of cardiac muscle to be lower and the changes of mechanism of inner membrane of heart adjustment. After the monkeys were suspended for 7-12 days, scientists found that reductions in the amount of functional capillary vessel on the section of soleus, gastrocnemius, and bicipital muscle of arm. Inside the muscle fibre there is scattered pellet and blocks of muscle cell quality that may cause barriers of blood dynamics and dropsy. Scientists also discovered the changes of viscera quality, brain quality, immunity function and, micro-circulation system in the experiments with animals in simulated weightlessness. These results of this research supply scientific evidence for people to understand and solve the medical problems in manned space flight.

Since 1987, 65 kinds of agricultural seeds and more than 370 varieties have been loaded by using backtrack satellites and high space balloons. Scientists conducted research on mutagenesis effects of crops in space environment to choose and cultivate many new varieties. For example, Gardening Office under Heilongjiang Agricultural Science Institute cultivated a new strain "satellite 87-2," one of the fruit weighed 175.45g, from the fourth generation of pimiento seeds loaded in space. The Institute of Plant and Ecology, CAS, cultivated a new strain "Plant Shanghai No.1" which has differential genetic signs of brown lists on the shell of spike, short plant, strong ability of tiller, and resistance ability of gibberellin disease. The Heredity Institute of Chinese Academy of Science and Agricultural Institute of Jiangxi province cooperated and cultivated big grains mutation stems which have stable genetic characteristics with a 93% increase in weight of a thousand grains from the generation japonica rice seeds of "Land reclamation and cultivation No.58" loaded on the satellite. The Institute of Botany, CAS did some research on the mechanism of plant incline to gravity through the root needle tissues and cells of seedling cypress loaded on the satellites. The results show that Ca^{2+} transferred from calcium storeroom and centralized to the cell membrane, cell organ and cell quality and calcium was regarded as second messenger system. It is also found in the experiment that space environments have much influence on root needle cells structure of cypress. The main representations are the biology effects of all the cells' separation of quality wall, increase of fat drops, accretion of vacuole, and thickening of the cell wall. Scientists found that there is a special cell organ having the same size as the core in the cell and its shape and properties remain to be studied. These research results are important in the theory of plant physiology and application of space technologies used in the field of crop breeding.

Pioneer species of aquatic living things formed tests for the manned space ship ecosystem. The Institute of Hydrobiology, CAS loaded a duality close controlled system with rotifer and micro-algae on a scientific recovering satellite. After the experiment and flight of 14 days, micro- algae made the light energy transform into chemical energy to supply the growth of the rotifer. CO_2 and waste excreted by rotifers are used by micro-algae and a closed circulation system was formed. Scientists also conducted research on space biology effects of chlorella and anabaena and found that the amount of starch grain reduced after the flight of the protein core. Activity of solid nitrogen of anabaena was enhanced and liposome reduced. Research included the genetic and chemical characteristics of space variation on algae trunk.

In 1987, The Micro-organism Institute of Chinese Academy of Science and relevant units have conducted some research on space biology effects through the micro-organisms like epiphyte, bacteria, and actinomyces aboard the scientific satellites. Observation results show that the speed of growth of some bacteria and actinomyces is quicken and the distilling of gemma bacillus produce gemma, companion cell crystal distill and release. Research on Micro-organism space biology effects was performed on the temperature-controlled micro-organism cultivation instrument in 1994. Among the loaded bifid bacteria scientists chose an excellent production plant whose ability of bearing perhydrol was improved in space, with fast growth and ability to resist radiation. The Institute of Plant Physiology and Ecology, CAS acquired a space plant variation from Cephalosporin and C amidase genic engineering, whose active of enzyme is increased by 4 .3 times. On the 17th recovery scientific satellite launched by our country in 1996, research on space biology effects on micro-organisms like antibiotics NIKKO production fungus plant streptomycin autogenetic solid azote alkaline bacteria, purple non-sulphur photosynthesis bacteria used to purify waste water and high grade epiphyte day lily and Pleurotu ostreatus has been made through biological all-purpose cultivation boxes. The results show the effects of space environment gene on some biological functions and characteristics of some plants, for example, the output of NIKKO mildew was increased by 13%~18%; livability of non-sulphur photosynthesis bacteria was higher than on Barth; the rate of reducing BOD was increased by 15.3%; the growth of mycelium of day lily was increased and the appearance of anlage was brought forward by 7 to 8 days with 16.2%-23.8% increase of input. In the space planting experiments of the second generation of Pleurotu ostreatus, the output was increased by 19.4%-27.3%. Transforming rate of edible fungus from straw was 40% and the efficiency of biology was as high as 129%. Scientists discussed the mechanism of variations in micro-organism space variation plants at same time of research as the Micro-organism space biology effects were made. In 1996 Jia Shifang and his collegue made analysis on plasmid DNA of bifid bacteria that has obvious biological characteristics of variation and they did not found the medical plasmid. In 1997 Mr. Luo Aiqun and associates analyzed a variation plant of NIKKO mildew production fungus and discovered that the variation had something to do with the conformation polarization gene. In 1998 Mr. Weng Manli,et al. discussed the mutation frequency of colon bacillus Lac- and Arg+ by the space environment gene and found that the mutation frequency of Lac- as high as 68 times that of the ground and the mutation frequency of Arg+ was increased by 10 times. The mutation frequency of uncontrolled genes in reversion body was increased.

Since 1988 Chinese scientists did protein crystallization experiments on the recovery satellites launched by our country by using COSIHA crystallization instruments developed by Germany for the first time. Two protein crystallization experiments were done successfully on the recovery satellites by using the instruments of pipe-shaped steam liquid pervasion TVD successively manufactured by China herself in 1992 and 1994. Through these growth experiments of space protein crystals, scientists not only observed the promotion effects on the forming of the core, the size of crystals, outside appearance and inner order by micro-gravity, but also acquired high grade protein crystals like bacteriolysin enzyme, acid phosphatide enzyme A2, and haemoglobin that grow in space and can be used for structure analysis. Excellent material has been developed for the research on biology macromolecules.

Over the past 40 years, many achievements have been made in the research of space life sciences, and these research results have demonstrated the huge potentiality of studying and developing biological resources by making use of the limitless space environment. Research and application of the effects of space biology, still awaits a theoretical breakthrough on theory that produces domino effect mechanism. We expect that there will be very fast speed in the development of theories and technologies of space life sciences accompanying activities of exploration the life outside the earth.

3. MAIN ACHIEVEMENTS OF SPACE SCIENTIFIC EXPERIMENTS ON MANNED SPACESHIP

China started the manned spaceship program in 1992. Up to now, four test spaceships have been launched successfully.

The main tasks of the science application system of Chinese manned spaceships are supplying payloads and relevant space-borne and ground-borne supporting system for space experiments, sounding of the Earth and space scientific technological tests, including micro-gravity fluid physics, space material, space life, space chronometer, and space physics environment exploration.

Manned spaceships consist of propulsive cabin, recovery cabin, orbit cabin and accessional sects in the front of the orbit cabin. The scientific application system payload is equipped in recovery cabin, orbit cabin and accessional sects. Systems of the spaceship are made up of subdivision systems of power, supply and distribution, system guidance (GNC) and controlling, propulsion, management of data, remote measurement and control, environment control and safeguard of lives and retrieval etc. Space ships are launched by CZ-2F carrier rockets.

Some results of payload and test items are summarized.

3.1 Shenzhou No.1 (launched on Nov. 20,1999 and returned on Nov .21)

3.1.1 Main payload of scientific application system of Shenzhou No.1

3.1.1.1 Detector of micro-gravity

3.1.1.2 Collector and editor of micro-gravity

3.1.1.3 Multi work position crystal growing stove

3.1.1.4 Crystal stove electronics controlled box

3.1.2 All the experiments the mission succeeded and reached the expected goals

3.1.2.1 System of supply and distribution of power of communal equipment recovery cabin worked regularly.

3.1.2.2 1553B terminal laid apogee worked regularly.

3.1.2.3 Whole course measurement of micro-gravity of Shenzhou No.1 and complete data was obtained.

3.1.2.4 Functions of electricity adding, changes of position and lifting and drawing were validated by Multi work position crystal growing stove.

3.2 Shenzhou No.2 (launched on Jan. 10,2001 and returned on Jan. 16)

The quantity and items of products of scientific application system loaded on Shenzhou No.2 were the same as these on Shenzhou No.1. But all the instruments were for the formal launching and underwent the formal orbit experiments.

3.2.1 Payload of scientific application system of space science amounts to 8 items:

3.2.1.1 Multi work position space crystal growing stove

3.2.1.2 Observation instrument equipment of space crystal growth

3.2.1.3 Crystallization instrument equipment of space protein

3.2.1.4 Equipment of space communal living things culture box

3.2.1.5 Space chronometer: High-energy radiation monitor of sun and universe celestial bodies

- Super-soft X-ray monitor and equipment
- γ-ray monitor and equipment
- X-ray monitor and equipment

3.2.1.6 Optics window module and equipment

3.2.1.7 Space environment monitor and equipment

- Detector of ingredients of atmosphere
- Detector of density of atmosphere
- Detector of track of solid

3.2.1.8 Surveying instrument and equipment of micro- gravity

3.2.2 Results of flight tests

Application systems on track worked efficiently during the independent flight of space ships and obtained complete remote measurement parameters and scientific data.

3.2.2.1 System of wide energy spectrum covering space chronometer detector composed of X-ray monitor, super-soft X-ray monitor and γ-ray monitor, processed the exploration of universe γ-rays and monitoring of high grade radiation of solar flare. After the contrast with data of global GRB associated observation network, scientists confirmed existence of more than 30 universal γ-ray and nearly 100 cases of sun hard X-ray solar flare.

3.2.2.2 Through the exploration of atmospheric ingredients, our country measured mass spectrum chart of space atmosphere within the scope of said orbit height for the first time; in the aspect of atmosphere density, global atmosphere density exploration preparation materials were acquired within the scope of spaceship moving orbit for the first time.

3.2.2.3 Supporting system on track of payload- communal equipment was completely examined and the work of whole course achieved complete success and laid a foundation for the design and production of products of follow-up space ship tasks.

3.2.2.4 Acceleration frequency spectrum of micro-gravity of all the spaceship incidents was measured during the independent flight of space ships through the measurement of micro-gravity.

3.3 Shenzhou No.3 (launched on Mar. 25,2002 and returned on Apr. 1)

Scientific application system of Shenzhou No.3 majored in scientific tests of optical remote sensing and micro-gravity, at the same time, monitoring and research of high layer atmosphere and the environment of earth was performed.

3.3.1 There were eight payload scientific application systems:

3.3.1.1 Equipment of spectrum instrument of formation of image with middle resolving power.

3.3.1.2 Cirrus sounder

3.3.1.3 Monitoring of the environment of the earth

- Receiving an giving off instruments of the radiation of the earth
- Spectrum monitor of solar ultraviolet radiation
- Monitor of solar constant

3.3.1.4 Survey of space environment and high layer atmosphere

- Detector of ingredients of atmosphere
- Detector of density of atmosphere
- Monitoring of particle radiation of recovery cabin

3.3.1.5 1 module of orbit cabin window of space ship

3.3.1.6 Multi work position crystal growing stove

3.3.1.7 Experiments of space life sciences

- Crystallization instrument of space protein
- Biological reactor equipment of cells

3.3.1.8 Micro-gravity detector

3.3.2 Results of flight tests

Application systems on track worked normally during the independent flight of space ships and finished complete remote measurements. Seven days later the orbit cabin separated from recovery cabin, and the latter was returned back successfully at the Amugulang pasture in the middle part of Inner Mongolia. The orbit cabin had been on track for application until work finished on August 5. The total flight experiment on track lasted for 127 days and all the equipment of the application system worked smoothly and achieved great success.

3.3.2.1 The whole course of work of the CMODIS worked normally on orbit and obtained 863 images, whose quality has reached the expected goal and trial products have been formed to supply to related users.

3.3.2.2 Complete probing materials have been acquired through the exploration of cirrus. Through comparing large areas of cirrus and GMS-5 material with synchronous observation materials of ground weather stations, scientists have found that main channels of data for upper air cirrus measured by detectors of cirrus is reasonable and the detector of cirrus can work efficiently. Expected test tasks have been fulfilled successfully.

3.3.2.3 Important observation results have been attained on monitoring of the environment of the Earth. Among them:

- Comparing the degree of solar ultraviolet radiation spectrum with the observation average value of International degree of solar ultraviolet radiation spectrum outside the Earth, the former and the latter are coherent within the scope of measurement errors. The brightness of Sun backward dispersion of ultraviolet radiation spectrum by the atmosphere is between the observations and minimum of the international atmosphere.
- Detector of Sun constant measured the data of Sun constant. It is coherent with the Sun constant data measured in the same period (March to September, 2002) in the world within 0.1%.
- Receiving and giving off instruments of the radiation of the Earth has worked for 2000 hours on orbit. The results from measured data show that whole wave bands radiation, reflected short wave sun radiation and the earth average long wave in radiation are identical with the foreign measurements.

3.3.2.4 Complete samples have been obtained for the scientific experiments of space material and results full of significance of scientific research have also been attained through the elementary analysis.

3.3.2.5 The growth rate of space protein crystal has reached 75%, 6 kinds of protein grew high quality crystals and the best capacity of diffraction of the crystals has reached the highest resolution power of the same kinds of crystals.

3.3.2.6 Cultivation results of space cells: four kinds of cells of mammal all survived and grew very well. Among them the big grain lymphocyte of humankind (NK92) grown in space was obviously better than that on ground and formed fairly big groups and blocks of growing. The rate of increase was about 12.5 times of that of the ground. The amount of secretion of albumen antibody of space anti root of Chinese trichosanthes was higher by (8%) than that in ground.

3.3.2.7 Large amounts of data have been acquired by upper atmosphere sound and some important results were attained after analysis and processing. For example, space atmosphere environment within the scope of the height of orbit periods was the atom oxygen layer. The lowest content of atomic oxygen was above 60% of total amount of each kind of ingredients of atmosphere and the highest could be about 90%. The mass spectrum chart of gas ingredients of the gas deflation layer near the surface of orbit cabin and elaborated exploration results of exceptional changes of upper air atmosphere ingredients caused by magnetic storm were both gained for the first time. And the phenomenon of latitudinal process movement of exceptional changes of upper air atmosphere main ingredients was also discovered.

3.3.2.8 Communal payload equipment was tested again and the advanced performances of the system, the stability and reliability of working were proved at the same time.

3.3.2.9 Other payload fulfilled tasks according to the expected goals.

3.4 Shenzhou No.4 (launched on Dec 30,2002 and returned on Jan 05,2003)

Scientific application system of Shenzhou No.4 majored in scientific tests of micro-wave remote sensing and micro-gravity, at the same time, comprehensive examination and research on space environment were conducted.

3.4.1 Payloads for scientific application system amounts to 6 items:

3.4.1.1 Multi-mode micro-wave remote sensor

3.4.1.2 Experiments of comprehensive precise orbit-tracking

- Equipment of GPS data receiving and processing
- Equipment of laser ranging: laser emitter

3.4.1.3 Equipment for tests of fluid physics of micro- gravity

3.4.1.4 Experiments of biological technologies

- Equipment of electric fusion of space cells
- Equipment of space electrophoresis meter

3.4.1.5 Examination equipments of comprehensive space environment

- Detector equipment of ingredients of atmosphere
- Detector equipment of density of atmosphere
- Detector equipment of high-energy proton and Heavy Ion
- Detector equipment of high-energy electron
- Detector equipment of low-energy particle
- Detector equipment of electric potential
- Detector equipment of single particle event
- Detector equipment of track of solid

3.4.2 Results of flight tests

Application systems on track worked efficiently during the independent flight of space ships and scientific experiments of micro-gravity were mainly done. After the 78th circle boot-strap examination tests of equipments left on track like micro-wave remote sensing were taken and complete remote measurement parameters and scientific data were obtained. Seven days later the orbit and recovery cabin separated and the latter was taken back successfully at the Amugulang pasture in the middle part of Inner Mongolia. The orbit cabin is left on track for application and now it is operating. Elementary analysis results after the callback of recovery cabin samples and experimental records are:

3.4.2.1 Experiments of electric fusion of space cells were successful and obtained experimental samples for the follow-up cultivation research and analysis.

- The rate of fusion of flora cells space experiment was 18.8% and that on the ground is only 1.65%. The rate of fusion was enhanced by over 10 times. After 162 hours of temperature-controlled preservation of the fusion of flora cells, including on track and callback and transport, altogether about 7 days, survival rate was 53.6% with that on ground was 38%. Obviously the rate of fusion in space is higher than on ground.
- The rate of fusion of animal cells in space experiment was about 11% and that on ground was 4%. The rate of fusion was enhanced by over 2 times. Survival rate of cells was lower than that on ground. Survival rate of

lymphocytes was about 43.48% of that on ground. Survival rate of marrow tumour was about 27.78% of that on ground. Owing to the long retention period on track after the fusion, the vitality of cells descended to some extent.

3.4.2.2 On track experiments of space separation pure chemistry, experimental research of biology macromolecule, and cells achieved success. Better light-checking curve was attained and the examination results of retrieved samples also proved the success of the experiments.

- The distance between the two peak values in the space experiment was increased by 38% and the distance in the bottom of ravines space experiment was reduced by 35%. It shows that the separation is better under the condition of micro-gravity than on ground and obvious advantages are possessed.
- Hb strips were narrower in space than on ground; the height of pinnacle was higher than that on ground. It shows the diffusion of samples and high resolving power.
- There were no empty tubes among the 30 collecting tubes. The average value of the collection amount of space was 37.3 and that on ground was 36.2Ml. It shows that the even characteristics of collection are almost the same in space and on ground.

3.4.2.3 Experiments of micro-gravity fluid were completely successful and got complete transfer of liquid drops and chart records of temperature field interference experiments. Designed programs of experiment instruments were reasonable with excellent functions and reached the expected demands. The data were of important value for research and experiments of micro - gravity fluid physics. Scientists will make analysis and research on relevant research subjects and put forward research report through the experimental material acquired.

3.4.2.4 All the monitoring work of comprehensive space environment was normal. Main exploration results initially obtained were the following:

- The distribution of high-energy charged particle on the orbit majored in electron and there were few proton and Heavy Ios. They mainly distributed over the South Atlantic Ocean. The effect of single particles that was caused by them was not serious and detector of single particles event still didn't monitor the reversal of single particles.
- Electric potential of static on the surface of space ship was not high. The lowest was minus 12 V and it was positive electric potential in the area of

illumination and was positive 20V in normal cases.

- During the tranquil period of the sun, low-energy particles were few in quantity. Its effect of charging to the surface of the space ship was weaker than the effect of photoelectricity of the sun.
- In the early morning of January 04, a small disturbance happened in space environment. The amount of charged particles didn't change obviously. But the disturbance aroused the increase of the density of orbit atmosphere by 1/3 than that of January 03.
- The data of ingredients of this orbit atmosphere was acquired through the monitor of atmosphere ingredients. The data acquired are being processed.

3.4.2.5 Experiments of multi-mode micro-wave remote sensing meter on track achieved an elementary success. Complete exploration dada was achieved successfully during the independent flight of spaceships. The system has been working smoothly up to now. Dispersion mode antenna operated for 51minutes and then stopped, but the other equipment of the system could work efficiently and get exploration data of wind field on the surface of ocean.

- Over one hundred thousand of packages of efficient data were obtained through many times of bootstrap of the five channels of the mode of radiation. The analysis of results of numerical value show that the radiation meter worked smoothly, bright temperature exported by five channels which were acquired through reversion became sensitive with the changes of aim of places and objects and there was obvious difference between the ocean with low bright temperature and the ocean with high bright temperature.
- High-grade mode worked for 13 times after bootstrap at the stage of independent flight of spaceship and of being left on orbit. The total time was four hours and three quarters and a large amount of data of echo and track was attained. The analysis of results of numerical value shows that the altimeter worked efficiently and the curve of echo was also normal. The height measured and uplifted Slope-Rising Edge and changes of shape of waves of FFT were reasonable. All these proved that the tracking methods of counting of the altimeter on the ocean were reliable and the altimeter had excellent performance.
- During the period of independent flight of space ships and dispersion mode the antenna was not opened. The state of working of launching and receiving

channels was examined by tests of addition of electricity. The working state of launching and receiving channels was normal, which was proved by the judgement of engineering remote measurement and levels which received noise. Antenna left on the orbit worked once successfully, total revolving time was 51 minutes, 47 seconds and 606 millisecond. The dispersion data acquired showed that the distribution of the surface of the ocean wind field was formed through reversion.

3.4.2.6 Working of experiments of comprehensive precision on fixed orbit was normal. Data of laser measuring distance fixed orbit initially acquired showed that the measurement precision of grade of centimeters could be achieved. It was also proved that it would be doable for the comprehensive precision track fixing by using UBS/GPS/SLR and it would be carried out in terms of engineering. The results of GPS difference and amendment of ionosphere have been gained and the data of comprehensive precision on fixed orbit is being processed.

3.4.2.7 Measurement of level of micro-gravity of the spaceship

Measurement of level of micro-gravity of the spaceship was normal during the independent flight of the space ship. During the whole course of work, varying data of statistical amount of micro- gravity, which was changing with the time, was attained. Working practice and speed increments of the engine of three times orbit maintenance course and of the course of changing of orbits were checked and data of 4ms sampling rate was supplied for the controlling of flight. Important events of the spaceship and payload were examined, such as the structure response caused by releasing of pressure of orbit cabin, various preparation actions before returning, the course of adjustment of posture, separation of orbit and recovery cabin, the course of applying the brake, the state of micro-gravity during the course of the electro-phoresis meter, fusion meter of cells, and experiments of fluid.

4. PROSPECTS AND DEVELOPMENT OF SPACE SCIENCE OF CHINA

In November 2000, the News Office of Chinese State Council issued the "Space White Book of China" for the first time. The three main tasks of Chinese space industry put forward in the white book are space technologies, space utilization and space sciences. This is the first time for our country to list the space science in the state documents as one of the main tasks of the national space program.

China National Space Administration places much emphasis on the development of space science of our country and is making the development programs and plans of exploration and research of space science during 2002 to 2020.

4.1 Prospects of Exploration and Research of Space Physics

In the first ten years of the 21st century, the strategic subject of the research of space physics of our country is the various chain course of the Sun-Earth space system and forecast methods of space weather numerical value. The main object of research is thorough understanding of the physical process of chain changes of sun-earth space, including the appearance of sun events and their course of spreading and evolvement in the interplanetary space ; the response course of disturbing of solar activities and disturbing of interplanetary by magnetospheric space storms, the response course of disturbing of solar activities, magnetospheric space storms and low layer atmosphere by ionosphere and middle and upper air, research on development and comparing of planet environment, establish dynamic mode of space environment, and develop forecast methods of numerical value of space weather.

From the years 2000 to 2020, the concept of the strategic idea for exploration plans is to cooperate on the research of space physics and to develop the exploration of sun-earth space and of the moon. Plans for exploration put forward initially are as the followings:

4.1.1 "Geospace Double Star Exploration Program" (aka "Double Star Program")

"Double Star Program"includes two small satellites: one is an equator satellite near the earth and the distance is 500 km when it is nearest to the earth and the distance is a 60,000 km when it is furthest to the earth with a dip angle of 28.5° . The other is a polar region satellite, the distance is 700 km when it is nearest to the earth and the distance is 40,000 km when it is furthest to the earth with a dip angle of 90° . Eight instruments of exploration will be placed on each satellite. These two satellites will cooperate each other and operate in the main activities area near the earth, which is still not covered by international exploration satellites of geospace and will form a unique system for innovative geospace plans of exploration. The main scientific and application object of the Double Stars is to explore and study the rules of spatiotemporal changes of midfield and particles and springing mechanism of magnetospheric space storms, to establish and develop the dynamic model and methods of physical forecast in space environment, and to enhance the space exploration technologies and scientific research through international cooperation. "Double Stars Program" cooperates with Cluster II of European Space Agency and forms constellation exploration plans with six-point exploration, which will be an important sun-earth space exploration program in the world in 21st century. The equator region satellite near the earth is scheduled to launch in December 2003. The polar region satellite near the earth is scheduled to launch in June 2004.

4.1.2 "Exploration plans of small satellites of Asia-Pacific cooperation"

This is a multi-task small satellite of multilateral cooperation of Asia-Pacific space technologies and application. And it is also a small satellite for the space scientific exploration and technological tests. Its orbit will be an approximate round orbit with a height of 796 km and a dip angle 98.6°. The main scientific object is to explore and study the influence on ionosphere and middle and upper air global structure by solar activities and magnetosphere disturbiance.

4.1.3 Space sun telescope

Sun telescope program designed and manufactured by China, includes five detecting instruments. The orbit parameter of the satellite will be round orbit with a height of 730 km and a dip angle 98.3°. The main scientific aim is to detect mechanics of magnetic fluid phenomenon of instant changes and stable state during the course of mechanics of magnetic fluid and make breakthroughs in sun physics.

4.1.4 The basis of the implementation of the above-mentioned exploration plans is to improve the scientific objects of exploration plans and the level of detecting technologies constantly and put forward the sun-earth space plans of exploration that will be implemented before the year 2015. The concept is to take part in the Chinese sun-earth space exploration plan "Plan of influence of sun by space environment of the earth" of the international cooperation plan "International Living With a Star".

4.2 Prospects and exploration of the Moon by China

Conceptual plans for the Moon exploration program and moon exploration in the near future have been brought forward and initially divided into three stages as "Environment, landing, and returning." The exploration plan of circling the moon orbit is scheduled to implement during the early 11th Five-year-plan. The development of chemical research of the Moon can be accelerated through the analysis and research of ingredients of the materials on the surface of the moon and the pre-research of the exploration plans of the Moon. Around the year 2015 it will be possible for our country to execute a soft landing on the Moon and lay a foundation for the utilization of the Moon.

4.3 The Prospect and Development of Chinese Micro-Gravity, Space Living Things and Space Material Science

China has built a ground research tower of micro-gravity and supplied the conditions for the experiments and research of micro-gravity. The implementation of plans of manned spaceships has promoted experiments and research for micro-gravity, space living things, and space material. Furthermore, scientists are seeking experiments and research on the International Space Station through international cooperation.

4.4 Prospects of Exploration and Research of the Mars of China

The key objectives of Chinese deep space exploration are the Moon and the Mars. On the basis of the exploration of the Moon, the exploration of the Mars will be scheduled. China is now ready to make programs and plans of exploration of the Mars and the exploration plans for circling the orbit of Mars probably will be implemented around the year 2020. Through the pre-research of the exploration of the Mars, the development of comparative planetology and the chemical research of the Mars can be promoted.

Acknowledgements: During the compiling of this article, I feel greatly honored to be provided the relevant material by : Ouyang Ziyuan academician from Guiyang Geography and Chemistry Institute of CASS, professor Zou Yongliao, Professor Li Chunlai, professor Liu Jianzhong, professor Liu Zhiheng from Biological Physics Institute of CAS, Department of Space Science and Application of CAS, professor Gu Yidong and professor Sun Huixian from manned spaceship application system.

Geological Sciences in China: Review and Prospects

Sun Shu Zhou Xinhua

Institute of Geology and Geophysics, Chinese Academy of Sciences

***Sun Shu,** geologist, Research professor at Institute of Geology and Geophysics, the Chinese Academy of Sciences; Member of the Chinese Academy of Sciences, Fellow of the Third World Academy of Sciences and Academician of International Euroasian Academy of Scinces, as well as Honorary fellow, Geological Society of London. His recent publications are as follows: "Mesozoic and Cenozoic sedimentary history of South China", "Indosinides in China and the consumption of eastern Paleotethys", "Tetonic facies map of China" and so on.*

Abstract: A brief review of the development and major achievements of geological sciences in China for the last 50 years has been made in this article; and it also presents the highlights of future development in this field for the coming 10-20 years.

1. INTRODUCTION

The ancient wise and nature-loving Chinese scholars pondered over the natural phenomena, which they faced and made up with some seeds of geological ideas.

Chinese geologists had started to engage in geology at the beginning of 20th century[1], when the modern natural sciences were just initiated in China. Over the last 50 years, geological education, survey, exploration, and scientific research have reached up to an unprecedented scope and made distinguished achievements[2-6]. Chinese geologists have taken part in a series of international cooperation projects and hosted a series of international conferences and symposia on various special geological topics held in China in last two decades. The 30th International Geological Congress held in Beijing in 1996 was crowned a success. It should be pointed out that Chinese geologists have actively participated in international scientific activities during the last years and will continue to develop new international cooperation in the coming years, as they have benefited from the Chinese government's policy of reform and opening to the outside world.

In the authors' view, three aspects of the strategy made the long-term development of geological sciences vigorous in China: the first one is to emphasize the geological education for training and bringing up talented people; the second is to place first priority to solving the problems of mineral resources, fossil fuels, and environment concerned with the national economy and the people's livelihood; and the third is to stress on the basic research for promoting academic standard.

The world population was quadrupled in the 20th century and will be possibly double in the 21st century. Thus, mankind will naturally be greatly concerned with the Earth with which we are closely associated, and aspire to know a variety of natural laws about the Earth and the inhabitability on the Earth. In this paper we would make a brief review on the development of geological sciences in China from the view point of the background of international geological sciences, with an emphasis on the development in the last half a century, especially over the past 20-30 years.

2. MORE PROFOUND UNDERSTANDING OF THE EARTH

Science and technology were rapidly developed in the whole world in the 20th century, as were the geological sciences.

In the first half of the 20th century, research on the Earth, mainly on the minerals, rocks, fossil fauna and flora, geological structures as well as various geologic phenomena and processes in the crust were carried out from different aspects of geology and, hence, many branching disciplines were constantly formulated and developed; the intercrossing and combination of geology with chemistry and physics brought about geochemistry and geophysics. The former studies elements, isotopes and radioactive dating, while the latter places its emphasis on study of the crustal structure and the structure of the whole Earth. The hypothesis of geosyncline became a dominant theory for explaining crustal structure and its development and evolution, although its theory of fixation of continents was challenged by the mobilism of the hypothesis of continental drift in 1915. With more understanding of the natural regularities of the Earth, many mineral resources and fossil energy were found and explored. Chinese geologists proposed geomechanics, polycyclic orogenesis, block-fauting tectonics, diwa tectonics and other hypothesis for explaining the tectonics of the China continent, even the problems concerned with global tectonics.

At the end of the1960s, based on the extensive global ocean investigation carried out after the Second World War, a series of new discoveries had resulted in the emergence of a theory of plate tectonics. It marks a revolution in the Earth sciences[7]. Since the end of 1980s, a scientific concept of the Earth system science has increasingly attracted great attention in development of geological sciences.

Over the last 30 years, plate tectonic theory of the lithosphere has been developed in China. Like in other parts of the world, it has changed many traditional concepts and knowledge of the Earth and has promoted many disciplines developing, intercrossing and incorporating with each other, such as structural geology, geotectonics, petrology, sedimentology, paleontology, stratigraphy, paleogeography, mineral deposits, oil-gas geology, geochemistry and geophysics, resulting in emergence of a series of new research fields. In China, multiple orogenic belts in different geological time, the Qinghai-Tibet Plateau with huge crustal thickness, various types of marine and continental sedimentary basins, and ultra-high pressure metamorphic belts provide rich and varied natural laboratories for study of plate tectonics. The geological evolution history of Chinese continent and its surrounding sea areas reflect a series of collision events among several cratons, numerous continental blocks and island arcs. Geologists have postulated a variety of concepts, such as small oceanic basins, archipelagic oceans, back-arc basins, terrenes, subduction-accretion and tectonic facies, which are important for understanding the evolutionary history of plate tectonics in China. The plate tectonic framework of China, including Tethyan, Central Asian/Paleoasian and Circum-Pacific tectonic domains, has become a research hotspot by Chinese and foreign geologists. Several tectonic models for China and whole Asian continent were set forth, the study on the continental accretion and the geodynamic processes are being carried out, and. the tectonic evolution of ultra-high metamorphic belt is being also studied by using deep continental drilling. Meanwhile, individual geological disciplines have scored great achievements, for example, many new minerals were recognized; the Lèping Epoch and its two stages named by Chinese geoscientists have been acknowledged by International Geological Time Scale; the global stratotype sections and points between Permian and Triassic Systems and between two stages of Middle-Ordovician were established in China. Significant advances were also made in studies of genesis of various types of rocks, isotopic geochemistry, organic geochemistry, Quaternary geology, sedimentary environment and sedimentary basins, paleogeography, paleooceanography, mathematical geology, paleobiogeography, and paleomagnetism.

Over the last 20 years, a series of new and major paleontological discoveries has attracted worldwide attention. These are highlighted by the discoveries of Chengjiang biota and Jehol biota. Since the discovery of the former in 1984, some new phyla, class and order were finely studied. In particular, the reliable discovery of the earliest pisces so far known, including *Myllokunmingia fengiaoa* and others, pushed the origin history of vertebrates back to the early Cambrian Period. The discovery of Chengjiang biota has invoked the impetus to probe into the origin of many metazoan texa and will be greatly helpful to making new breakthrough in the evolutionary theory of organism. The Jehol biota has been studied for more than 70 years. However, during the last years, early birds, feathered dinosaurs, primeval mammals, and the oldest angiosperm were discovered in West Liaoning area, which became a hotspot attracting attentions of Chinese and foreign scientists. The fossil of early birds found in the region is the most abundant in the world, while the discovery of feathered dinosaurs close to birds has strongly supported the hypothesis of origin of birds from dinosaurs.

During the last years, Chinese geologists have stressed the study of interaction between the various subsystems inside the Earth system. On the one hand, stress has been put on the interaction between the crust/mantle and mantle/core for the study on the processes of plate subduction and exhumation, magmatism, metamorphism, mantle plume and geochemical cycles by examining the elements, isotopes, minerals, rocks and geological structures. On the other hand, the geologists have also put stress on study of interactions between lithosphere-hydrosphere-atmosphere-biosphere for the study of sedimentation, sea-floor hydrothermal activity, biogeochemical cycle, environmental changes, and evolution of organism through examination of sedimentary rocks, fossils, loess, lacustrine deposits, karstic deposits and deep sea sediments. Study on the Phanerozoic lithospheric evolution in China has revealed the dilamination and over one hundred kilometer thinning of the lithosphere in the Mesozoic-Cenozoic, and their possible geodynamic background and consequence on the transition of tectonic framework in some regions have been discussed. It also has indicated the crucial significance of underplating since the Paleozoic and the important role of juvenile crustal material in lithospheric evolution and magmatism in some regions. Study on loess-paleosol sequences has revealed an evolutionary history of paleo-monsoon in East Asia. A deep-sea isotopic sequence over the last 30 Ma set up firstly in the western Pacific Ocean has revealed a series of characteristics of climate evolution in the lower-latitude ocean.

Over the past 100 years,great progress of geological sciences has been achieved in understanding of the Earth, as well as in understanding of the landmass and sea areas of China. However, we have to admit, or acknowledge that there still are so many unknowns. The reason is that the Earth has its long evolutional history and the geological processes interweaved with physical, chemical and biological processes are fairly complex. The products of the Earth' s surface processes and the processes in the Earth's interior can occur or may be superimposed on each other at one place and the geological phenomena of different times can also be composited at one site, but mankind could not reproduce these processes under laboratory conditions. Therefore, development of geological sciences in China requires its branches not only to study various phenomena and processes, but also, to develop its research on Earth system from the view of complex system and major attention must be paid to the influences of environmental change induced by human activities. The more the understanding of various natural processes and their interrelations of the Earth on which we live, the higher the capability of our controlling and predicting the natural laws are, and hence, the better the effectiveness of our management of the Earth is[8].

In the coming 10-20 years, to strengthen basic theoretical research of geological sciences in China, it seems that the following aspects must be stressed [5,6,8-12]:

2.1 Continuing to push forward in-depth development of individual disciplines of geological sciences

The development of geological sciences is composed of the achievements of its all disciplines. General advance in geological sciences cannot be taken without progresses in its individual disciplines.

All geological disciplines should pursue basic theoretical research with strategic significance and keep abreast of their individual current frontiers and research hotspots in time. Of cause, some freely probed subjects out of the scopes of frontiers and hotspot can also create their specific bright and unexpected outcomes. It should be pointed out that all geological disciplines have significantly enlarged their development fields in the last years.

2.2 Advancing interdisciplinary and multidisciplinary researches aiming the science of earth system[11,12]

Many natural phenomena are of complex systems, while many practical needs of society are often of systematic engineering. Thus, treatment and solution of complex objects by means of multidisciplinary and interdisciplinary research approaches are an important requirement to development of science and technology. As described above, to understand many natural laws requires knowledge of Earth systems and to solve many practical problems cannot ignore the interaction between the Earth's subsystems.

The interdisciplinary and multidisciplinary researches are inferred to intercrossing and integration between all geological disciplines with those concerning the solid Earth and liquid Earth as well as other relevant sciences. The top priority would be placed onto following fields[6,13,14].

- Lithospheric dynamics of China and Asian continent;
- The Earth's environment, life process and biogeoscience;
- The formation and evolution history of marginal seas of Asian continent;
- Energy flow and mass flux between Asian continent and West Pacific Ocean at middle-lower latitudes and their evolution in geologic history;
- Fluids in the crust;
- Low temperature geochemistry;
- Environmental geochemistry;
- Interactions between the surficial crust, atmosphere, hydrosphere, and bio-

sphere, and their products and evolution;

- Interaction between the Earth's crust and mantle and between the mantle and core;
- Mantle convection;
- High-resolution paleoclimatology and global environmental changes;
- Multidisciplinary and interdisciplinary researches on mechanism for earthquake occurrence and earthquake prediction;
- Geology and geophysics of the Antarctic and Arctic;
- Modern geological mapping and high-resolution geochemical mapping of the China continent and its offshore sea floors.

2.3 Development of Blank and Weak Disciplines

As early as the end of the 1950s, Chinese scientists, as published in a variety of articles, expounded a developing trend of Earth sciences as "to explore the surrounding planets, the Earth's interior and the offshore seas." However, various reasons caused the development to be entirely unsatisfactory. For example, as regards to probing the offshore seas, the research on coastal seas of China remains to be further strengthened although it has been carried out for several decades and major progresses have been made; whereas the research on deep ocean area has just started. China was not able to take part in deep ocean investigation, such as DSDP, in 1960s -1980s, resulting in revolution in earth science. Up to date, this historical gap is still one of the constraints on the nowadays development of earth sciences in China[14]. At the beginning of 1990s, the Division of Earth Science, Chinese Academy of Sciences again proposed"to explore the surrounding planets, the Earth's interior and the offshore seas." Later, China joined the Ocean Drilling Program and i mplemented its Leg 184 in the South China Sea area. At the beginning of 21st century, it is the right time to propose again to explore the Earth's interior, the oceans and the surrounding planets. It would contain the following research topics[6,14].

- Structure, composition and processes of Earth's deep interior;
- Experimental study at high-temperature, high- and ultra-high pressure;
- Experimental and theoretical studies of geothermodynamics;
- China continental scientific drilling;
- Research on oceanic lithosphere (including paleo-environment, deep-sea hy-

drothermal fluids and metal deposition, deep biosphere and earthquakes, etc.), mid-ocean ridges and subduction zones with active participation in the Integrated Ocean Drilling Program(IODP);

- Research on Moon and Mars.

3. ENSURING AND GUARANTEEING FOR SUSTAINABLE DEVELOPMENT

In China, and perhaps in the whole world, at the mention of "geology", people would immediately link it to "search for mineral" and "search for oil". It indicates that geology is by no means an ivory tower but is a science directly related to economic development and is close to people's daily life.

During the 20th century, minerals and fossil energy played an irreplaceable role in global economy. Minerals are the raw materials for many industries, coal always ranks an important place in energy, uranium ore produces electric power delivered to common families, and oil is named as black gold. The consumption of mineral resources rose steeply in the 20th century. At the beginning of the 20th century, only 20 elements were used, but at present time, 92 elements are utilized. In 2000, energy production was increased about 5 times as against 1950 and more than 20 times as against 1850[15].

In order to provide geologic background information for exploring various mineral resources required for economic construction, it is necessary to develop regional geological investigation. A regular regional geological survey and mapping work on scale of 1:200,000 had started in China in 1955. By 1997, the surveying and mapping of 88.07% of China continental area on scale of 1:200,000 were performed and 98.20% of the area was mapped on scale of 1:1,000,000. Thus, geological and tectonic maps of China and of Asian-European Continent were published. At the same time, a large amount of regional gravity survey, aerial geophysical survey, remote sensing, and regional geochemical prospecting was performed.

Over the past 50 years, totally more than 200,000 mineral deposits and ore occurrences have been found by development and application of ore-forming theories, geophysical and geochemical explorations, remote sensing, drillings, and other prospecting techniques. The proven reserves were only of two mineral resources in 1949 and increased up to 157 mineral resources by the end of 2000. The potential gross value of reserves of 45 major mineral resources are ranked the third place in the world and the second place in their output values in the last years[16]. The mineral resources having their proven reserves are coal, tungsten, antimony, titanium, rare earths, niobium, tantalum, tin, molybdenum, vanadium, lithium, bismuth, beryllium, magnesite, barite, graphite, gypsum, bentonite, etc. Metallogenic theories, such as theories of metallogenic series, strata-bound ore deposits, regional metallogenesis and super-large ore deposits, have been formulated. In 1949, oil production was only about 120,000 tons. This production level had kept for a considerable long time. Since 1955, oil prospecting has been developed on unprecedented large scale. A great breakthrough was made in finding of a world class oil field, the Daqing Oil Field in Songliao Basin in 1959. Later, oil-gas fields were successively found in the vast China land and its sea areas. By 1998, oil production exceeded 160 million tons in the whole country and production of natural gas reached 27 billion m^3 in 2000. China's oil and gas production has ranked the fifth place in the world, and the oil exploitation and refining engineering have became one of the mainstay industries in China. Petroleum geological and geophysical practices gave rise theories of oil generation and reserving in continental facies, and a set of techniques for oil exploration in continental facies was developed. It must be pointed out that part of oil-gas fields in western China is of marine facies origin.

With the population growth and the economy development, the global demand for metals is estimated possibly to be doubled in the 21st century, which could not be provided only by metal recovery and reuse. It is necessary to exploit many mines. According to current supply levels, the proven oil and gas reserves could be yet exploited for 37 and 66 years, respectively[17]. Thus, we must continue to find new oil reservoirs in the exploited oil areas, and to develop exploration for oil and gas in new areas, such as in deep water area of the seas and to study the possibility of methane hydrate as a new energy.

Minerals and fossil energy products have entered into global markets. China has become a member of the World Trade Organization and hence can adjust the demands for products by import and export. In general, per capita possession of the proven minerals and fossil energy reserves is not high in China. Moreover, in the first half of the century, China will build a well-off society into a moderately developed society. Therefore, in view of tremendous demand for natural resources in China itself and in world markets, it is fairly necessary to find out many more resources in China territory, meanwhile, China must be involved in international resource exploration projects.

In the last 50 years, Chinese geoscientists have carried out extensive geological research and scored tremendous achievements in solving geological problems encountered in construction of engineering projects, in surveying groundwater resources for providing water needed in daily use and industries in arid regions, and in development of prevention from and prediction of geologic hazards. It has also promoted the development of the relevant disciplines, such as engineering geology, hydrogeology and seismic geology.

A large growing population and unprecedented development of the economy in the 20th century caused global environmental crisis. A book *Silent Spring,* published in 1962, first described an ecological crisis caused by pesticides and insecticides used in agriculture. Later, a series of environmental hazards convulsed and shocked mankind with over-pasturing grassland and over-deforestation that resulted in water and soil loss and desertification; industrial water and solid wastes brought toxic, deforming and carcinogenic materials into lakes, rivers and soils; fossil energy combustion and polluted atmosphere induced diseases even decease of mankind; and in long-term respect, greenhouse gas inducing global climatic warming will cause the global environment to dramatically change. All these environmental problems have a serious tendency to threat many species and human existence. Geological sciences in China, as in other parts of the world, had paid more attention to our ambient environment just in the late of the 20th century and will hold the environment and the mineral resources in equally importance place in the 21st century.

A notable sustainable development concept was suggested in a book of *Our Common Future in* 1987 and it now attracts more and more attention. The Chinese government had worked out *China's Agenda 21* in the early 1990s, and has designed a sustainable development strategy as a national policy. To achieve the target of sustainable development, it is necessary to provide society with exploitable natural resources and to reasonably utilize, save, and preserve all the resources, so that the descendants could have the resources in sustainable utilization. We should stimulate faster economic growth to raise the living standard of the people and also well preserve, control, and optimize the environment and ecology, so that the descendents could have a suitable and survival environment. To achieve the target of sustainable development, we should make the social and economic development fully coordinate and harmonize with environment and ecology, so as to promote economic development and to treat well the environment and to restore the damaged environment in time.

Geological sciences in China in the 21st century should pay attention to the national development and people's blessedness, to make a due contribution to whole mankind, and to ensure the sustainable development in the fields of resources, environment and disaster reduction[18]. Thus, we must put the stress on the concrete scientific researches:

3.1 To Find out, Explore and Exploit Fossil Energy and Mineral Resources

- Distribution regularities and the availability evaluation of fossil energy, mineral resources and geothermal energy;
- New geological, geophysical, geochemical prospecting and remote sensing methods for searching out and exploring the buried mineral resources;
- Formation and evolution of mineral deposits, oil and gas systems and their controlling effects;
- Techniques for exploitation and utilization of resources;
- New techniques for preservation and reasonable utilization of resources;
- International strategy of exploration and utilization of mineral resources and energy.

3.2 To Survey, Exploit and Preserve Ground Water Resource

The importance of groundwater resource attracts increasing attention. People in arid-semiarid regions have deep perception to it. People in northern China have suffered the worry of water shortage. The groundwater is a treasure, which must be appropriately and reasonably utilized. In doing so, we should study and develop following topics and techniques:

- The evolutional regularities and theories for adjustment of and control on groundwater resource in big delta areas and water-consuming industrial areas in eastern China;
- Aquifers, groundwater systems and their circulation in relation to surface, ecologic and meteoritic waters;

- Management model for sustainable utilization of water resources;
- Techniques for preservation of groundwater resource and prevention of groundwater from pollution, and
- Techniques for controlling ground subsidence induced by pumping groundwater.

3.3 Prediction and Reduction of Geologic Hazards

- Mechanisms, prediction and prevention of landslides and debris flows;
- Relationship between active structures and earthquakes in densely populated areas, such as urban, coastal and other areas;
- Systematical investigation and observation on landslide and debris flow potential risk zone and frequent hit areas;
- Risk evaluation and disaster reduction measures to geologic hazards in major cities and densely populated areas.

3.4 Study on Geological Foundations and Safety for Engineering Construction and Engineering Planning

- Stabilities of soil and rock base for buildings and engineering facilities;
- Geological safety for exploitation of mines and oil fields;
- Development regularities and consequence evaluation of long-term slow geologic processes in urban areas, along main traffic lines and in major engineering project areas;
- Stability of major engineering facilities under the condition of earthquake occurrence.

3.5 Protection and Optimization of Environment

This is a research field, where a special attention should be paid.

- Transport and diffusion regularities and techniques for control on and treatment of prevention of pollutants, especially persistent organic and toxic pollutants;
- The effect of sea level rise induced by climate warming on the coastal regions of China;
- Monitoring, prevention and treatment of environmental pollution in extraction and production processes of energy and mineral resources;

- Principles and techniques for environmental restoration of the exploited mines;
- Principles and techniques for disposal of solid castoffs, including nuclear waste; and
- The relationship between geological environment and ecology, agriculture and human health.

3.6 Development of Geological Engineering and Techniques[19]

- New geological, geochemical and geophysical engineering and techniques for exploration by using GPS, RS and GIS;
- New techniques for mineral processing;
- New kinds of ores and their processing techniques;
- Monitoring and treatment of delayed geochemical hazard ("chemical timing bomb");
- principles and techniques for underground disposal of carbon dioxidedischarged from industries;
- geoecological principles and techniques for remediation of rock desertification areas;
- The genesis of gas hydrates, its exploration and exploitation techniques;
- Techniques for *in-situ* extraction of copper, uranium and other minerals;
- Urban geological, geophysical and geochemical integration techniques, and
- Coastal zone management engineering and techniques.

4. CONCLUSIONS

It is usually believed that geology is characterized by a close connection of basic research with applied study. To enhance the knowledge of the Earth and to solve the issues of resources and environment which mankind faces for long time, i.e. to probe the natural mystery and to actively serve the society are the intrinsic power for development of geological sciences. With the aid of high-tech and new techniques, such as tomography of the Earth, Earth observation, precise analytical measurement and computation, experimental modeling, etc., the modern geological sciences will be fully vigorous and have its development opportunities in the 21st century, so it can provide vast space for "curiosity"and creative research. Chinese geologists would make great efforts for the course of development of modern geological sciences.

Acknowledgements: The authors are gratefully indebted to Wang Hongzhen, Li Tingdong and Ma Fuchen for the constructive suggestions and comments made for the manuscript; and to Su Jilan, Wang Pinxian and Yin Hongfu for the modifications made for the English version.

REFERENCES

1 Huang Jiqing, A brief discussion on the main achievements in geological science in China over the past 80 years and the orientation of further efforts. Geological Review, 1982, 28 (6): 515-523 (in Chinese)

2 Li Siguang, The upswing of geology following Chinese sciences. Renmin Ribao, People's Daily, Sept . 21, 1952; Collection of Papers by Li Sigunag, 1952, 8, 270-286 (in Chinese)

3 Zhu Xun, Struggle for flourishing geological science and technology in China: Report at 70th Anniversary of Geological Society of China. Geological Review, 1992, 38 (6) (in Chinese.

4 Wang Hongzheng, 50th Anniversary of Geological Science in China. Beijing: China University of Geological Science Press, 1999, pp. 331 (in Chinese)

5 Wang Hongzheng, Zai Yusheng, You Zhendong, Shi Baoheng, Yang Weiran, and Yang Guangrong, A brief review of Geological science in China for 50 years. Geological Review, 2000, 46 (1):1-7 (in Chinese)

6 Sun Shu,. Geological science in China in the past, present and future. Geological Review, 2002, 48: 576-584 (in Chinese with English abstract)

7 Moore, E. M. and R.T. Twiss, Tectonics. W.H. Freeman and Company, New York. 1995, pp. 415

8 US National Science Foundation Directorate for Geosciences,:NSF Geosciences Beyond 2002: Understanding and Predicting Earth's Environment and Habitability. Full Report (NSF 20-27). Last Modified. 2002, http://www. nsf. Gov/search 97 cgi/ vtopic

9 US National Research Council, Solid-Earth Sciences and Society. National Academy Press. Washington, D.C.1993, pp.346

10 UK National Environment Research Council, Strategy for Earth Science and Technology, 2002, http:// www. NERC. Gov

11 Ministry of Science and Technology, National Natural Science Foundation Commission,. Report of Basic Science Development. 2001, pp. 367 (in Chinese)

12 National Natural Science Foundation Commission, The Research Fields Priority-Supported by National Natural Science Foundation in the 10th Five-Year Plan. Beijing: Atomic Energy Publishing House, 2001, pp.87 (in Chinese)

13 Research Group on Strategy of Earth Science Development in China, Geoscience Department of Chinese Academy of Sciences,. Some Problems in Strategy of Earth Science Development in China, from Great Geoscience Nation toward Powerful Geoscience Nation. Beijing: Science Press, 1998 (in Chinese)

14 Research Group on Strategy of Earth Science Development in China, Geoscience Department of Chinese Academy of Sciences, Earth Science: Review and Prospects in the Transition between Centuries. Jinan: Shandong Education Publishing House, 2002, pp.210 (in Chinese)

15 Zhang Jinghu, World resources and environment.Cultural university of China Press, 2002, pp. 238 (in Chinese)

16 Song Ruixiang, Report of Mineral Resources in China. Beijing: Geological Publishing House, 1997, pp.308 (in Chinese)

17 Gries R. Who dares, wins. Geoscientists, 2002, 12(6):4-11

18 Tian Fengshan,2000,Geological sciences and the sustainable development of China,Lecture at the 31st International Geological Congress,Brazil

19 Hsu K.J., A reform in earth science education after a revolution in earth science, Episodes, 1997, 20(3) :151-157

Global Change and Arid Environments in China

Liu Dongsheng*

Institute of Geology and Geophysics, Chinese Academy of Sciences

Liu Dongsheng *, Born in November 22, 1917 in Shenyang, China. Bachelors degree in Geology, National Southwest Associated University, Kunming, China (1938-1942); Honorary D. Sc. at Australian National University (1987) and honorary Ph. D. at Lingnan University, Hong Kong (1995); Junior Geologist and Paleontologist for the Geological Survey of China (1945-1949); Senior Geologist with Ministry of Geology, China (1949-1953); Professor, Institute of Geology, Chinese Academy of Sciences (1953-present); Secretary General of the China Association for Sciences and Technology (1982-1986); Director, China Science and Technology Museum (1982-1987); Director, Xi'an Laboratory of Loess and Quaternary Geology(1985-1992). Academician of Chinese Academy of Sciences since 1980; Academician of The Third Word Academy of Sciences since 1991; Academician of the International Eurasian Academy of Sciences since1996; President, China Quaternary Research Association (1981-1998); President, International Union for Quaternary Research (1991-1995); Member of Scientific Steering Committee of PAGES (1985-1999). Major awards include the National Natural Science Prize of China (four times in 1982, 1988, 1991 and 2000), Chen Jiageng Award for Natural Sciences (1989), Distinguished Award from the China Green Prize of Environmental Science and Technology (1993), Ho Leung Ho Lee Prize in Earth Sciences (1995); Tyler Prize (2002).*

The author of this paper is the laureate of the 2002 International Taylor Environment Prize.

Abstract: The modern environmental distribution pattern of China was progressively developed from early Cenozoic era. The Miocene is a critical period during which a profound change in the environmental structure occurred. In late Miocene-early Pliocene, the distribution pattern of the arid zones in China was much similar to that in today. During Pliocene, the arid zone extended eastwards and became the world largest desert in mid-latitudes. Surround the desert region, extensive and thick loess deposits have been developed. In the last 2.6 Ma, climate fluctuated in great amplitude, and the winter monsoon and continental aridity were significantly intensified. The "harmony of man with nature" became a core strategy for the management of environmental desertification of western China. As the critical obstacles for improvement of environmental quality differ from region to region, specific policies should be practiced individually. Adjustment of mutually beneficial relations between major factors in an ecosystem should be addressed and the policies for controlling human activities in type and scale are also crucial.

1. INTRODUCTION

In the 1950s, a series of social effects from environmental pollution shocked the whole world, and environment science came to the forefront to address that dire peril!In the following 50 years, significant breakthroughs have been made in raising awareness towards these environmental problems. In 1962 American biologist Dr.Rachel Carson published her epoch-making book "Silent Spring", calling on the broad masses of the people to make their choice at this intersect, either to take this final opportunity to protect our environment, or to pay no attention to an ever increasing deterioration of our environment. The World Environment Conference held in Stockholm in 1972 made the viewpoint of "We have only one earth" strike root in the hearts of the people throughout the world. The ever increasing environmental crisis has thereby aroused great concern for the whole world for the first time.

With the deepening of our understanding of environmental problems, the concept of environment has been further developed from "polluted environment","environment of life", "ecological environment" to "environment of existence". From macroscopic rules and microscopic mechanisms of environmental change, scientists and engineers engaged in various fields of research have done much research and practical work on the protection of Earths environment and the improvement of environmental quality, hence making great contributions to the probe of regional and global environmental changes, assessment of environmental quality variations, regional division and environmental planning, implementation of environmental projects, and policy-making toward environmental problems.

In the 1980s, "International Geosphere-Biosphere Program "(IGBP) started, and global change became a hot spot of concern by scientists, especially environmental scientists, throughout the world. Meanwhile, due to increasing deterioration of the environment and exhaustion of natural resources, readjusting economic activities toward improvement of the Earth's environment has been a focus in policy-making for numerous nations and social communities. After the "Conference on Environment and Development" sponsored by the United Nations was held in 1992, "the concept of entirety and mutual dependence of the Earth's environment" has been commonly accepted by political circles, ordinary people, and academic circles. The "21st Century Agenda" and the "Theory of Sustainable Development," especially the"World Summit Conference on Sustainable Development " held in South Africa in 2002 shed the first light on the requirements for coordinated development of economy and environment .

Aridification is one of the major global environmental problems. In human history, droughts, floods and earthquakes are three principal natural disasters that are hard to combat. The occurrence of drought not merely has special characteristics of regional distribution, but also affects the inter-annual fluctuation of climate. If in the years when arid regions are attacked by droughts, the consequences would be hard to imagine, as can be described by a phrase"snow plus frost". Under the scientific outline of"Global Change", understanding of the environmental evolution characteristics and maintaining a better environmental quality for arid regions will be of special scientific value and great significance with respect to sustainable development. In 2001 the four major scientific organizations (IGBP, IADP, WCRP, DIRER) jointly proposed in the Netherlands that air (CO_2), water and food would be regarded as a common research objective, and this gave great impetus to the study of global arid environment. This is also an important opportunity for Chinese scientists to make great contributions and achieve good results in their work on sustainable development.

2. ARIDITY PROBLEMS AND CHINA'S ARID AREAS

2.1 Aridity of Natural Environment Caused by Fresh Water Shortage

Widespread in Asia, Africa, Australia and South America are arid and semi-arid regions. For example, the African Continent, where one-third of the land area has an average annual precipitation of less than 200mm because of its specific geological environment and geographical position. As a result, the total area of arid regions in the African Continent ranks first over the world, and the African Continent is usually regarded as the "Thirsty Continent". Since the 20th century, drought calamities on the African Continent have become more and more serious. In the twenty years starting from 1968, there have occurred big droughts on the African Continent, which set records in recent history with respect to their long duration. These droughts would be considered as the most serious calamities encountered by mankind in the recent history of Africa. From late 1960s to early 1970s there had occurred the most serious droughts for five consecutive years on the African Continent. Late in the 1980s, droughts became more and more serious. During the years of 1982-1984 there also occurred the most serious droughts, which had never been encountered in the past one hundred years, for three consecutive years, affecting 24 African countries and 40% of the total population in the whole of Africa, causing a casualty of more than 50×10^4 persons, making more than 6 million people become destitute and homeless, and two hundred million people to survive on the line of starvation and death. During the years of 1992-1993, big droughts had attacked 11 countries in the southern part of Africa, resulting in a 50% decrease in grain production and more than half the population suffered from starvation. More than 1000×10^4 tons of grain have to be imported annually[1].

In the arid regions of China there is a popularized saying "Drought strike in two out of the three yeas" and "at least, a small drought occurs within three years and a big one in ten years" . Although aridification can lead to environmental problems such as soil desertification and sand storm, the most direct consequence of aridification will be the loss of lives and poverties of local residents. For instance, during the three years of 1928-1930 big droughts had taken place in the whole province of Shaanxi, accompanied with febrile diseases, causing a casualty of more than 2.5 million people and making more than 6 million people become destitute and homeless. As another example, in the year of 1995 spring and summer droughts attacked the vast area of northwestern China, directly causing great losses in agricultural production. The drought-stricken area of Gansu Province reached 6500×10^4 Mu, and 360×10^4 persons and 160×10^4 animals had no sufficient water to drink; more than 140×10^4 Mu of summer grain crops had no harvest, or a little harvest[1].

More importantly, some arid and semi-arid regions are just the regions where many third-world countries are located, and the backwardness of economic development in these third-world countries would significantly affect their capabilities of combating natural disasters. Therefore, research on arid environment would be the important responsibility to be shouldered by third-world scientists. We are very gratified that the Third-World Academy of Sciences has long been concerned with this important environmental problem and is now trying to find effective means to handle it.

2.2 The Distribution of China's Arid Regions

Most arid regions are distributed in the northwestern part of China, for example in Xinjiang, Qinghai, Gansu, Ningxia, Shaanxi and Inner Mongolia, where the average annual precipitation is lower than 200mm for a number of years. So they are the typical arid or semi-arid regions, with a total area of 345 $\times$ 10^4km^2, accounting for 39% of the total land area of China, and a population of about 8178 $\times$ 10^4, only accounting for 7.3% of the total population of China. The average population density is less than 27 person/km^2. In this sense, Northwest China is a terrain with a vast territory but a sparse population.

Northwest China, the area to the west of the Helan Mountains is an inland arid region (253 $\times$ 10^4km^2), and to the east of the Helan Mountains a semi-arid grassland region (29 $\times$ 10^4km^2) , and to the west of the northern trunk stream of the Yellow River is a semi-arid and semi-moist region(63 $\times$ 10^4km^2)(Fig.1).

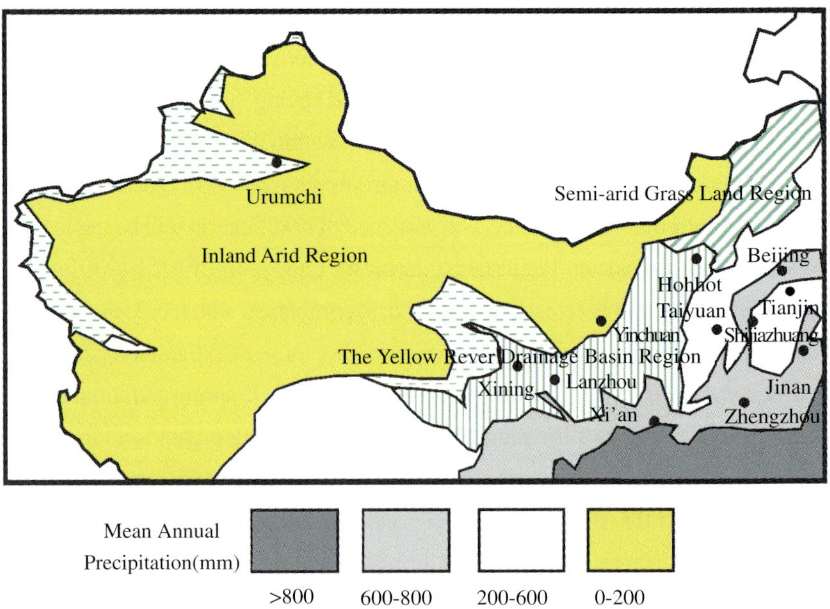

Fig. 1 Sketch map showing the regional division of arid environments in the northwestern part of China.

Although these three types are all characterized by varying 6 degrees of aridification, they are different in landscape due to their different natural frameworks.

The inland arid region comprises a number of basin units that are surrounded by mountains. With the exception of the Qinghai Lake, vast deserts are usually developed in the center of the basins. Some streams originated from precipitation on the mountains or melted ice and snow and found a home in deserts, or wetlands among the deserts. Deserts, gobies and oases were formed, depending on the constraints of water resources. Natural or man-made oases are important habitat for human communities. So, as viewed from the angle of environmental type-regions, the population is distributed densely, but in a scattered manner regionally. The semi-arid grassland region has a slightly higher annual precipitation, which shows an increasing tendency from west to east. In the region there are distributed only some seasonal streams. Grassland and wetland are the principal landscape types for human communities to live and multiply, as well as their economic activities, so the population is of scattered distribution. The semi-arid/semi-moist region includes mainly the source region of the Yellow River, the Loess Plateau and sandy areas distributed along the Great Wall. The total population on the Loess Plateau is now over 8000 $\times 10^4$ and the population density exceeds the average level of the whole country.

Northwest China's arid/semi-arid regions are not only the regions where water resources are severely lacking, but also the regions where water resources are distributed extremely unevenly. In the mountainous regions there is much rainfall, but in those extremely arid regions the annual precipitation is less than 10mm. The water resource amount per capita in the whole region is 1781m^3. Although the water source amount per capita comes up to 80% of the total water amount shared per capita in the whole country, water resources there are distributed extremely unevenly from one place to another. In many places the water resource per capita is less than 30%of the per capita in the whole country. The Yellow River Valley within the bounds of Ningxia Autonomous Region (217m^3 /year per capita), the Wei River Valley in Shaanxi Province (326m^3 /year per capita), the Huangshui River Valley in Qinghai Province (618m^3 /year per capita) and the Shiyang River Valley in Gansu Province (761m^3 year/ per capita) can be taken as typical examples in this aspect.

3. FORMATION AND EVOLUTION OF CHINA'S ARID ENVIRONMENT

3.1 The Formation and Evolution of China's Arid Framework

China'smodern environmental framework has been developed since Cenozoic time. The Paleocene (65-54 Ma B.P.) environment inherited the Late Cretaceous fundamental environmental framework. Under the control of the planetary wind system, there was formed an E-W-trending arid zone at about 18°N-35°N(Fig.2), in which there were developed large quantities of salts and gypsum deposits. Beyond the arid zone there developed coal seams with a lot of animal and plant fossils. During the Eocene (54-38 Ma B.P.), the environmental framework was generally similar to that of the Paleocene, but the arid zone was slightly shifted northwards (Fig. 2b). During the Oligocene(38-24 Ma B.P.), the climatic belt was still distributed latitudinally, but the arid region in the southeastern part was obviously moistened (Fig. 2c). The suitable increase of precipitation and humidity marked the appearance of southeast summer monsoon in an embryonic form.

During the Miocene (24-6.5 Ma B.P.) great changes took place in China's environmental framework. Many coal-bearing basins appeared in the regions of Southwest China, and the environment was transformed from arid to moist (Fig. 2d), indicating the formation of southwest summer monsoon. Meanwhile, the abundant coal seams and forests also indicate the precise intensification of southwest summer monsoon. At that time, the Qinghai-Xizang Plateau and the Himalayan Mountains began to lift to a certain level, thereby hampering the transfer of the moisture carried by southwest summer monsoon. As a result, the geographical framework was initially created, which is approximately equivalent to the commencement of China's arid zone at present time[2,3,4]. During this period of time, there occurred the event of aridification in East Asia, i.e., the formation of loess. In recent years a loess-paleosol sequence developed since 22Ma has been found at Qin'an, Gansu Province, which indicates the origination of Asian inland deserts and Asian monsoon circulation system. The alternative appearance of loess and paleosol layers is also indicative of variations in intensity of winter and summer monsoon circulation.In particular the inland aridification in Asia reached its maximum during 15-13 Ma and 8-7 Ma. On the whole, the aridity is relatively constant. Although the persisting cooling on a global scale during the Cenozoic and the closure of the subtethys Sea seems to have played an important role in the formation of inland deserts, the rise of the Qinghai-Xizang Plateau during the Early Miocene is the main factor leading to the origin ation and evolution of the inland deserts (Guo et al., 2002). The sand-dust record in the Indian Ocean[5] also provides strong evidence for the occurrence of serious aridification on the African Continent, while the oceanographic record shows that the Indian Ocean summer monsoon was drastically intensified during 8Ma B.P.[6].

The distribution pattern of arid zones in China during the Late Miocene-Pliocene (6.5-3.4 Ma B.P.) (Fig.2e) is still closer to the present framework, and during the Pliocene (3.4-2.6 Ma B.P.) the arid zones expanded eastwards (Fig. 2f), leading to the development of the largest middle-latitude desert in the world. Around the desert were developed loess deposits of great thickness. The main portion of the *Hipparion* red clay, deposited to the east of the Liupan Mountain is also composed of aeolian deposits, recording the evolutionary history of Asian inland aridification on a large scale during 6-2.6 Ma. The maximum aridification appeared during 6.2-5.0 Ma B.P., then tending to decrease till the obvious intensification during 3.6 and 2.7 Ma B.P. The intensification of aridification during 3.6Ma is consistent with the rise of the Qinghai-Xizang Plateau, but the evoltuionary history of aridification is highly cosnsistent with the process of development of the polar ice cover since its formation during 6-7 Ma B.P., as reflected by pelagic oxygen isotope records and Arctic ice raft deposits. So the formation and evolution of the Arctic ice cover and the rise of the Qinghai-Xizang Plateau are both important factors (or driving forces) leading to the development of Asian inland aridification during that period of time. The influence by both the factors on the formation and development of Sibrian high pressure and the hampering of warm/wet currents by the plateau are the main paths leading to aridification[7,8].

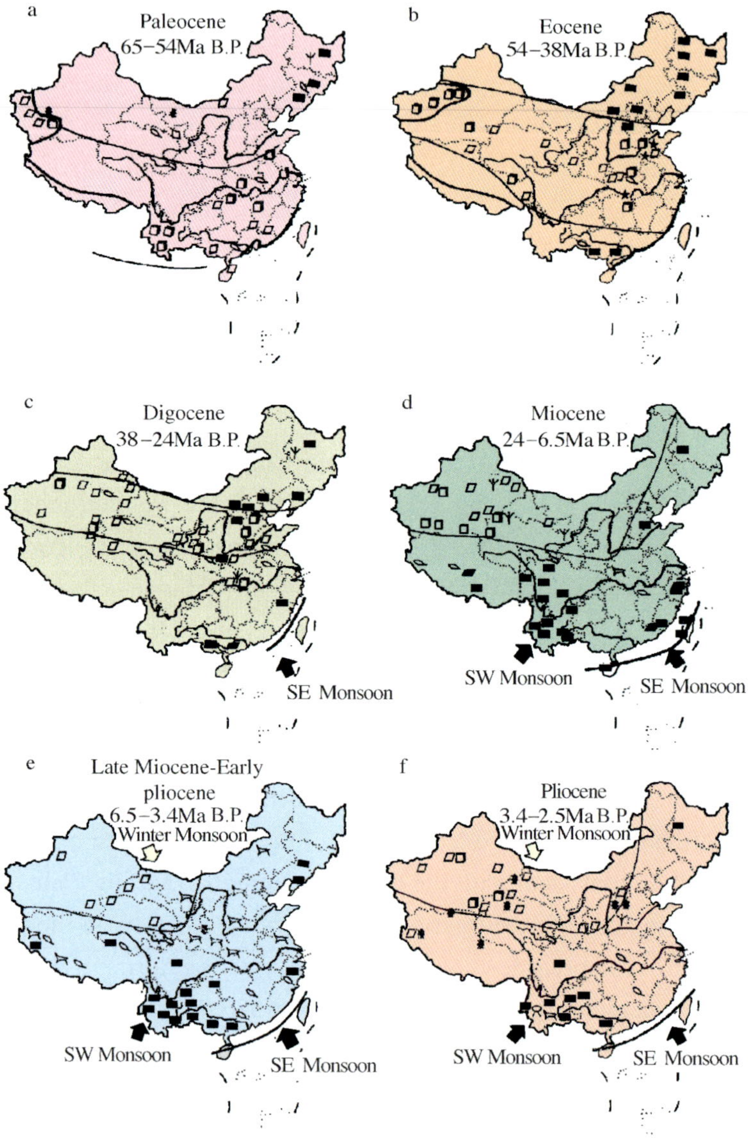

Fig. 2 Sketch map showing climatic and environmental changes in China during various geological periods of the Cenozoic

3.2 China's Loess Accumulation and Aridification History of Asian Inland since 2.6 Ma B.P.

During the Quaternary, global climate and environment experienced periodical fluctuations. These fluctuations have also been recorded completely in the loess-paleosol sequence strata formed as a result of alternation of arid climate and relatively moist climate, which, as an ideal information-carrier, together with pelagic sediments and polar ice cores, has become the three important databases of Quaternary global climate change.

The Loess Plateau covers an area of about 4.4×10^5 km^2. Loess deposits in the areas of Shaanxi and Gansu provinces are thickest, exceeding 300m in thickness. Loess deposits on the Loess Plateau tend to decrease both in thickness and in grain size from northwest to southeast, showing the typical features of aeolian deposits. The bottom boundary age of loess deposits is deduced to be 2.6 Ma B.P. Underlying the complete vertical loess profile are Tertiary red clay accumulates with a continuous transition and conformable contact to the loess. The Luochuan Loess Section in Shaanxi is one of the standard loess sections in China. In the loess section there are two kinds of stratigraphic units which have significant differences in both color and texture. One is represented by loess layers, usually grayish-yellow in color, homogenous in composition and of no precise texture; the other is paleosol layers, red in color, with obvious soil texture and soil-genesis stratigraphic sequence[7]. Loess-paleosol stratigraphic sequences can be distinguished in loess sections of different regions. As for any major loess and paleosol unit, comparison can be made at least among more than three sections[9]. Such continuity and completeness of China's loess deposition provide a possibility to understand the history of Quaternary paleoclimatic change. Detailed analysis of variations in grain size, composition and susceptibility of the loess section, its plant silicates, spore-pollen assemblage, snail fossils, ^{13}C-^{18}O and ^{10}Be isotopes, and other environmental indices and the establishment of a high-resolution time scale for the environmental evolution sequence have laid down the sound basis for a better understanding of the long-term aridification history since 2.6 Ma B.P.(Fig.3).

the existence and development of mankind itself. The restoration of harmonious integration of human and nature is the key for the coordinated development of humans and nature [14].

In terms of the results of scientific research, the Chinese government has formulated the state policies for "the protection of natural environment", "sustainable development" and"harmonious coexistence of humans and nature". These policies are of great significance in deepening people' sunderstanding of the importance of environmental protection and of the dependence of economic construction on science. China National Environment Protection Bureau made investigations into ecological environment in the western part of China, conducted the regional division of ecological functions and strengthened the environmental assessment of major engineering projects and environmental prevision in construction. Consequently, outstanding achievements have been acquired in ecological environmental protection.

In particular, the implementation of "West China's Development Strategy" and the formulation of the policy toward the coordination of Northwest China's development and ecological environmental construction have pointed out the way for the sustainable development of society and economy in northwestern China' s arid/semi-arid regions. The "Western China' s Development Strategy "is the key for the future development of China.

4.3 The Trains of Thoughts Concerning Sustainable Development in China's Arid/Semi-Arid Regions

Although the classical philosophical principle of "integrating universe and earth in one" has laid down the important ideological foundation for sustainable development in China's arid/semi-arid regions, there are significant differences in regional environments among China's arid/semi-arid regions. So, it is necessary to formulate different strategies of sustainable development in accordance with regional environmental characteristics. Two important aspects are: ① to bring into full play the adjustment and maintenance relations of ecological systems among different environmental zones; and ② to scientifically adjust and control the type and scale of human activities.

4.3.1 To bring into full play the adjustment and maintenance relations of ecological systems among different environmental zones.

Regional spatial analysis showed that different environmental structure types would influence each other within one region. Analysis of the geological agents leading to variations in environmental quality indicated the existence of interactions between economic activities and natural environment. Also there exist mutually dependent relations between the influence and the interaction, giving rise to the zonal structures of regional environmental ecological systems[15]. Following such zonal structures is the key to ensure sustainable development in a given region [16]. The aim of implementing the strategy of "integrated construction of urban and rural environment" is to solve the problem of how to realize regional coordination between economic development, rational disposition of natural resources and environmental protection[17,18].

Northwest China's inland arid region, semi-arid grassland region and semi-arid/semi-moist region show significant differences in zonal structure of regional environments because of their different natural frameworks and environmental types. Taking the arid inland region made up of a number of basin units surrounded by high mountains as an example, each of the basin units constitutes a relatively independent environmental domain, in which oases or residential quarters are the essential parts of zonal structures. Due to intensive interactions between human activities and natural elements, the environmental ecosystem would become fragile. The mountainous regions adjacent to the basins are the shielding parts of zonal structures, where natural elements occupy the dominant position, and they have strong capabilities of combating external disturbance. So, the environmental ecosystems are relatively stable there. Between the zonal structures mentioned above is the transitional zone, which serves as the channels for the transfer of materials and energy among the environmental zonal structures and as the bridge for the mutual adjustment and maintenance of ecological systems. As for the arid/semi-arid regions, the most important function of the transitional buffering zone is to transfer water resources. In the semi-arid grassland and Loess Plateau regions there should exist respectively numbers of relatively independent and mutually associated environmental domains with zonal structures.

4.3.2 Types and scales of human activities adjusted and controlled by scientific means

As for the development and utilization of a certain region, thc loading capability of its natural environmental system should be first and foremost taken into consideration. Under given circumstances pertaining to human activities, the loading capability of the environmental natural system refers to the integrated effects of geology, land, forestland, living organisms, hydrology, climate and other factors [14]. In the study of such loading capabilities the types of human economic and social activities are taken as the fundamental objects. If the type of human activity matches, or is in concordance with, the characteristics of the natural environmental system, the loading capability of natural environmental ecological system can be brought into full play. As an environmental ecological system in an arid/semi-arid region, no matter what type of human activity would be, hydrological factors (water amount and water quality) would have an important bearing on the loading capability of environmental ecological system.

To understand the loading capability of the environment, the assessment of the environmental capability of pollutants is always required to be conducted in consideration of the production and purification of pollutants in the processes of human activities. The environmental capability refers to the maximum amount of environmental pollutants loaded by a certain regional environment in case no damage is done to human health, human existence and ecological system. This is the concept specially proposed for environmental management and controls over the total amount of pollutants and their concentrations. Under the environmental conditions of arid/semi-arid regions, environmental pollutants have specific capability with respect to their dilution, diffusion and purification, and so the assessment of their environmental capabilities is of special significance.

Since the ancient times, residents in the arid/semi-arid regions have paid much attention to the problems of how to adjust themselves to the natural environment under arid/semi-arid climate conditions. In cases where the productivity was relatively backward and the pressure of population was not so serious, it is possible to maintain the concordance of humans with nature and acquire great achievements in economic and social development. The essence of such concordant development is to adjust the economic and social activities of mankind itself so as to make proper use of the loading capability of natural ecosystem. In the future for sustainable development of arid/semi-arid regions the type and scale of human activities should be scientifically adjusted and brought under control.

4.4 Establishment of the nationwide (global) integrated ecologically environmental safety system

China's arid/semi-arid regions only represent one of the various environmental types throughout the country. Different environmental-type regions have their own characteristics and their own special environmental problems. The lack of water resources in arid/semi-arid regions is a social manifestation of aridification, and aridification is the natural root of the lack of water resources. Although "water resources and aridification " is the outstanding environmental featrure of arid/semi-arid regions, which would cause the problems of desertification and sandstorm, other environmental- type regions also have other environmental features and environmental problems. For example, in the karst mountainous regions of southwestern China they have encountered the environmental problem of rocky desertification caused by water erosion; in the economically developed regions of eastern China there has been the problem of water pollution due to release of industrial wastes. To coordinate the integrity and dependence of the nationwide living environment and the general procedure between economic development and environmental improvement, it is quite necessary to implement the strategy of "integrated construction of environment in eastern and western China" [14] so as to establish the nationwide integrated ecologically environmental safety system.

In addition, as stated previously, under the influence of specific climatic factors there would occur consecutive droughts in arid/semi-arid regions, and other kinds of environmental disasters would occur under specific hydrological and climatic conditions in other environment-type regions. All these go to show that in the construction of the nationwide integrated ecologically environmental safety system the measures of predicting, preventing and alleviating various kinds of sudden environmental calamities should come first.

Of course, northwestern China should be regarded as an indispensable part of the world. To coordinate the procedure of global economic and social development, especially to enhance the ability of combating droughts under urgent conditions, it is of great significance to establish "the nationwide integrated ecologically environmental safety system." This may be not only a subject the Third-World scientists are considerably concerned with, but also one of the important fields for international collaboration, especially for long-term collaboration among the Third-World scientists.

Acknowledgements: The author participated in the project "Strategic Research on Water Resource Disposition, Ecological Environmental Construction and Sustainable Development in Northwestern China" sponsored by the Chinese Academy of Engineering. The data concerned in this paper are cited from the above study. Thanks are due to Wan Guojiang, Han Jingtai, Han Jiamao, Guo Zhengtang and Hou Juzhi for their involvement in the preparation of this manuscript.

REFERENCES

1 Liu Jichun. The Chronicle of Human Disasters (1900-1999), No.1, The Visitation of Providence

2 Liu Tungsheng. Geological Environments in China and Global Change. 30th International Geological Congress, Beijing, 1996, 1:15-26

3 Liu T S and Guo Z T. Geological environment in China and global Change. Selected works of LIU Tungsheng, Science Press, Beijing, 1997, 192-202

4 Liu Tungsheng, Zheng Mianping, Guo Zhengtang. Initiation and evolution of the Asian monsoon system timely coupled with the Ice-sheet growth and the tectonic movements in Asia. Quaternary Sciences, 1998(8): 194-204

5 Guo Z T, Ruddiman W F, Hao Q Z, Wu H B, Qiao Y S, Zhu R X, Peng S Z, Wei J J, Yuan B Y, Liu T S. 2002: Onset of Asian desertification by 22 Myr ago inferred from loess deposits in China. Nature, 416, 159-163

6 Hovan S A and Rea O K. Post-Eocene record of eolian deposition at sites 752, 754 and 756, easterm Indian Ocean. Proc. ODP, Scientific Results, 1991, 121: 209-240

7 Prell W L and Kutzbach J E. Sensitivity of the Indian monsoon to forcing parameters and implications for its evolution. Nature, 1992, 360: 647-652

8 Guo Z T, Peng SZ, Hao Q Z, Biscaye P E, An Z S and Liu T S. Late Miocene-Pliocene development of Asian aridification as recorded in an eolian sequence in northern China. Global and Planetary Changes (in Press).

9 Shackleton N J, Pisias N G. Atmospheric carbon dioxide, orbital forcing, and climate. In Sundquist, E T and Broeker W S (Eds.), The Carbon cycle and atmospheric CO2: natural variations Archean to present. Geophysical Monograph, 1985, 32:412–417

10 Shackleton N J, Berger A, Peltier, W R. An alternative astronomical calibration of the lower Pleistocene

timescale based on ODP Site 677. Trans. R. Soc. Edinbugh, Earth science. 1990, 81:251–261

21 Shackleton N J, Hall M A and Pate D. Pliocene stable isotope stratigraphy of Site 846, edited by Pisias et al, Proc. Ocean Drill. Program Sci. Results, 1995, 138: 337–355

32 QING Xi-tai. Taoist Thought of Ecological Ethics and its Realistic Significance. Journal of Sichuan University (Social Science Foliation), 2001(1): 39-43

43 Wan G J. A Preliminary Discussion on Regional Environmental Research. In: Advances on Environ mental Quality Research, Guizhou People□s Publishing House, Guiyang, China, 1985 32-47

54 Wan G J, Chen Y C and Xu Y F. The Division of Environmental Protection in Beijing-Tianjin-Bohai Bay Area. In: The Environmental Evolution and the Way of Development and Protection in Beijing-Tianjin-Bohai Bay Area (Ed. by Wan G J), Science Press, Beijing, China, 1989, 157-164

65 Wan G J. The Environmental Construction Merging into an Organic Whole in the Cities and the Countryside. Chinese Science News, 1989, (44): 3, 16/June

76 Wan G.J. and Pu H.X.. The Studies of Environmental Strategic for Economic Development in Southwest China. Science Press, Beijing, China, 1991, 121pp

87 Wan G J. The Geochemical Principles for Environmental Quality. Environmental Science Press, Beijing, China, 1988, 216pp

98 WAN Guojiang. The Integration of East and West China. Green Strategy, (Ed. by Li Zhengdao and Zhou Guangzhao), Qingdao Press, 1997, 340-344

Atmospheric Science in China

Zeng Qingcun Zhao Sixiong

Institute of Atmospheric Physics, Chinese Academy of Sciences

Zeng Qingcun, *research professor of Institute of Atmospheric Physics (IAP), Chinese Academy of Sciences (CAS), member of CAS, foreign member of Russian Academy of Sciences and fellow of the Third World Academy of Sciences. Born in Guangdong Province, May 4, 1935. He now is the vice president of China Association for Science and Technology and director of the International Center for Climate and Environment Sciences (ICCES/IAP, CAS). Main research fields: geophysical fluid dynamics, general circulation of atmosphere and oceans, theory of numerical weather prediction, climate dynamics, climate system model, climate numerical modeling and prediction, environmental ecosystem dynamics, natural cybernetics and atmospheric remote sensing. More than one hundred publications and articles, they are: Principles of infrared remote sounding of the atmosphere,The Physical-mathematical bases of numerical weather prediction, etc..*

Abstract: In this paper, the history and present state of meteorology and its extended field—atmospheric science — in China is briefly described. Then, several of the most important fields among them are introduced, for example, the research of climate and global climate change, disastrous weather prediction, atmospheric sounding, remote sensing, atmospheric physics, detection and prediction of air pollution, artificial modification of weather, natural cybernetics, and other application studies.

1. SURVEY

There is a five thousand-year-old history in China and a lot of meteorological events and climate facts have been recorded in Chinese historical books. Some famous historical politicians, strategists or scientists who knew meteorology well, mentioned some meteorology regulations and made climate and weather predictions, for example, the application of the 24 solar terms in agriculture and the dense fog prediction in the Yangtze River in winter. At the beginning of the 20th century, Chinese scientists had already begun to use instruments to observe the meteorological elements, analyze the weather maps and compile a picture of the regional climate. At the present time, the world's most dense surface and upper sounding network is in China. China has developed her own meteorological satellites, various meteorological radars, meteorological data collection and communication systems, and a warning and prediction system. Weather prediction, climate prediction and ocean-meteorological prediction have served as the routine operational program in China, East Asia and the West Pacific Ocean.

China has the Chinese Meteorological Administration (CMA) and various provincial meteorological bureaus. In the Chinese Academy of Sciences, CMA and universities, there are outstanding research organizations and universities, where education in meteorology and atmospheric science can be provided. In addition, the China Meteorological Society has many members, and in China, about 2,000,000 people are employed in the field of meteorology.

A wide cooperation exists between China and other countries. China can provide training in meteorology for foreign graduate students in masters and Ph.D. programs and for visiting scientists on short-term visits.

2. STUDY OF CLIMATE AND GLOBAL CLIMATE CHANGE

There is a very long history of climate research in China, especially regarding its formation, variability, and drought and flood prediction relating closely with economic and social development. The prediction of ENSO and its influence on the variability of the East Asian monsoon is conducted in China. The Chinese Academy of Sciences has developed the extraseasonal dynamic climate prediction system (CP- II), which has successfully predicted the summer drought and flood tendency (patterns) in China since 1998, and was provided to the operational department as a real-time prediction system.

China has paid more attention to historical climate change and global climate warming and their relation with environment change (namely, global change). Ancient climate maps have been compiled and duplicated, and the tendency and characteristics of the various regions of East Asia for more than six thousand years have been discovered. The research shows that sudden changes occur in the climate process. For example, two sudden climatic changes can be found in China during the 20th century, and the phenomena were closely associated with global climate.

China actively participates in the international climate research projects, including WCRP, IPCC and IGBP and has made important contributions in this respect. The International Centre for Climate and Environment Sciences (ICCES), CAS and T-RRC of START especially has been engaging in this kind of research. They can provide opportunities for Ph.D.study and for short-term training of young scientists from Third World countries.

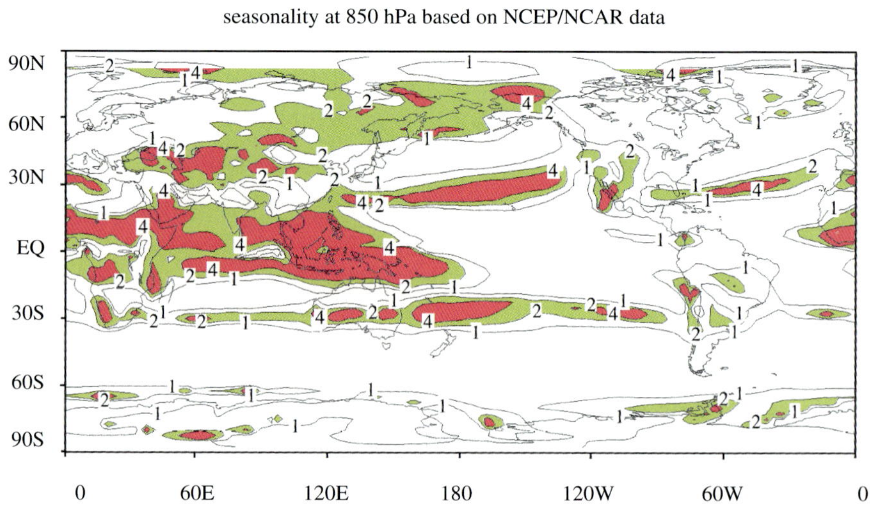

Fig.1 Global monsoon system

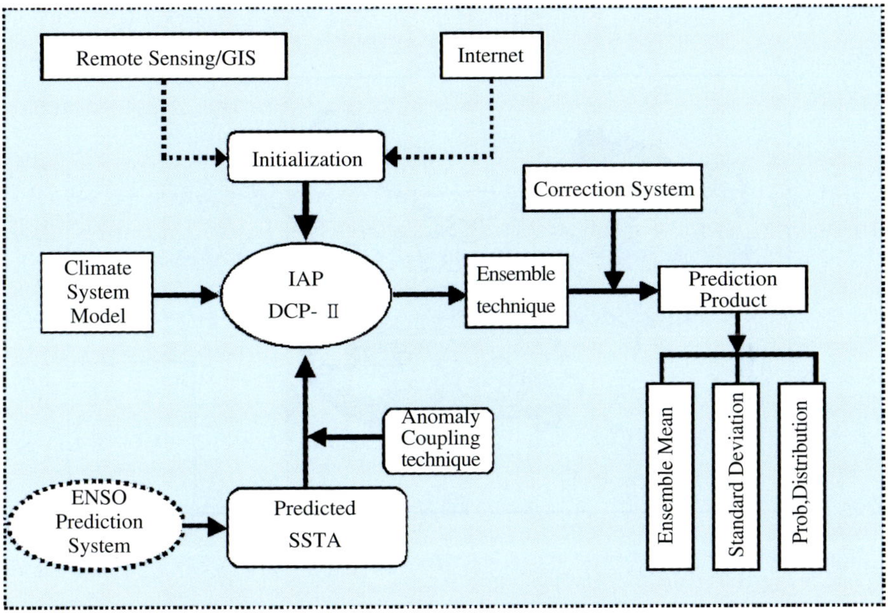

Fig.2 IAP DCP flow chart

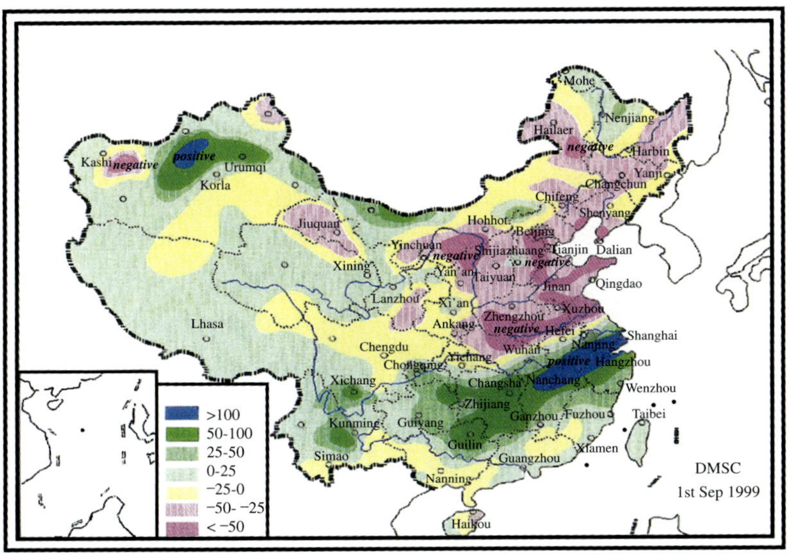

(a) Observed percentage summer rainfall anomalies over China for year 1999

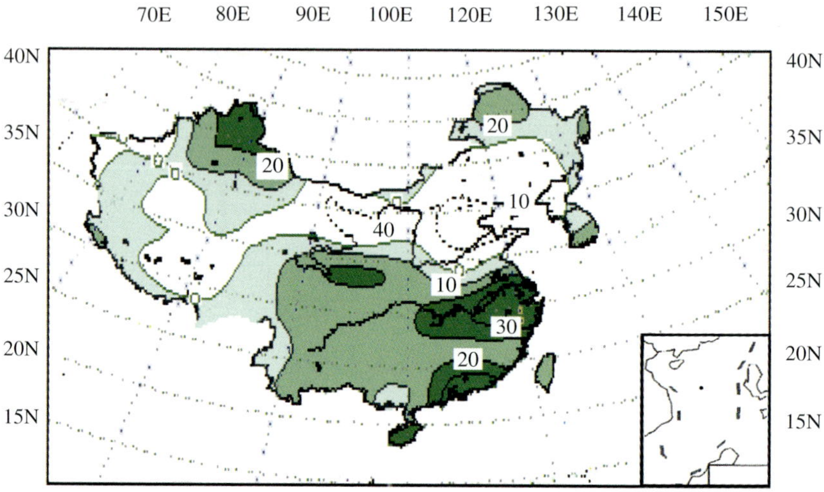

(b) predicted percentage summer rainfall anomalies over China

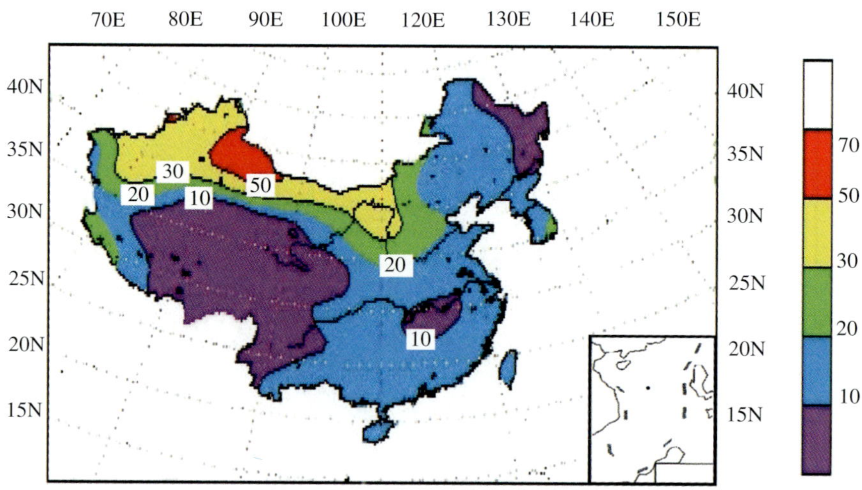

(c) The standard deviation

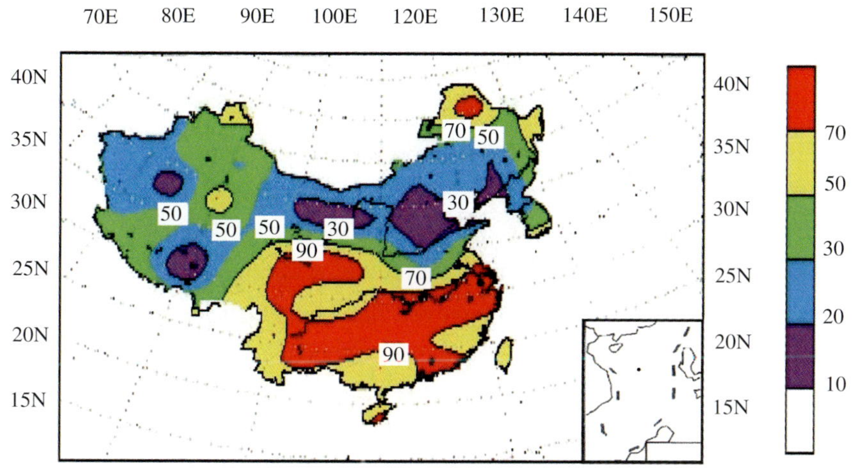

(d) probability distribution of positive rainfall anomlies of the ensemble rainfall

Fig.3 Precipitation prediction and verification in June-August, 1999

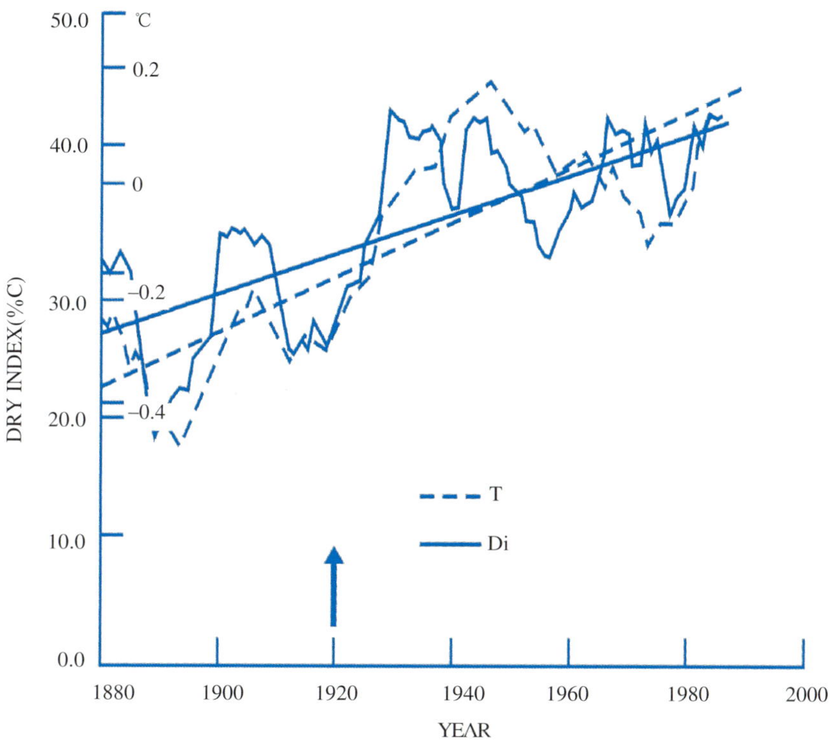

Fig.4 Sudden changes of climate during the 20th century

3. DISASTROUS WEATHER PREDICTION

China attaches great importance to prediction research of weather patterns and weather phenomena that produce meteorological disasters, especially heavy rainfall, typhoons, dust storms, cold waves, heat waves, hailstorms, thunderstorms and so on, excepting routine short range and middle range weather prediction. China has applied and developed some relevant observational systems, warning systems and numerical prediction models and methods. However, the mechanisms are still not clarified because of the sudden occurrence and quick development of some disastrous weather systems. So, sometimes unsuccessful predictions are still made. Therefore, further study on disastrous weather mechanisms and prediction theories should be conducted extensively. In addition, urban applied meteorology and radar meteorology have been studied and services provided routinely during recent years.

Fig.5 FY2 Satellite image for a heavy rain case

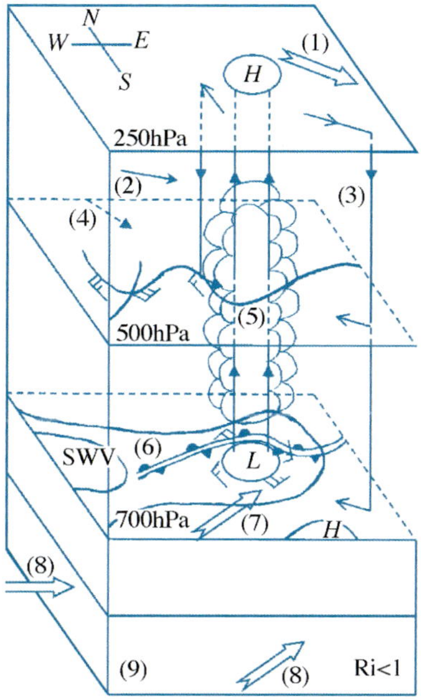

Fig.6 Physical model of sudden heavy rainfall in July 1998

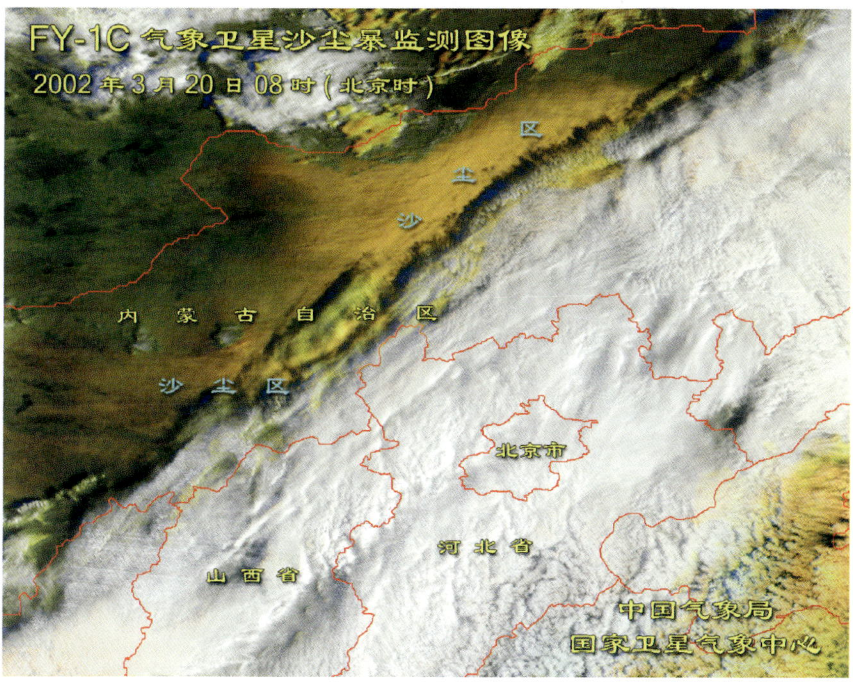

Fig.7 FY2 Satellite image for a dust storm case

4. ATMOSPHERIC SOUNDING, REMOTE SENSING AND ATMOSPHERIC PHYSICS RESEARCH

The most conventional meteorological instruments (upper air sounding and surface usage) used in China are of our own manufacture, for example, special observational instruments including various radars, lidars, sodars, Doppler radars, and so on, and special equipment, such as that for research of radiation, the planetary boundary layer, cloud macrophysics processes, aerosol characteristics, atmospheric trace gas content, and so on. Chinese scientists have designed and launched China's own meteorological satellites and also various related remote sensing instruments (including infrared and microwave sensors).

Institute of Atmospheric Physics (IAP) scientists have concentrated on the study of cloud and rain characteristics, artificial modification of precipitation, and the principles and technology of suppressing hailstorms. The study on both active remote sensing and passive has been done as well. The technology has been used extensively in practice.

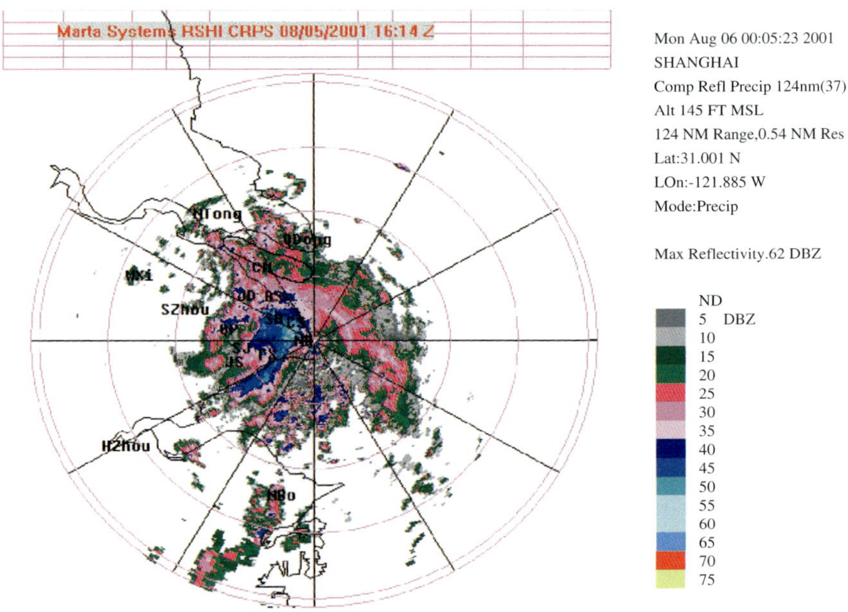

Fig.8 Doppler radar

Fig.9 Laser radar

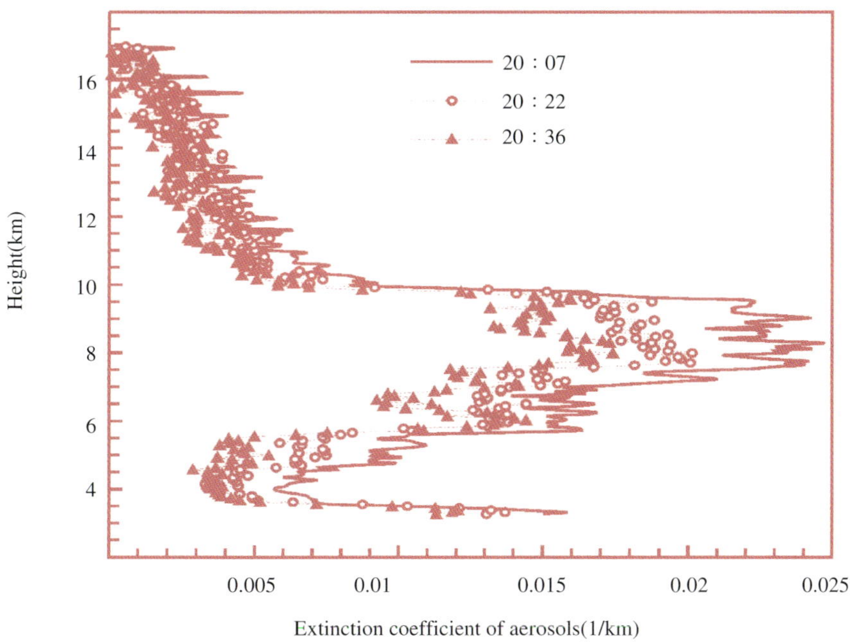

Fig.10 Distribution of dust and aerosals

Fig.11 The 325m meteorological tower for planetary boundary layer research

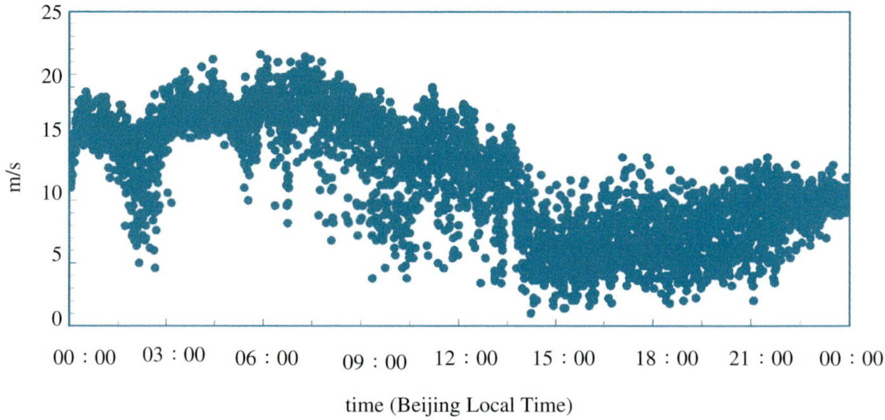

Fig.12 The gust of wind provided by the 325m meteorological tower

Fig.13 VHF radar at the Institute of Atmospheric Physics (IAP)

5. DETECTION AND PREDICTION OF AIR POLLUTION

Since the 1970s, the detection stations and network for air pollution monitoring have been set up in succession for the observing and controlling of various pollutants in the atmosphere, especially in urban areas. There are stations of background pollution observation and their network. Even five years ago, an air quality bulletin was being distributed in some of China's large cities, and short range prediction of air pollution was also being provided.

6. ARTIFICIAL MODIFICATION OF WEATHER AND NATURAL CYBERNETICS

The artificial change of the local environment, and then the improvement and deterioration of the climate in a local area have been studied earlier. And many practices and a lot of experiences in this aspect have been obtained in China. However, theoretical study has been conducted only in recent years.

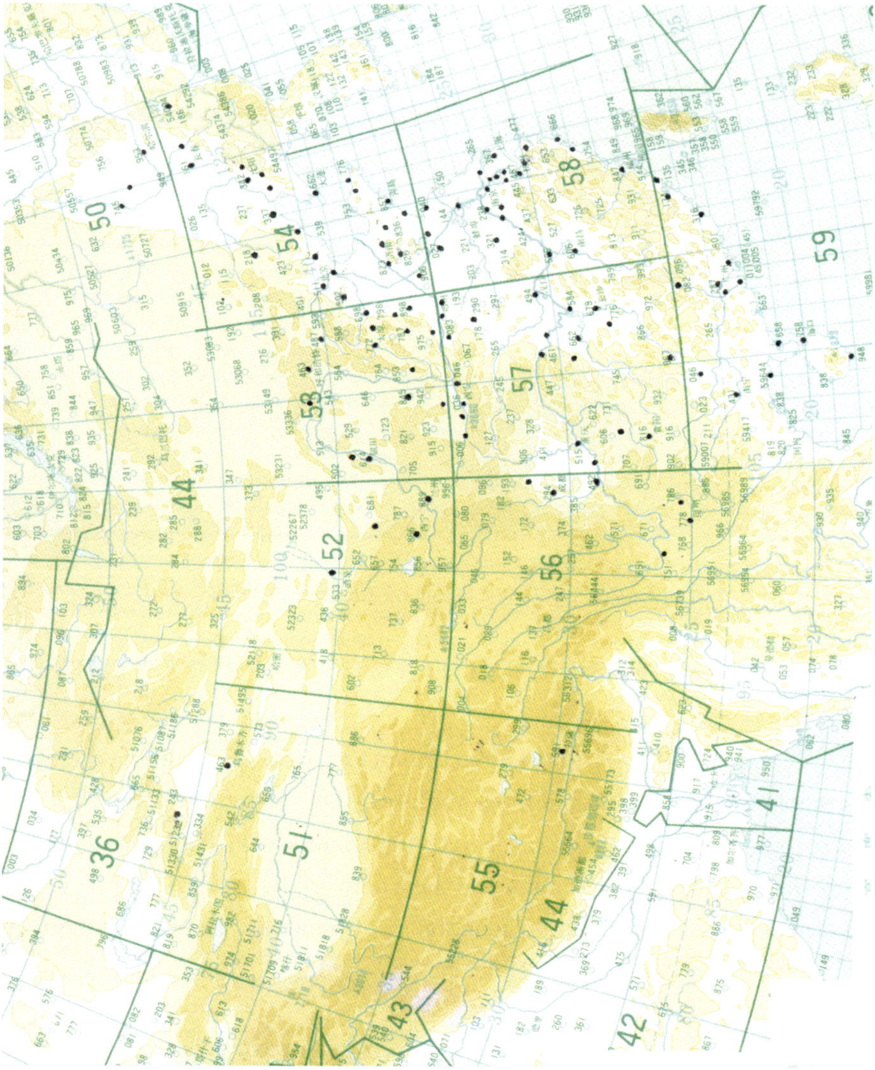

Fig. 14 Observation Stations for air pollution

Fig. 15 Observational equipment for CH_4

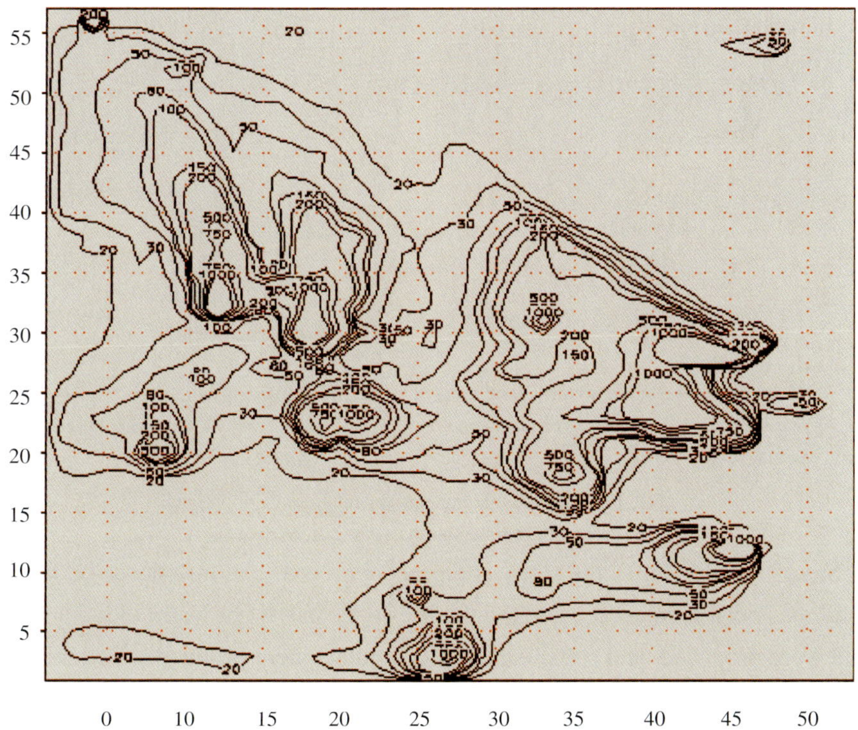

Fig. 16 Prediction for air pollution in urban areas

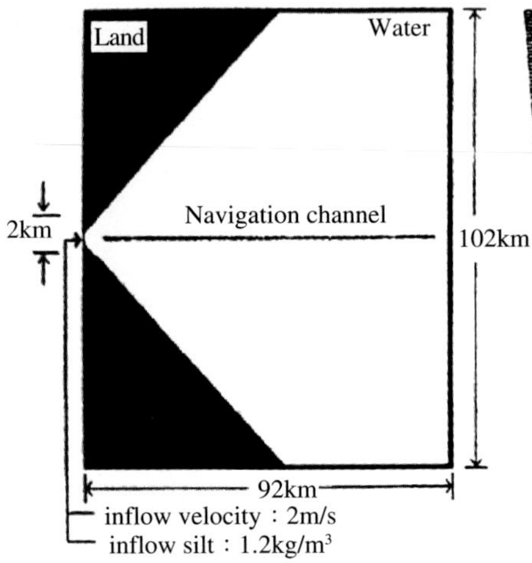

(a) Model domain, open boundary conditions and positions of Navigation channel.

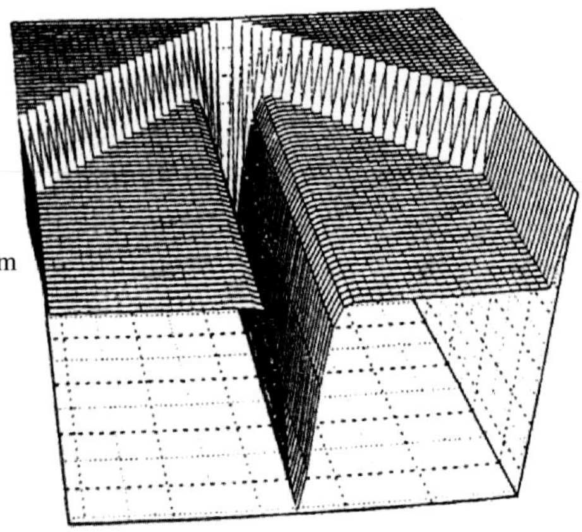

(b) A 3D shapshot of ideal bottom topography from which model run 40-year simulation to generate initial condition (shown in (c)) for this study.

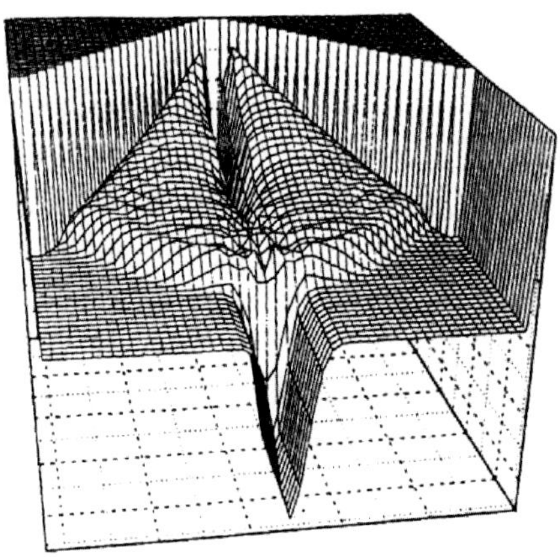

(c) A 3D shapshot of resulting bottom topography of 40-year simulation. This is used as initial condition in optimal control experiments.

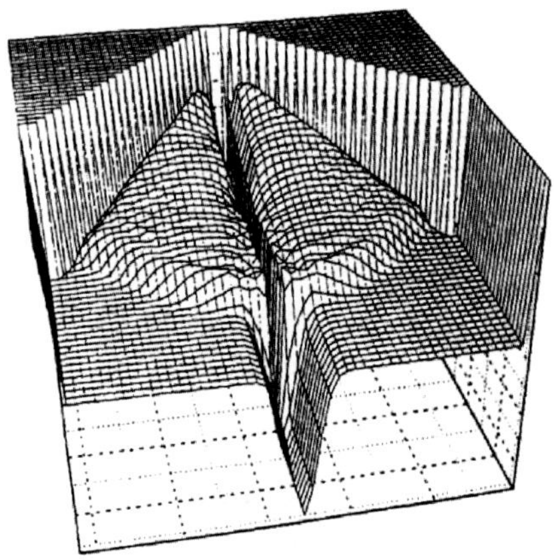

(d) A 3D shapshot of new bottom topography after optimal dredging is applied.

Fig. 17 Some computing experimental results of optimal control of sedimentation in navigation channels

Since the late 1950s, the study and practice of artificial influences on fog and rainfall has been focused on, and the theories for catalytic of warm cloud and cold cloud were proposed, and some catalytic promoters were produced. The scientific verification is still not enough, even though large amounts of research work have been done.

In the 1990s, Chinese scientists put forward the general principle for combining artificial modification of weather and natural environment, and then they developed natural cybernetics. Therefore, they successfully put together the research mechanism and the operational practice.

7. THE OTHER APPLICATIONS RESEARCH

Research has shown the existence of the relationship between human activities and meteorological conditions. Therefore various relevant applied meteorological problems are proposed according to different regions, groups of people, and different requirements, for example, urban meteorology, forest meteorology, medical meteorology, and so on. These above-mentioned studies and operational services have been developed extensively.

Acknowledgement: The CAS and CMA provided the necessary data. Thanks for their kind assistance. In addition, I would also like to be grateful to Xie Feng, Lin Zhaohui, Sun Jianhua, Bei Naifang, Qi Linlin, Lu Daren, Zhao Yijun, Ren Zhenhai, Wang Mingxing, Hu Fei , Zhu jiang and Pei Shen for successively providing the figures in this paper.

Advances in Ocean Science

Su Jilan   Yuan Yaochu

Laboratory of Ocean Dynamic Processes and Satellite Oceanography
Second Institute of Oceanography, State Oceanic Administration

Su Jilan, *is a Physical Oceanographer. Born in Youxian, Hunan Province. Graduated from National Taiwan University in 1957. Received MS from Virginia Polytechnic Institute, USA in 1961, and Ph.D. from University of California at Berkeley, USA in 1967. Elected fellow of Third World Academy of Sciences in 1994 and foreign member of the Russian Academy of Sciences in 1999. Member, Executive Committee of the Scientific Committee on Oceanic Research, ICSU (1988-1992). Chairman, Intergovernmental Oceanographic Commission, UNESCO (1999-2003). Research professor and director (1994-1999), Second Institute of Oceanography, State Oceanic Administration. Director, Laboratory of Ocean Dynamic Processes and Satellite Oceanography, State Oceanic Administration.*

Since 1980, Su's research interests have been focused on circulation dynamics of coastal oceans and estuaries. In particular, he paid special attention to the influence of the Kuroshio on the coastal ocean circulation of the Yellow Sea, East China Sea, and South China Sea. He was the Principal Scientist (Chinese side) of the China-Japan Joint Research Program on the Kuroshio (1986-1992). He has collaborated with oceanographers from the Taiwan Province in carrying out several extensive observation of the South China Sea. Since the early 1990s, he has been promoting, collaborating with

Chinese fishery oceanographer, and the study of ocean ecosystem dynamics in China. He was a Co-Principal Scientist of the major project of the Natural Science Foundation of China on the ecosystem dynamics of the Bohai Sea, which paid attention, among others, the recruitment of the Penaeus chinensis. He was also the Principal Scientist of a key project of the National Natural Science Foundation of China on the role of the secondary Changjiang River plume in the sediment transport in the Hangzhou Bay. He was elected Member of the Chinese Academy of Sciences in 1991.

Abstract: Selected advances in China in some areas of ocean science are reported, covering the period since 1980 but with special emphasis on the last 10 years. In the physical oceanography, we concentrate on the research results on the western boundary currents of the North Pacific, circulation of the China Seas, and the Antarctic Ocean. Findings of the ocean ecosystem dynamics study in the Bohai Sea, Yellow Sea and East China Sea, and Taiwan Strait are discussed. In the paleo-oceanography we focus on the results from the recent drillings of Leg-184 of the Ocean Drilling Program at the South China Sea. Finally, achievements from the recently completed first phase of the ocean technology component of the National High Technology Research and Development Program are presented briefly.

China began to build up research capability in ocean science in the 1950s. A national multi-disciplinary study of the coastal waters of China, called the Comprehensive Survey of the Coastal Seas of China, was implemented in 1958-1960. Even during the difficult years of 1965-1977 large-scale studies in marine geology and marine geophysics were conducted. Since 1978, China has resumed its attention in all areas of ocean science, including actively engaging in international cooperation with many countries.

Ocean science spans a wide range of areas. There have been already several reviews on the advances of ocean science in China over the last 50 years (e.g., see ref. 1-3). In this short review, we shall briefly highlight findings and achievements in a few areas, namely, physical oceanography, ocean ecosystem dynamics, paleo-oceanography, and the ocean technology. The period since 1980 is covered, with special emphasis on the last 10 years.

1. PHYSICAL OCEANOGRAPHY

The nature of the ocean environment dictates that the physical oceanography forms the foundation of all the fields of ocean science. Consequently, like many coastal states, China has paid most of its attention to, in addition to marine biology, physical oceanography. In this section, we shall discuss the results of selected research projects in the western boundary currents in the North Pacific, circulation of the China Seas, and the Antarctic Ocean.

1.1 Mid and Low Latitude Western Boundary Currents in the North Pacific

As the North Equatorial Current flows westwards in the North Pacific Ocean, it bifurcates around the 13° N near the Philippines. The northward branch forms the beginning of the Kuroshio and the southward branch the Mindanao Current. Here we shall review the findings by the Chinese scientists on the Kuroshio, the western boundary current east of the Ryukyu Islands (called the Ryukyu Current in the following), and the Mindanao Undercurrent beneath the Mindanao Current.

1.1.1 Kuroshio and the Ryukyu Current

Both the Kuroshio and the Ryukyu Current are western boundary currents of the North Pacific Subtropical Gyre. The Kuroshio flows northwards east of the Luzon Strait and then enters the East China Sea along the east shore of the Taiwan Island. It then follows the shelf break northeastwards until it deviates to the east, exiting through the Tokara Strait. The Kuroshio exerts strong influence on the marine environment of the South China Sea, East China Sea, and Yellow Sea. The Ryukyu Current, which has a complicated interaction with the Kuroshio, flows northeastwards along the east of the Ryukyu Islands.

During the 1970s, Kuroshio studies by the Chinese scientists were based on data from the Cooperative Study of the Kuroshio (CSK) sponsored by the Intergovernmental Oceanographic Commission. From the mid 1980s, data collected by the Chinese scientists themselves were also used for the study of the Kuroshio and the Ryukyu Current, including cooperative projects such as the China-Japan Joint Research Program on the Kuroshio (JRK) and the China-Japan Cooperative Study of the Subtropical Circulation System.

Mainly because of the topographic effects of the passageways along its path, the total volume transport of the Kuroshio changes greatly between neighboring seas. Its transport was found to be around 45 Sv (1 Sv = $10^6 m^3 s^{-1}$) east of Taiwan, of which about 25 Sv was estimated to enter the East China Sea due to the blocking of the shallow ridge between Taiwan and Yonakuni Island. Part of the Kuroshio Surface Water was also found to originate from the seas southeast of the Ryukyu Islands. The Kuroshio Intermediate Water inside the East China Sea was found to have augmented from the Philippine Sea through the deep trough over the submarine ridge south of Okinawa. All is extra volume input resulting in an annual average transport of about 27 Sv for the Kuroshio in the East China Sea.

East of Taiwan, other than the branch of the Kuroshio that entered the East China Sea, sometimes the Kuroshio was found to have an eastern branch flowing northeastwards to the area east of the Ryukyu Islands. However, this eastern branch was not found during two different observations in 1997, a strong El Niño year. This suggests the existence of at least two current patterns for the Kuroshio east of Taiwan.

The Ryukyu Current was in fact discovered in the 1960's. However, there was always doubt whether the current has a significant steady volume transport. The Chinese scientists found that the Ryukyu Current is indeed a steady northward western boundary current with two cores. One of the cores is centered at the subsurface with a rather large volume transport. This subsurface core may provide the principal source for the subsurface transport of the Kuroshio south of Japan. There appears to be three sources of water, not necessarily all concurrent, for the Ryukyu Current, namely, an anticyclonic recirculation gyre southeast of Okinawa, a westward current from the east at mid latitudes, and the eastern branch of the Kuroshio east of Taiwan.

An abyssal western boundary current was also found to flow steadily southwestwards all the year round below 3000 m east of the Ryukyu Islands. As the deep basin of the Philippine Sea is connected to the North Pacific through only several deep trenches, this current was overlooked in the classical theory on global abyssal currents.

1.1.2 Mindanao Undercurrent

In the mid 1980's China actively participated in the ocean-atmosphere interaction experiment over the Tropical Pacific. Several observations were conducted between 7°N -8°N in the seas east of the Philippines during this experiment.

Analysis of this set of data showed the existence of a strong northward current, named the "Mindanao Undercurrent " by Chinese scientists, beneath the southerly Mindanao Current. This undercurrent is about 150 km wide and has a total transport around 10 Sv. Normally it has two cores. The nearshore core, centered at around the 900 m depth, is about 1000 m thick, while the offshore core, centered between 300-400 m, is only 500 m thick.

1.2 Circulation of the China Seas

The China Seas, located west of the Kuroshio, include the Bohai Sea, Yellow Sea, East China Sea, and the South China Sea. They are under the strong influence of the Kuroshio and the northeast monsoon. The hydrography and circulation of these seas have distinct seasonal characteristics. The Comprehensive Survey of the Coastal Seas of China in 1958-1960 has provided a general view of these characteristics.

1.2.1 Circulation in the South China Sea

The South China Sea, with an area of about $3.5 \times 10^6\, km^2$ and complex bathymetry, is the largest marginal sea in the western Tropical Pacific. Through several straits, it is connected to the Philippine Sea, East China Sea, Sulu Sea, Andaman Sea, and Java Sea. It has a basin-scale circulation imbedded with active meso-scale eddies. Since the 1980's China has conducted many scientific studies in this basin.

The comprehensive study in the late 1950's discovered a northerly current in the shelf break area south of China between Hainan Island and the offshore area of Shantou. It was found to be persistent even under the northeast monsoon. Chinese scientists named it the "South China Sea Warm Current ". Its existence has been reconfirmed in subsequent observations since the 1980's and several possible generation mechanisms for this current have been offered.

The Kuroshio, flowing by east of the Luzon Strait, has a significant average influx to the South China Sea. However, the manner of its transport into the South China Sea is largely unknown. All the existing synoptic survey results, mostly by Chinese scientists, do not support the intrusion of a branch, or a loop, from the Kuroshio, although an anticyclonic eddy with core characteristics of the Kuroshio frontal water was identified in one of the surveys. Offshore of the South China Sea Warm Current a persistent year-round southwesterly current has been discovered.

Its volume transport is between 4-7 Sv and several generation mechanisms for this current have been suggested.

Meso-scale eddies have been frequently observed in synoptic surveys. Among these eddies, the most noteworthy ones are the strong anticyclonic eddies often found, respectively, east of the Dongsha Islands and offshore of southern Vietnam.

The South China Sea is at the northwest end of the warm pool of the Equatorial Pacific Ocean. The SST of the South China Sea was found to have a large positive anomaly five months after an El Niño event happened in the East Pacific. During an El Niño event the summer circulation in the South China Sea is characterized by the strengthening of both the cyclonic gyre in the north and the anticyclonic gyre in the south, while the winter circulation shows a general weakening of the cyclonic gyre over the entire basin.

1.2.2 Shelf Circulation in the East China Sea

The East China Sea has a wide epicontinental shelf. Circulation over the shelf is influenced by the Kuroshio, the monsoon and the discharge from the Changjiang River.

The comprehensive study in the late 1950s also discovered a year-round northerly coastal current along the coast of the Fujian and Zhejiang provinces, called the "Taiwan Warm Current." Even during the strong northeast monsoon, it persists under the surface layer. The origin of this current has always been one of the focal interests of the Chinese oceanographers. Some suggested it to be a branch from the Kuroshio northeast of Taiwan, while others regarded it as a current originated from the Taiwan Strait. Analysis of the observations in the 1980s demonstrated that in either summer or winter both origins contribute to the Taiwan Warm Current. It in fact has two branches, one nearshore, and the other offshore. Their flow patterns have distinct seasonal characteristics. Their generation is a result of complex interaction between the sea surface elevation fields associated with, respectively, the subarctical and the subtropical gyres, as well as the shelf intrusion of the Kuroshio northeast of Taiwan.

The year-round northerly current in the Taiwan Strait connects the South China Sea Warm Current and the nearshore branch of the Taiwan Warm Current. It is likely driven principally by the difference of the sea surface elevations at the north and south ends of Taiwan, associated with the subtropical gyre. The annual average transport has been estimated between 1-2 Sv.

The Changjiang River is the world's third largest river in discharge. As early as in the 1960s Chinese scientists discovered that in summer the Changjiang River plume turns northeast towards the Cheju Island, instead of turning southward along the coast as predicted in the classical theory. In the 1980s, Chinese scientists noticed a good correlation between this summer behavior of the Changjiang River plume and the amount of the river discharge. They further suggested that both the Taiwan Warm Current and the southerly summer monsoon also helped this "anomalous" plume behavior. In the 1990s, they found that the bottom pool of the East China Sea Dense Water, formed in the winter, also exerts important influence on the spreading of the plume water.

1.2.3 Circulation in the Bohai and Yellow Seas

The Bohai and Yellow Seas form a semi-enclosed sea. The northward Yellow Sea Warm Current, characterized by its warm tongue, can be identified in winter all the way from the southern end of the Yellow Sea to the Bohai Sea. In summer, a basin-wide cold-water pool, called the Yellow Sea Cold Water Pool, occupies the bottom layer.

It was shown in the early 1980s that the Yellow Sea Warm Current was basically a wind-driven subsurface current flowing northward along the deep trough of the basin. In summer, it is rather just a weak part of the Taiwan Warm Current steered by topography. Its influence is limited to the southern end of the Yellow Sea. The Yellow Sea Cold Water Pool is bounded by tidal fronts near the coast. It was demonstrated that in its interior there is upwelling in the surface layer and downwelling below the thermocline. Its circulation is cyclonic in the upper and middle layers, but anticyclonic near the bottom layer.

1.3 Antarctic Ocean

Antarctic research by Chinese scientists began in the mid 1980s. Soon afterwards China set up two Antarctic stations consecutively, first the Great Wall Station on the Antarctic Peninsula and then the Sun Yat-sen Station by the Prydz Bay. Recent researches found that both the deep and bottom waters in the Bransfield Strait have originated from the shelf area of the Weddell Sea. For the Prydz Bay five water masses have been identified, namely, the Antarctic Summer Surface Water, Prydz Bay Shelf Water, Amery Ice Shelf Water, Modified Circumpolar Deep Water, and Prydz Bay Bottom Water. It was found that the potential temperature of the bottom water at the Amery basin has the same value as that of the Antarctic Bottom Water just off the Prydz Bay. Further analysis of the trace element deuterium suggests that the Antarctic Bottom Water just off the Prydz Bay is likely formed from mixing of the Circumpolar Deep Water and the Prydz Bay Shelf Water.

2. OCEAN ECOSYSTEM DYNAMICS

Study of ocean ecosystem dynamics requires close collaboration between several disciplines such as marine chemistry, marine biology, fishery oceanography, and physical oceanography. Since the mid 1980s there have been three major studies related to ocean ecosystem dynamics in China. Their study areas are, respectively, the upwelling system at the Taiwan Shoal, the early life habitat of *Penaeus chinensis* in the Bohai Sea, and the anchovy spawning grounds in the Yellow Sea.

2.1 Upwelling System at the Taiwan Shoal

Taiwan Shoal, situated offshore of southern Fujian, is an important fishery ground. It is a typical subtropical coastal upwelling ecosystem driven by the summer monsoon. Chinese scientists first developed a method for *in situ* measurement of iron at different valence states. With this technique, diurnal variation of iron at double valence states in the upwelling ecosystem could be understood. It was found that the biogeochemistry of carbon and nutrients in this ecosystem are closely related to the changes, over various time scales, in the marine environment at the Taiwan Shoal. With the construction of a model for the biochemistry of the essential life elements in this ecosystem, relationships between the environmental factors and the fishery catch were established.

2.2 Early Life Habitats of *Penaeus Chinensis* in the Bohai Sea

Penaeus chinensis, an annual-cycle prawn, is the only penaeid prawn species found in temperate high latitudes. It is also the only species that has a migrating life cycle, spawning mainly in the Bohai Sea in late spring and over-wintering in the Yellow Sea. *Penaeus chinensisi*, being of high economic value, was an important fishery stock for the fishery industry in Northern China. However, since the 1990s this stock has experienced a very sharp decline in catch.

Since the 1980s, the environment of the early life habitats of *Penaeus chinensis* has experienced significant negative changes. Although large fluctuation of the number of the spawning stock determines directly the recruitment stock, changes in the habitat environment were found to have pronounced effects on the recruitment as well. In addition to the increasing amount of land-based pollutants discharged into the coastal waters, other anthropogeneous activities, such as damming of the rivers, drastic withdrawal of river water and rapid extension of the river mouth delta because of human interference, have had equally detrimental impacts on the habitats. These activities have seriously altered the physical and chemical processes in the habitats that are favorable to the successful recruitment of *Penaeus chinensis*. Sharp decline in the recruitment has resulted in the dwindling of the stock of the species.

2.3 Anchovy Spawning Grounds in the Yellow Sea

Study of the critical processes of the anchovy spawning grounds in the Yellow Sea is part of an ongoing project. During the warm season, an important pathway for the transport of coastal water into the interior of the Yellow Sea was found to be through the thermocline. This is so because of the double-celled structure of the vertical circulation of the Yellow Sea Cold Water Pool (see discussion in the Physical Oceanography section).

Calanus sinicus is the major prey for adult anchovy. It was found that this species has an interesting over-summering strategy. When the water temperature is above 25℃, the activity of several digestive enzymes will drop to a very low level. Consequently, in summer *Calanus sinicus* is found only in the center of the Yellow Sea Cold Water Pool or other area with stable low temperature.

3. PALEO-OCEANOGRAPHY

Studies of paleo-oceanography in China started in the early 1980s and have focused mostly in the South China Sea since the 1990s. In the spring of 1999, the Ocean Drilling Program completed Leg-184, exploring the Asian Monsoon through Drilling in the South China Sea. Leg-184 drilled 17 cores at six stations in waters of 2000-3000 m depths. Altogether, a total of more than 5000 m of high quality sediment cores have been recovered,with the longest one being over 850 m. One of the cores has over 30 million years of sedimentary record since the opening of the South China Sea. Its isotope record, showing a continuous stratum for the past 26 million years in the late Cenozoic, is the only ocean record from a single sediment core. Another core of about 500 m length shows a Quaternary deposit accumulated over the last million years. This is the highest sedimentation rate known today for a deep-sea deposit. In the following, a few recent findings are discussed, although active research is still going on with these cores.

Fluctuation of carbon with periods coinciding with glacial-interglacial cycles was found, with the carbon maximum preceding the onset of glacial period. This fluctuation pattern was then also found in cores at other sites globally. This suggests that long-term changes of tropical processes, such as monsoon and El Niño, due to fluctuation of solar radiation received at low latitudes as a consequence of the Earth's orbital motion may cause carbon fluctuation in the global ocean. This carbon fluctuation has an important effect on the global climate. This poses a serious challenge to the prevailing thinking that only the high latitude region of the North Atlantic Ocean determines the glacial-interglacial climate change. In other words, it is not only the physical processes of the polar ice that determines the global climate. As two of the unstable regions of the climate system, both the warm pool at the western Equatorial Pacific and the high latitude region of the North Atlantic Ocean play equally important roles in the climate changes.

From the continuous late Cenozoic record of cores at many different sites, it is evident that both the East Asia Monsoon and South Asia Monsoon were strengthened concurrently at two periods, namely, 7.6-6.5 Ma and 3.2-2.2 Ma before present. However, the core records from the South China Sea showed that the winter monsoon was obviously stronger for the paleo East Asia Monsoon. Thus, both the fluctuation periods due to Earth's orbital motion and the paleo-monsoon proxy records for the East Asia Monsoon are different from that for the South Asia Monsoon. The records of the paleo-monsoon from the deep-water cores of the South China Sea compare well with those from the loess cores from Central China. Both set of cores can be used to calibrate the climate change history of China.

The cores from Leg-184 cover the entire history of the opening up of the South China Sea. It showed that a deep basin already existed at the beginning phase of the rift. Analysis of microfossil also demonstrated the deepening process of the basin. It has yielded the time period for the large-scale tectonic movement, known to have occurred, at between the early and late Tertiary.

Another ongoing project, Evolution of the Marginal Seas next to China, has obtained results on the final expansion phase of the East Basin of the South China Sea and the characteristics of the rift boundary of the northern part of the South China Sea. In addition, some preliminary information on the resource distributions of oil, gas, and methane hydrate in the South China Sea and East China Sea has been delineated.

4. ADVANCED OCEAN TECHNOLOGY

In the late 1980s, China initiated the National High Technology Research and Development Program. Ocean technology was not included at that time. In the late 1990s, ocean technology was included as the eighth area under the program. Under the ocean, technology area three offices were established to oversee three aspects of the ocean technology, namely, marine environment monitoring, marine biological technology, and seabed exploration and ocean resources exploitation. In the following, we shall highlight some of the achievements of this area in the first five years of its implementation.

4.1 Marine Environment Monitoring

A prototype acoustic correlation current profiler was built. In this prototype, key components such as emission signal design, return signal processor, transducer array, and others were developed. The profiler works in water up to 110 m depth. It can sense five layers across a depth of 50 m with an accuracy of $\pm$ 0.05 ms^{-1}. Another prototype of shore-based high-frequency radar system to measure currents and waves was also built. For this prototype key components developed were phase controlled antenna array, broadband transmission/acquisition switch, and return radar signal sampling and processing. Effective range for measuring ocean currents and wind directions is 150 km, and 120 km for the waves and wind speeds.

4.2 Marine Aquaculture

Optimal conditions to induce triploids and to culture the larvae have been found for oysters, nacres, scallops, abalones and prawns. In addition, sex control methods in cultural animals have also been developed. These techniques have been successfully tested in pilot projects conducted at several aquaculture grounds.

A series of crucial processes for closed seawater-recirculation plant for aquaculture have been developed and applied to a demonstration project. These processes include high-efficiency filtering system, high-efficiency nitrogen removal, high-percentage recirculation, and automatic water quality monitoring and control.

4.3 Topographic and Geomorphologic Survey of the Seabed

A high-precision topographic and geomorphologic seabed survey system has been built around multi-beam data. This system has met the international standard for demarcation. It can monitor and process the multi-beam data in real-time and optimally correct errors caused by rolling and pitching of the sensor. Tidal effects can also be removed. It has a positioning capability that can achieve a real-time *in situ* accuracy of less than 10 m error within a range of 800 km. Furthermore, it has an interactive digital map generation subsystem for the topography and geomorphology of the sea floor. In addition, it is capable of processing images obtained by either optical or acoustical means, including converting the relative positioning data of the towed body to geodetic coordinates.

4.4 Major Offshore Gas Field Drilling

Drilling through a deep stratum (beyond 2000 m below the seabed) under high-temperature and super-hydrostatic pressure has always been a world-class challenge. A suite of techniques have been developed to partially meet this challenge. A high-resolution seismic method was developed, achieving a 7-10 m resolution of stratigraphic record at 2500-4000 m below the seabed. A series of oil-based drilling fluids and a series of well-completion fluid capable to control corrosion were produced for high-temperature and high-pressure conditions. Methods for prediction and monitoring of stratigraphic pressure were developed with excellent results. A system was also developed for testing high-temperature high-pressure gases with excellent safety control. All these techniques have been applied to exploration drillings through a gas field under high-temperature and super-hydrostatic pressure condition at the Ying-Ge Sea.

REFERENCES

1 Chinese Academy of Sciences. 2003 Report on the High Technology Development in China. Science Press, 2003, Beijing. (In Chinese)

2 Liu Guangding (Chief Editor). Chapter 5: Marine Sciences. In: Chinese Academic Canon in the 20th Century. Fujian Education Press, 2002, Fuzhou, China. (In Chinese)

3 Su Jilan, Jin Xianglong, Yuan Yaochu, Hu Minghui and Xiang Jianhai. Ocean Science in China at the turn of the Century. In: Progress in Science and Technology and Discipline Development in China (Chief Editor: Zhou Guangzhao), Vol. 1, Chinese Science and Technology Press, Beijing, 1998, 105-110. (In Chinese)

Earthquakes and Seismic Hazard in China

Chen Yong Qi Cheng

China Seismological Bureau

***Chen Yong**, graduated from the University of Science and Technology of China in 1965, majoring in Geophysics. From 1970's he worked on rock physics under high temperature and high pressure. Prof. Chen Yong was the director of the Institute of Geophysics (1983-1985), the deputy director-general of State Seismological Bureau (1985-1996), vice president of World Data Center -D (1986-1994) and the member of executive committee of International Seismological Center (1989-1997). He is the vice president of Chinese Geophysical Society and the vice president of Chinese Seismological Society. Prof. Chen Yong has published more than ten textbooks and monographs, of which "The Tangshan great earthquake of 1976-An anatomy of disaster", "Rock physics", "Seismic Hazard and risk" and "Fractal Geometry" have had a significant influence both in China and abroad. Thanks to his contributions to the study of rock physics and seismic risk assessment, he was elected the Member of Chinese Academy of Sciences in 1993.*

Abstract: Earthquakes have a long history as natural disasters for humans. Large earthquakes cause many great calamities, and no other natural phenomenon can cause such deadly destruction in such a large area within such a short time as does a large earthquake. Humans have accumulated more and more knowledge about earthquakes, and have introduced of a set of related subjects such as seismology. Earthquakes are not only a natural hazard but also a useful tool that can be employed to study the inner structure of the earth. Earthquakes and seismological methods will play more important roles in the development of the society. As the society modernizes and the global urbanization increases, more and more people live in cities with social wealth concentrated in cities, which makes seismic disasters change from architectural hazard to social hazard. With the approach of the digital age, newly developed technologies in the fields of Earth observation, information processing and communication will provide new opportunities for earthquake study.

1. INTRODUCTION

As a primary natural disaster, earthquakes have a long human history. The first documentation of an earthquake appeared in China in 1831 BC. The impression of earthquakes is usually that of disaster. In history, some major earthquakes gave rise to many great calamities. No other natural phenomenon can cause such deadly destruction in such a large area within such a short time as does a large earthquake (Fig. 1).

Fig.1 The Great Wall dislocated by an earthquake

Ancient Chinese attributed earthquakes to the abstract "imbalance of yin and yang" which was pure imagination in the pre-science age. True scientific understanding of earthquakes began with the advent of the "earth movement instrument" created by Zhang Heng in Chinese East Han Dynasty in 132 AD. This ancient instrument was constructed based on an understanding of the nature of earthquakes, that is, a ground movement spreading from a distant place that had not been known by western scientists until the 18th century. The appearance of the "ancient seismograph" and the scientific idea it was based on represent the beginning of seismology. Modern seismology started at the end of 19th century when precise and sophisticated seismic instruments came forth. From the first day of appearance, seismology has developed along two directions. One is to understand earthquakes; the other is to make use of earthquakes.

2. UNDERSTANDING EARTHQUAKES

As a natural phenomenon, there are major earthquakes and minor earthquakes. Major earthquakes can cause strong ground shakes making hills and houses collapse. Minor earthquakes are only measured by sensitive instruments and are not perceived by humans. The energy released by an earthquake is often used to represent the intensity of the event, and named by seismologists the earthquake magnitude. The widely used magnitude is the Richter magnitude, which is scaled by the common logarithm of the amplitude of seismic waves recorded by seismometers. The relation between the energy E of seismic waves and the magnitude M of the corresponding earthquake can be expressed as follows (the energy E is measured in ergs):

$$\log E=11.8+1.5M$$

It can be seen clearly from the above equation that the energies released by earthquakes of different magnitudes vary greatly. Generally speaking, with the magnitude increasing 1 unit, the total seismic energy released will be multiplied by about $10^{1.5} \approx 31.6$ times, that is to say, the seismic energy of an event with magnitude 2 is 31.6 times that with magnitude 1, and with magnitude 3, the seismic energy reaches $10^{1.5+1.5}=1000$ times. As a result, the majority of energy is released by the major earthquakes, although the minor events are more numerous.

What order of magnitude are the energies of earthquakes? For comparison, if the atomic bomb dropped to Hiroshima by USA in 1945 (equivalent to 20,000 tons of standard TNT explosive) had been buried 10 km deep under the ground and exploded, it would be equivalent to an earthquake of magnitude 5.5. The "great event" that occurred in Tangshan of China in 1976 was equivalent to 2800 such atomic bombs exploded simultaneously underground. Without question, the energies of earthquakes are tremendous. We can sort various natural phenomena according to their energies. Fig.2 is an energy diagram expressed in ergs. For example, the energy of lightning is about 10^{16} ergs. Now the largest energy we have known is about 10^{32} ergs that was produced when a celestial meteor or comet of 10 km in diameter collided with the Earth with a velocity of 20 km/s around 70 million years ago. This terrible collision produced vast ashes and made the Earth become a dark world. Some scholars believe that it is this catastrophe that caused the extinction of dinosaurs. In such a comprehensive energy diagram earthquakes (denoted by red dots in Fig. 2) lie approximately in the middle part and the energy of a great earthquake is about 10^{23} ergs. It is obvious that the effect of earthquakes to the human society and the whole nature is considerably large, which is the first understanding of earthquakes we have made.

The second understanding of earthquakes is that the spatial distribution of earthquakes is neither completely regular nor completely random, that is, regularity

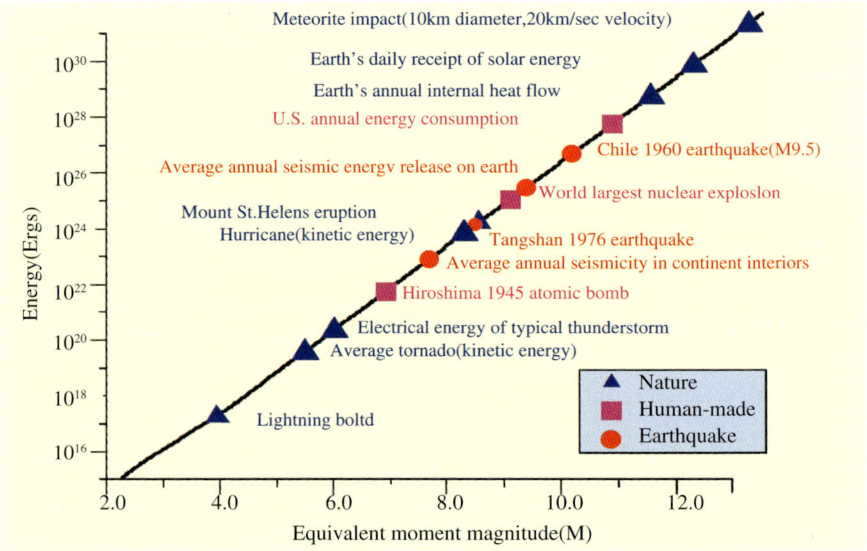

Comparison of the energy released by earthquakes and other energy in the nature

Fig.2 Sorted by energy of various natural phenomena

and randomness coexist in seismic activity. In the global hypocenter map (Fig.3), earthquakes are mainly located in three seismic zones. About 70% of earthquakes occurred in the circum-Pacific seismic zone including many well-known seismically active areas, such as Japan, Taiwan and San Andreas Fault in California, etc. The second seismic zone is the Eurasia seismic zone from the Mediterranean to the Himalayas. In this region the earthquake distribution is relatively sparser compared with the first seismic zone and about 15% of earthquakes took place here. The third seismic zone is the mid-ocean seismic zone along the ocean ridges, in which about 5% of earthquakes happened. Such a belt-distribution of global seismicity can be interpreted by plate-tectonics theory. The theory says that the lithosphere is composed of some rigid plates. While the interiors of the plates are relatively stable, many tectonic actions such as subduction, collision and shear, etc. are taking place at the boundaries. It is just the interactions between different plates that cause the development and occurrence of earthquakes, therefore, the majority of earthquakes located around the plate boundaries. As a result, most of the earthquakes all over the world take place in these three seismic zones. However, there are still 10% of the events not distributed regularly. On the contrary, they are located in some places far from the plate-boundary seismic zones. These earthquakes are called "intra-plate earthquakes". A famous example is the large events frequently taking place in the New Madrid seismic zone in USA which is within the interior of the North-America plate. In China most of the earthquakes are intra-plate earthquakes for which the mechanics of occurrence is still a scientific mystery.

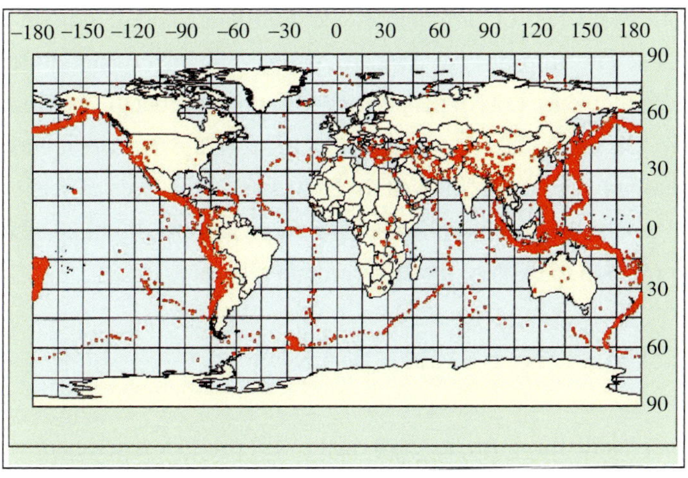

Fig. 3 Map of global earthquake distribution and plate tectonics

The third understanding of earthquakes we have is about the occurrence frequency of earthquakes. Earthquakes can be major or minor, but the question is how frequently do the major ones and the minor ones take place? We have known that there are more minor events than major events, the bigger the magnitude, the fewer the earthquake occurs. There are some proportional relations describing such kinds of occurrence frequency. That is, the ratio of the earthquake number of magnitude 9 to that of magnitude 8 is equal to the ratio of magnitude 8 to magnitude 7, and also equal to the ratio of magnitude 7 to magnitude 6, and so on. It is interesting that there are many other natural phenomena that also appear like this. For example, heavy floods are much less encountered than light ones following the similar magnitude-frequency relation as in the case of earthquakes. Another example is stars. The ratio of number of first order stars to that of second order stars and the ratio of that of second order stars to that of third order stars are equivalent. These ratios are proportional constant and have power exponent relationships, which may be a common rule for various natural phenomena. Such a rule has been found very early in seismology — the famous Gutenberg-Richter relation (G-R relation). In this relation, $N(M)$ denotes the number of earthquakes of magnitude M and above, then there is a power exponent relation between $N(M)$ and M :$\log N(M)=a-bM$(a, b are constants). From this relation the ratio $N(M+1)/N(M)=$ constants can be inferred easily.

We have accumulated lots of knowledge about earthquakes, but there are still many seismological mysteries left for us to understand, such as, the formation mechanics and the development process of earthquakes, etc. Earthquake prediction, in particular, is still a global scientific problem. Even so, we have made considerable progress in utilizing the earthquakes. Lots of people may feel strange about this because in common people's eyes the earthquake is undoubtedly a root of the trouble of endangering the safety of human lives and social wealth. However, in the eyes of seismologists and exploration geophysicists the earthquake is a bright lamp to illuminate the inner Earth and a sharp weapon to explore the underground resources.

3. UTILIZING THE EARTHQUAKES

As vibrations inside the earth, an earthquake will produce a series of waves named seismic waves when it occurs. Seismic wave is the only type of wave we have known at present that can penetrate through the earth. It can propagate to all the directions through the underground media and can be received by seismographs on the surface. The waveform recorded by seismographs involves the information of the features of epicenters, earth media and seismographs. By analyzing the seismic records on the surface we can not only know the location, depth, magnitude of the earthquake but also extract the information of the underground media, since the velocity of seismic wave varies with the medium it travels through. Seismic waves can be reflected, refracted and diffused at the interfaces. Analysis of the reflected and refracted waves makes it possible for us to understand the deep structures in the interior of the earth. In 1996 a Chinese scholar Dr. Song Xiaodong in USA found that the Earth's inner core is rotating faster than the planet itself (Fig.4). This finding has advanced greatly the understanding of the earth and was appraised as one of the ten most important pieces of scientific news in USA in that year.

Seismic exploration is another important method to utilizing seismic waves. Its history can be traced back to the mid 19 century. In 1845, by making use of the seismic waves generated by an artificial explosion, Robert Mallet had tried to measure the velocity of elastic waves in the crust. During the World War I, the two

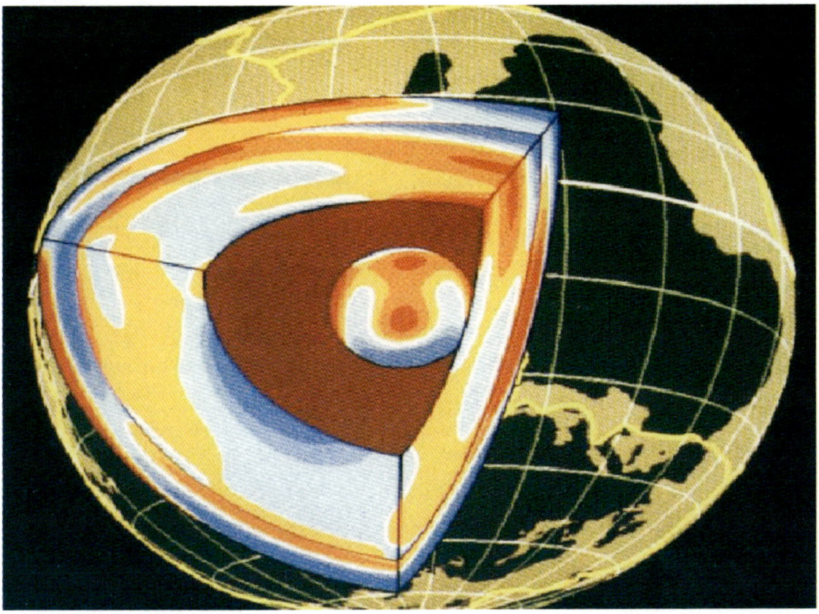

Fig.4 A Chinese scholar Dr. Song Xiaodong in USA found that the Earth's inner core is rotating faster than the planet itself

fighting sides had used the seismic waves produced by the recoil of cannons to accurately locate the place of the other. All these examples are considered as the buds of seismic exploration. With the high precision and resolution that other geophysical exploration methods can not achieve, seismic exploration is one of the most effective methods used in the exploration of petroleum and many other mineral resources. Mineral resources are often labeled by some special tectonic characteristics, for example, it is only in the certain kind of enclosing structures that petroleum and natural gas can be created and reserved. When seismic waves reach such structures, it will be reflected and refracted. Through detailed studies of the signals recorded on the surface, the tectonic features, depth and configuration of the underground structures can be inferred, which provides accurate positioning results for the following exploration work.

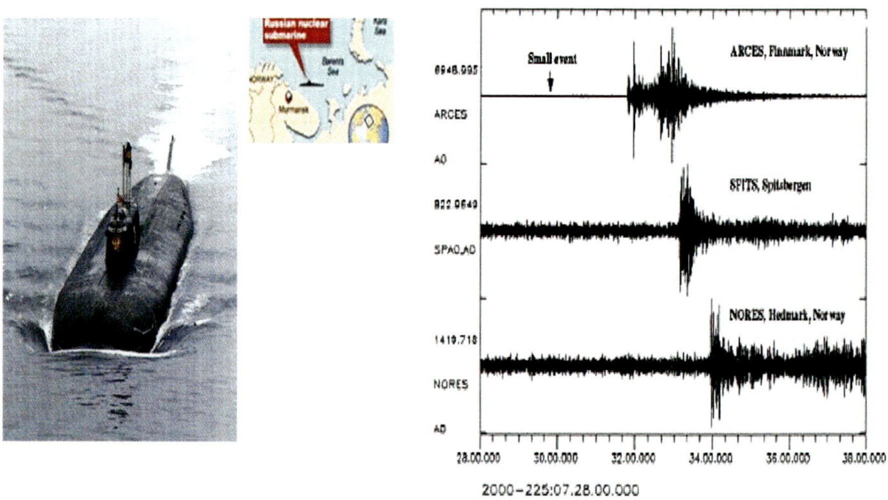

Fig. 5 Using seismic wave to find out the cause of the sinking of the Russian Kursk submarine

Earthquake utilization lies in many aspects. For example, when the Russian Kursk submarine wrecked in the Barents Sea in October 2000, nobody had thought to seek help from seismologists. However, it is the seismologists who made the discussion about the cause of this accident come to an end in January 2001. The seismograph recorded on the Baltic Sea showed the vibration generated by the Kursk submarine explosions, which provides a reliable explanation for the tragedy (Fig. 5). It is believed after data analysis that a torpedo accidentally induced this tragedy when the submarine was on the surface of the sea, and then several torpedoes exploded deep under the sea. However, at the beginning the Russian government attributed this accident to a collision of an unknown foreign submarine. Another example, Fig. 6, shows the seismic waves recorded at the station Palisades 34 km far from the World Trade Center when the center was collided by the planes and collapsed. From the records, the whole process of this event can be known clearly [1].

As a matter of fact, the utilization of earthquakes goes further than the abovementioned. In recent years great improvement has been achieved with seismological methods in the prediction of volcano explosions. Studies on reservoir-induced earthquakes can provide theoretical bases to guarantee security for large reservoirs. For example, in the Three Gorges Dam Project of China, the research of earthquake disasters in this region is one of the principal contents for the engineering feasibility analysis. Monitoring mining earthquakes is one of the basic methods for ensuring the safety of mines. In addition, seismological methods are also used in planet exploration. The large-scale structures in the planet interiors can be revealed through studies of the free oscillations of the planets. In a word, earthquakes should be understood in two ways. On one hand, it brings terrible calamities to the human beings; on the other hand, it can be a kind of benefit if we know how to utilize it.

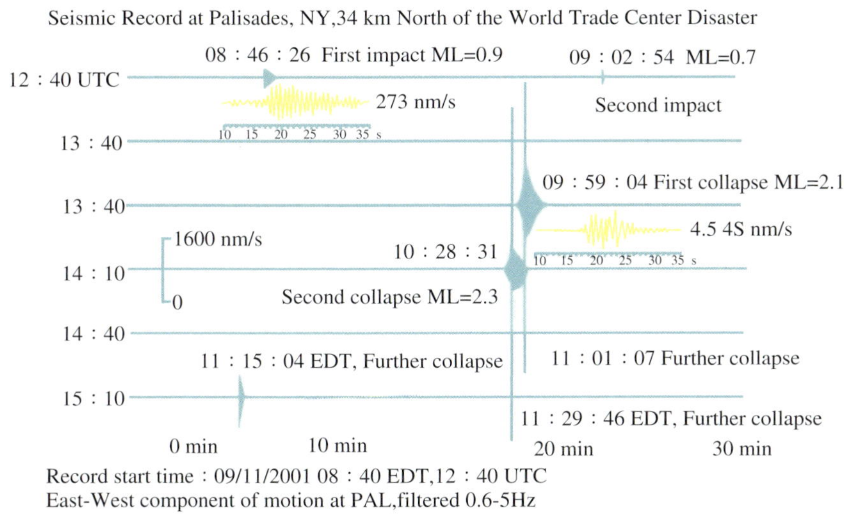

Fig. 6 A station 34 km far from the World Trade Center recorded the whole process of the 911 event

4. SEISMIC DISASTERS AND URBANIZATION

In recent years, great destructive earthquakes occurred one after the other and resulted in great disasters in USA (1994), Japan (1995), Turkey, Taiwan (1999) and India (2001). People may ask a question of whether we have entered an active seismicity period now? Statistic studies show that the average number of earthquakes occurring every year around the world didn't change obviously in the past ten years. But why are there so many news reports about earthquakes? Why are seismic losses so heavy? The answer is that although the natural phenomena don't change much, the human society has changed dramatically during such a short time.

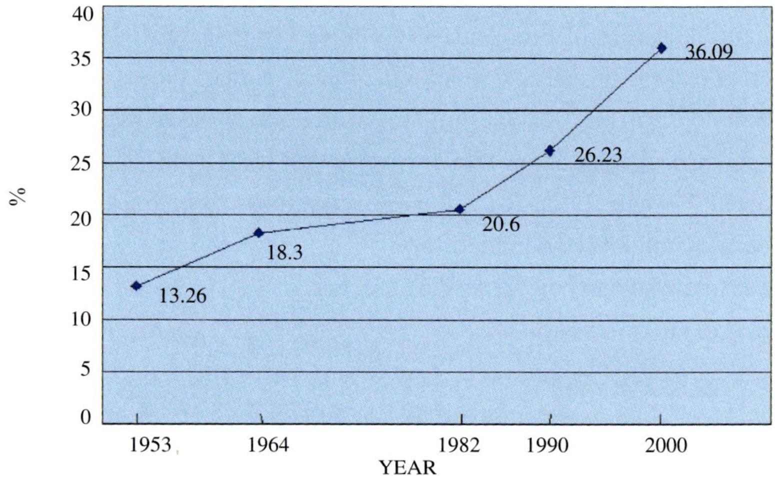

Fig. 7 The tendency of population urbanization in China

In the 1950s only 10% of people live in cities in the developing countries, and less than 30% of people lived in cities all over the world. In the next 50 years the tendency of population urbanization became more and more intensive (Fig. 7). Now there are about 50% of people living in the urban areas that occupy less than 1% of the earth surface. Population urbanization results in the high concentration of social productive force and social wealth to such limited terrains where high buildings stand in great numbers, lifeline engineering is extremely intricate, and communication, finance, and transportation networks are distributed widely. From disaster prevention point of view, all these make the urban areas more vulnerable by disasters. Once being attacked, even a small or moderate disaster can cause heavy losses, not to speak of the extremely destructive disaster of earthquakes. Not only direct losses in the cities are so heavy, but also a series of chain actions can be induced such as fires, poison gases plagues, etc. that will also cause tremendous losses.

With the social wealth described in terms of GDP (gross domestic product), our statistical results indicate that disaster-caused losses will grow much more quickly with the development of economy and the increase of GDP [2]. The correlation of earthquake-related economic losses L and GDP (millions US \$) is:

$$L = a + b \times \text{GDP} + c \times \text{GDP}^2$$

In this equation, a ,b and c are constants:

$$a = 4 \times 10^7$$

$$b = -5000$$

$$c = 2.5 \times 10^{-1}$$

Therefore, with the development of economy, the severity of seismic disasters will be paid more and more attention to. How to reach the purpose of earthquake resistance and disaster reduction will become a common problem that governments of various countries have to face.

The emphasis of earthquake resistance and disaster reduction should be laid on cities. In order to do the work well, earthquake sciences (seismic hazard analysis), engineering sciences (vulnerability analysis) and social-economy sciences (the structure, total amount and distributions of wealth of society) should be combined together to mitigate the losses of seismic disasters to the least degree.

Earthquake prediction is of great importance for disaster reduction. However, it hasn't reached a high level in science and technology at present. Once false predictions are made, great confusion and heavy losses will be inevitable. Therefore, it is important for seismic hazard mitigation to improve largely on disaster prediction. This work mainly involves hazard assessment, engineering defense, early prediction, etc.

In the 1990s, with the progressing of International Decade for Natural Disaster Reduction (IDNDR), seismic hazard quantification became more pressing, and some problems in the traditional hazard assessment methods unfolded gradually as well. For example, the extremely exhaustive information from geology, architecture and economy required by the traditional methods make it difficult for hazard assessment to be implemented widely all over the world. Under such a situation, Chinese seismologists developed a new kind of methodology [2-5] for seismic hazard quantification in which seismicity data are used instead of potential epicenter data to perform seismic hazard analysis and GDP data instead of building classification statistics to conduct seismic risk estimation (Fig. 8). The theoretical model based on is simple and rational, the data needed are relatively easy to obtain, and analysis approaches are practical and feasible. As a result, these methods passed the expert evaluation organized by IASPEI (International Association of Seismology and Physics of the Earth's Interior) and IAEE (International Association of Earthquake Engineering) in 1997, and were recommended to be applied all over the world. Using these methods the first global seismic hazard map and seismic risk estimation map were worked out and the assessment of future seismic hazards was conducted for the whole Central America in 1998.

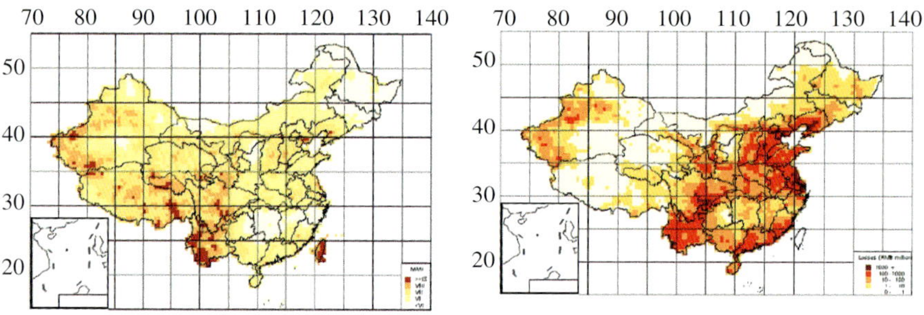

Seismic Hazard Map in China with exceeding probability of intensity 10% in the next 50 years.

The expected physical losses caused by earthquakes in future 50 years in the mainland of China.

Fig. 8 Seismic hazard map and seismic risk map for China

5. THE NEW AGE OF SEISMOLOGY

With the rapid development of computer technology and various observation technologies, recently the field of seismology has changed enormously. The continuous emergence of various new technologies, new concepts and new methods has brought new progress and prospects to seismology.

In the early 1980s China began to set up the China Digital Seismic Network (CDSN). The newest Capital Circle network was built up in 2001(Fig. 9). The Capital Circle area is about 150,000 square kilometers centered in Beijing ranging from N38.5° to N41.0° and from E114.0° to E120.0°. In 2000 its population was approximately 100,000,000 and the GDP reached 100 billion dollars. This area is a key place of politics, economy and cultures in China. In the last 10 years its GDP grew at an average speed of 8% in each year and the tendency of population urbanization became more and more intensive day by day. If geological disasters take place, the after-effects are difficult to estimate. The construction of the capital circle network improves the capability of seismic detection greatly, making the seismic records of this area accumulate very quickly.

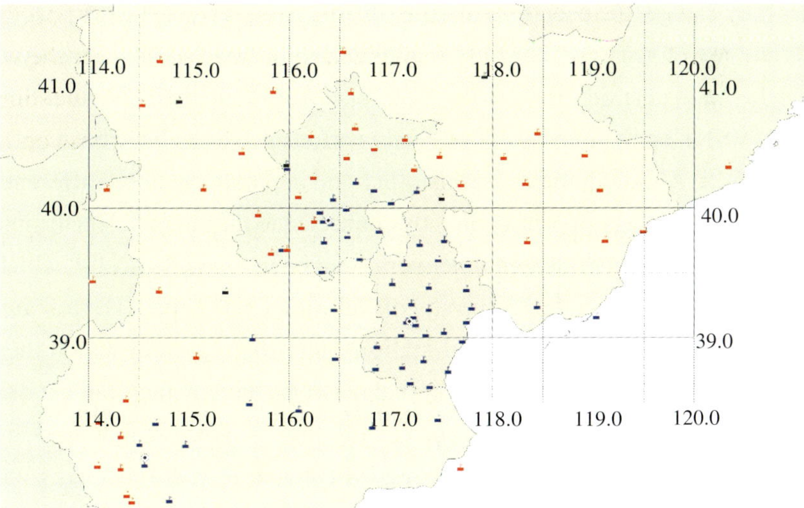

Fig. 9 Digital seismic stations distributed in the Capital Circle area
107 stations,59 short period seismometers (color in blue);
43 broad band seismometers (color in red); 5 very broad band seismometers (color in black)

Nowadays with the rapid development of computers and internet, humans are entering an information age. In the information tidal wave seismology also changed greatly. Seismic data acquirement and delivery became more and more convenient. In the past, it would take several months for seismologists to get the newest data from the global network. Now common people can get the real-time global seismic data freely from the internet at home. In recent years the Live Internet Seismic System (LISS) and the Near Real-Time System (NRTS) have been developed as parts of the global seismic network (GSN). LISS first collects all the data from all over the world, then processes these data and puts them on the internet quickly, and updates them constantly. Therefore, people in any part of the world can get the newest seismic activity information from the internet. For example, if an earthquake occurred several minutes ago in some place, people can not only see the waveforms and propagation situations of the seismic waves but also truly sense the earthquake personally on the internet. The seismic data, which can only be accessed by a few seismologists in the past, can be acquired by everyone now. Seismology is not mysterious any more in common people's eyes.

In the future it won't be necessary for seismologists to know completely how to operate seismic stations and how to collect and deliver seismic data. From the internet they can get the pseudo-real time seismic data, such as the original digital records and waveforms, etc. The only challenge they will face is how to analyze and process the mass data on the internet. They also can set up their "Virtual Seismic Network (VSN)"on the internet. Now many various networks have been set up all over the world. Although their responsibility and apparatus may be different with each other, all these networks have already connected to the internet. Via the internet different real-time data of different networks can be used synthetically, which has been tested successfully at the University of California, San Diego. For researchers, they only need to choose the data of stations near the region in which they are interested on the internet and set up a virtual network, thereby they can get the best observation effect for the study region. Though this kind of research working has just begun, its development prospect and influence to earth sciences will be inestimable.

REFERENCES

1 David Jackson, Seismic recordings at Palisades for the events at World Trade Center (WTC) on September 11. EOS, Transactions, American Geophysical Union, 82, 47, Nov.20, 2001, 565

2 Chen Yong et al., Researches on human activities, natural disasters and active tectonics, Quaternary Research, 2001(21):313-320

3 Chen Yong et al., Vulnerability analysis in earthquake loss estimate. Natural Hazards 2001(23):349-364

4 Chen Yong et al., Seismic hazard and risk analysis: a simplified approach. Beijing: Science Press, 2002

5 Chen Yong et al., Seismic hazard analysis and earthquake disasters prediction, Beijing: Seismological Press, 1999

6 Zhou Guangzhao, Zhu Guangya editor in chief, Jointly Walking Towards Sciences — series reports of hundreds of academicians on sciences and technologies, Beijing: Xinhua Publishing House, 1997

7 Chen Yong editor in chief, Earthquake observation in digital age, Beijing: Seismological Press, 1998 (The reader can find more related contents in the following web sites:www.usgs.govn and www.iris.org)

Remote Sensing Application and Digital Earth

Chen Shupeng Guo Huadong

Institute of Remote Sensing Applications, Chinese Academy of Sciences

Chen Shupeng, *an academician of the Chinese Academy of Sciences (1980-), fellow of the Third World Academy of Sciences (1992-), and academician of the International Eurasia Academy of Sciences. He is an honorary director of the Institute of Remote Sensing Applications, Chinese Academy of Sciences (1985-), former director of State Key Laboratory of Resources and Environmental Information System (1985-1993), and a member of China Environment and Development International Cooperation Committee (2002-). For a long period, his research interests have been in the fields of geography and environment, including remote sensing studies of natural resources and urban environments, flood disaster evaluation information systems, biomass estimation, and monitoring of coastal environment and global change. He is the author of "Exploring the Geosciences" (6 volume) (1992-2003) and many other books. He has received many awards including the Second Class and the Third Class Prizes of National Natural Science Award, the First Class Medal of International Eurasia Academy of Sciences in 1999, and the American Geographical Society's OM Miller Cartographic Medal in 1998.*

Abstract: Remote sensing in China began in the 1970s, and has undergone several distinct periods of its history from the initial introduction, the early experimentation, to the present innovative development phase. This paper presents an overview of significant achievements in China's capacity building of spaceborne and airborne Earth observing systems, remote sensing ground satellite receiving station networks, image processing systems, and other technology achievements. An introduction is provided of the fruitful results yielding basic theory for remote sensing research, remote sensing applications in global change and oceanographic investigation, land resources reconnaissance, environmental protection, disaster monitoring, urban planning and engineering construction, remote sensing archaeology, and reconstruction of ancient environments. This paper also illustrates the achievements in the construction of localized Digital Earth strategies in China, including national spatial data infrastructure, "Digital Province", "Digital City","Digital River" , and the Digital Earth Prototype system and other aspects. The mutual relationships among remote sensing, Digital Earth, and Earth Information Science are also discussed.

1. INTRODUCTION

Remote sensing, developed with the space age of the 1960s, is a new field in Chinese science. It represents a suite of advanced technologies and is a driving force for humankind's space programs for observing Earth systems and acquiring spatial information about the Earth surface. Remote sensing science has the capacity to actively observe and monitor large land areas accurately, comprehensively, and repeatedly.

Remote sensing in China began towards the end of 1970s. Preliminary review indicates that, in the past 30 years, China had over ten satellite remote sensing receiving stations in operation, more than 160 remote sensing agencies, and approximately 400 companies providing geographic information related business services and dozens of universities providing advanced technical training in the subjects of remote sensing and GIS.

Remote sensing in China has undergone approximately 30 years of development from its introduction and absorption in the 1970s to the present day export of techniques and experts. The practical experiences from the early historic periods up to the modern era of innovation in remote sensing science and technology in China can be briefly summarized as follows.

- Early remote sensing applications promoted the advancement of remote sensing science and technology, and introduced integrated airborne remote sensing experiments to support the continued development of satellite applications and development of application satellites.

• Development of visible and near-infrared remote sensing sensors was followed by far-infrared, multi-spectral, and hyperspectral sensors. Further development activities included microwave remote sensing sensors, and conducting multi-band, multi-polarization applications. Construction of multi-band, all weather and all time Earth observing information systems accompanied the gradual improvement of information acquisition capacity.

• The development of satellite application platforms began with an experimental satellite, progressed into operational meteorological satellites, resources satellites, and ocean satellite series as well as a small satellite constellation, to realize the goal of "Earth and space all in one for" "application satellites and satellite applications".

• Pioneering promotion for the localization of Digital Earth applications and contributing to the common sharing of global information built on the Earth observing technology system, which combines global positioning system, geographic information system and network communication system, and linked within a comprehensive "Digital Earth" strategy.

• Continuous development of new fields in remote sensing. By the end of the 20th century with focus on the natural resources, environment and energy issues in Earth science research was conducted on water-carbon cycle, chlorophyll and net primary productivity, and attention started to be paid on life science applications. In the 21st century, remote sensing applications included archaeology and demographic spatial analysis,pioneering new initiatives for space science in the human and social science studies.

• On the basis of a significant body of scientific knowledge gained from applied experiments and under the inspiration of demands for social dimensions, and some key laboratories at national and ministry level were established, and calibration sites and field experimental stations were built. Emphasis was placed on the research into characteristic spectra of electromagnetic waves, the transmission rule and its verification for remote sensing information, and the fusion of remote sensing information. These research efforts have laid a preliminary platform for the integration of remote sensing technical systems and the development of a new generation of application software as well as for the foundation of Earth Information Science.

2. CAPACITY BUILDING OF REMOTE SENSING INFORMATION ACQUISITION

2.1 Spaceborne Earth Observing System

China's space program has made noticeable progress in the world over the past 30 years. By the end of 2002,50 satellites and 4 spacecrafts had been launched. According to the prediction of the white book "China's Space Program", China will continuously develop and launch about 30 different satellites, in the tenth five-year-plan period, for communications, meteorology, ocean, mineral resources, navigation, astronomy, environment and disaster monitoring, as well as scientific space probing satellites. This constellation of satellites will strengthen directly or indirectly the Chinese capacity for acquiring Earth observing remote sensing data and information.

It is foreseen that in about ten years, a durable and stable period for operational satellite observing system will be established, consisting of meteorological satellite series, resources satellite series, ocean satellite series as well as environment and disaster monitoring constellations. Thus, dynamic global monitoring of land, atmosphere, and ocean can be implemented as an operational national program. In autumn of 2003, it is planned that the Chinese astronauts will be onboard the spacecraft "Shenzhou-5", realizing the dream of Chinese people for traveling in the space. At that time, new manned remote sensing sensors will be applied to Earth observation aboard the "Shenzhou-5". Details regarding the acquisition capacity of Earth observing remote sensing data from orbiting satellites are given below.

2.1.1 "Fengyun" Series Meteorological Satellite

China has built a"Fengyun (FY)" meteorological satellite series with polar-orbiting and stationary-orbiting satellites as well as a data application system. Four "FY-1" series satellites were launched in 1988, 1990, 1999 and 2002 respectively. The effective sensors of FY-1A and FY-1B satellites are 5-channel visible and infrared radiometers. The FY-1C satellite has increased the number of channels to 10, therefore, enhancing the Earth observing capability to acquire 4-channel global coverage data with a resolution of 3.1km once a day. The "FY-2" series are the Earth stationary-orbiting meteorological satellites. The FY-2A and FY-2B satellites were launched in 1997 and 2000 respectively, which can provide a scene of Earth imagery covering one third of the globe every half an hour. The second-generation polar-orbiting"FY-3" series meteorological satellites have been arranged into the developing program of 2002 to 2020, which will greatly enhance the Earth observing and space probing capabilities.

2.1.2 Resources Satellite Series

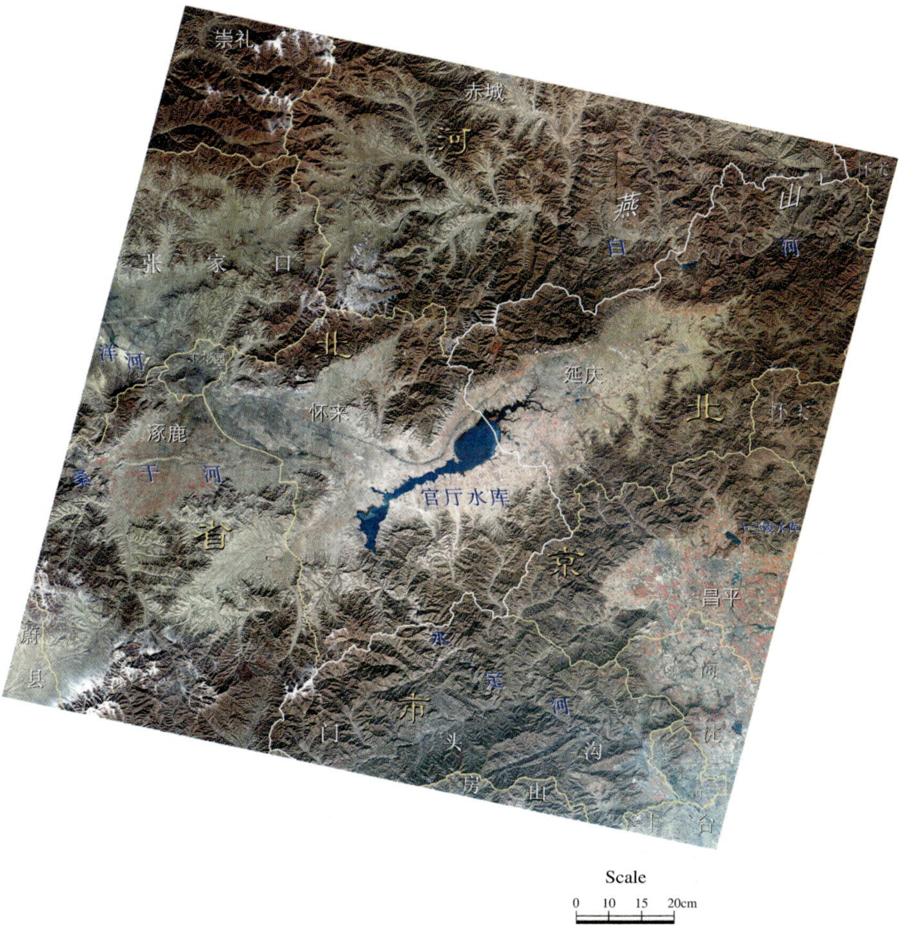

Fig. 1 CBERS Image Map of Guanting Reservoir Area in Beijing

In October 1999, the China-Brazil Earth Resources Remote Sensing Satellite (CBERS) was successfully launched. The satellite made a landmark example of space high-technology cooperation within developing countries. The satellite's effective load of sensors includes a 5-band CCD camera with a resolution of 20m, a 4-band infrared multi-spectral scanner with a resolution of 78m, and a 2-band wide field-of-view imager with a resolution of 256m. After three years of operation, China has acquired remote sensing images for over 80% of China, as well as of neighbouring countries and regions. The relevant agencies have acquired and archived over 320,000 scenes of images (Fig1). It is planned that in 2003 CBERS-2 will be launched with considerable improvement in the resolution and quality of its sensor systems.

2.1.3 Haiyang Satellite Series

On May 15, 2002, China launched an ocean color satellite ("Haiyang" HY-1) together with a FY-1D satellite on the same rocket. The HY-1 satellite has a 10-channel ocean color scanner and a 4-channel CCD imager, used for detecting ocean color and temperature, evaluating fisheries, forecasting fishing seasons, monitoring ocean pollution, providing information about estuary sediments, sea-shore ecology, and sea ice, etc. In addition, China will develop the Ocean Dynamic Environment Satellite series (HY-2 series satellites) for acquiring all-weather ocean wind field, sea surface height and sea temperature, through microwave detection to reach the support of disaster mitigation and prevention. An Ocean Environment Integrated Monitoring Satellite series (HY-3 series satellite) is also going to be developed for acquiring synchronous information of the ocean color and dynamic environment.

2.1.4 Earth Observing Small Satellite

In June 2000, the "Tsinghua-1" small satellite, developed by Tsinghua University in cooperation with Surry University of the United Kingdom, was successfully launched. The satellite has three CCD cameras operated at visible and near-infrared bands, which are able to make optical imaging of the Earth with 40m resolution along a 40 km imaging swath. This light weight satellite can be used in the fields of mineral resources investigation, environment and disaster monitoring, military reconnaissance, hydrological and geological surveying, and meteorological observation.

2.1.5 Environment Satellite Program

At present, China is speeding up the development of satellite systems for environment and disaster monitoring. It is planned that a small satellite constellation consisting of two optical satellites and one SAR (synthetic aperture radar) satellite be launched before 2005 and another small constellation comprising four optical satellites and four radar satellites be launched before 2010. These small satellite constellations will be developed to enable all weather and all time monitoring of the environment and disaster events.

2.1.6 "Shenzhou" Spacecraft

In March 2002, the third "Shenzhou" spacecraft (SZ-3) was orbited with a medium resolution imaging spectrometer (CMODIS) onboard. The CMODIS, orbited at a height of 343 ± 5km above sea level, has a ground resolution of 400-500 m, two day repeated coverage and a 650-700 km imaging swath, with 34 bands, over a spectral range from 0.4-12.5μm (Fig 2). In December 2002, the fourth "Shenzhou" (SZ-4) spacecraft carried a multi-mode microwave sensor system comprising a microwave radiometer, a radar altimeter, and a radar scatterometer. This experiment marked the first world record that three microwave observing sensors operated under the same monitoring system, and the first time adopting a scanning antenna to measure wind direction and wind speed. The acquired data will play an important role in understanding wind field, ocean wave dynamics and the exchange of ocean-atmospheric energy, as well as analysis of ocean disasters and fishery resources.

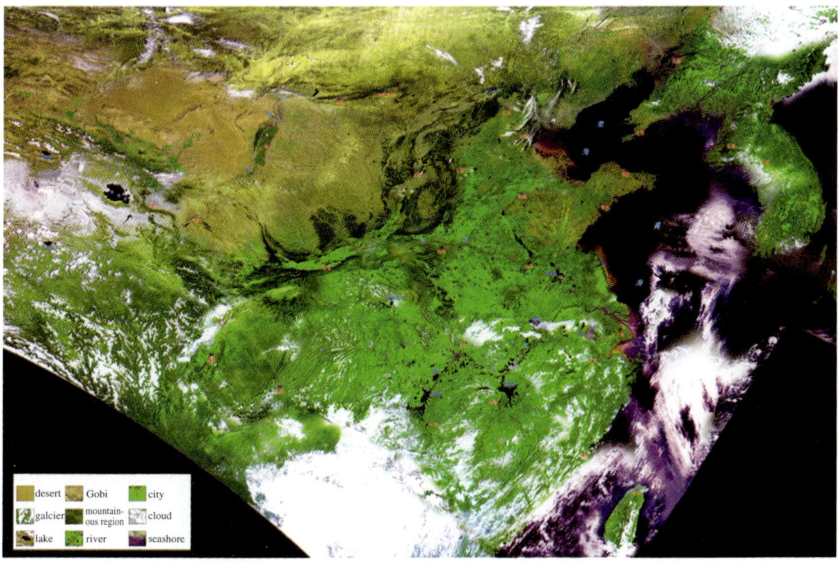

Fig. 2 Shenzhou-3 CMODIS Mosaic Imagery

2.2 Airborne Earth Observation and Experimental System

China is continuously improving its capacity to construct airborne Earth observing systems. It should be noted that the systems, developed under the direction of the Subject of Information Acquisition and Processing Technology as part of the National High-Tech Research and Development Program (863-308), consists of five new remote sensing sensors, including an Operational and Modular Imaging Spectrometer (OMIS), a Push-broom Hyperspectral Imager (PHI), a Plane Array

CCD Digital Camera, a 3D imager, and an L-SAR. These sensors provide high spatial and spectral resolutions, 3-D imaging, all weather, and all time imaging capabilities.

The operational and modular imaging spectrometer system (OMIS), which combines imaging technology and spectrum technology, simultaneously images the same geographic area along continuous spectral bands. The technical characteristics of the OMIS system are as follows. The OMIS has a wide band coverage; there are 128 detecting bands distributed in the range of 0.46-12.5μm at the atmospheric window. The instrument has a scanning field-of-view over 70°.The scanning system, imaging system, and spectrometer system are designed as individual modules, enabling alternation of two working modes at 128 bands and 68 bands. The GPS system can acquire the positioning data of the image, and standardized image products can be produced for specific mineral and natural resource applications (Fig 3).

A R : 2.08-2.35μm G : 916.8nm B : 608.2nm

B R : 3.0-5.0μm G : 888.0nm B : 608.2nm

Fig. 3 OMIS Imagery of Changping Area, Beijing

The push-broom photo-electricity remote sensing sensor (PHI) is becoming an important asset for the new-generation Earth observing system, due to its high spectral resolution, high sensitivity, non-mechanical moving parts and other performance characteristics. The successfully developed PHI contains 244 bands, with a spectral range of 0.40μm -0.85μm, and less than 5nm spectral resolution, a FOV of 42°, and a signal to noise ratio greater than 100.

The core technology for the high resolution plane array CCD digital camera system is a fully digital plane array CCD detector with 4096 by 4096 pixels. The detector together with specifically developed large FOV, large aperture, high resolution and low aberration optic system constitutes the main body of the aerial photogrammetric camera. The CCD camera includes a 3-axis gyratory stabilizing platform, data storage system with high-speed and large capacity, GPS positioning and receiving system, and other parts to make it an aerial photogrammetric camera system with digital, high resolution, and good suitability characteristics.

The 3-D Imager contains innovative characteristics of real-time or near real-time 3-D image production capability. It is an integrated system, and its information acquisition system is composed of the sub-systems for the scan imaging technique, laser distance measuring technique, GPS technique, and attitude measuring technique. An information processing system consists of software for directly positioning of the targets and synchronously producing the matched geo-coded image and DEM. Its major functions can synchronously generate terrain images at one time, or provide contour lines and orthographic images. Secondary products include 3-D demonstrations, thematic, and surveying map production (Fig 4).

The L-SAR operational system possesses a left and right antenna, which can be switched over from left to right for changing the observing directions during the imaging process. There are two polarizations enabling the acquisition of multi-polarization radar images. Two imaging modes can be selected for either narrow swath at high spatial resolution (resolution 3 × 3m) or wide swath at a lower resolution. It has the capabilities to record raw data and real time imaging to satisfy different application demands. In the design of the system, interferometric SAR imaging capability was considered, and experiments had been made for interferometric SAR imaging so as to acquire 3D information of the Earth surface. This system has been used in real-time monitoring during the devastating flood in 1998

Fig. 4 Imagery of Shanghai Pudong Area Acquired by 3-D Imager

(Fig. 5), and provided information for rapid decision-making support for disaster mitigation. It is also an ideal test-bed for developing the Chinese SAR satellite.

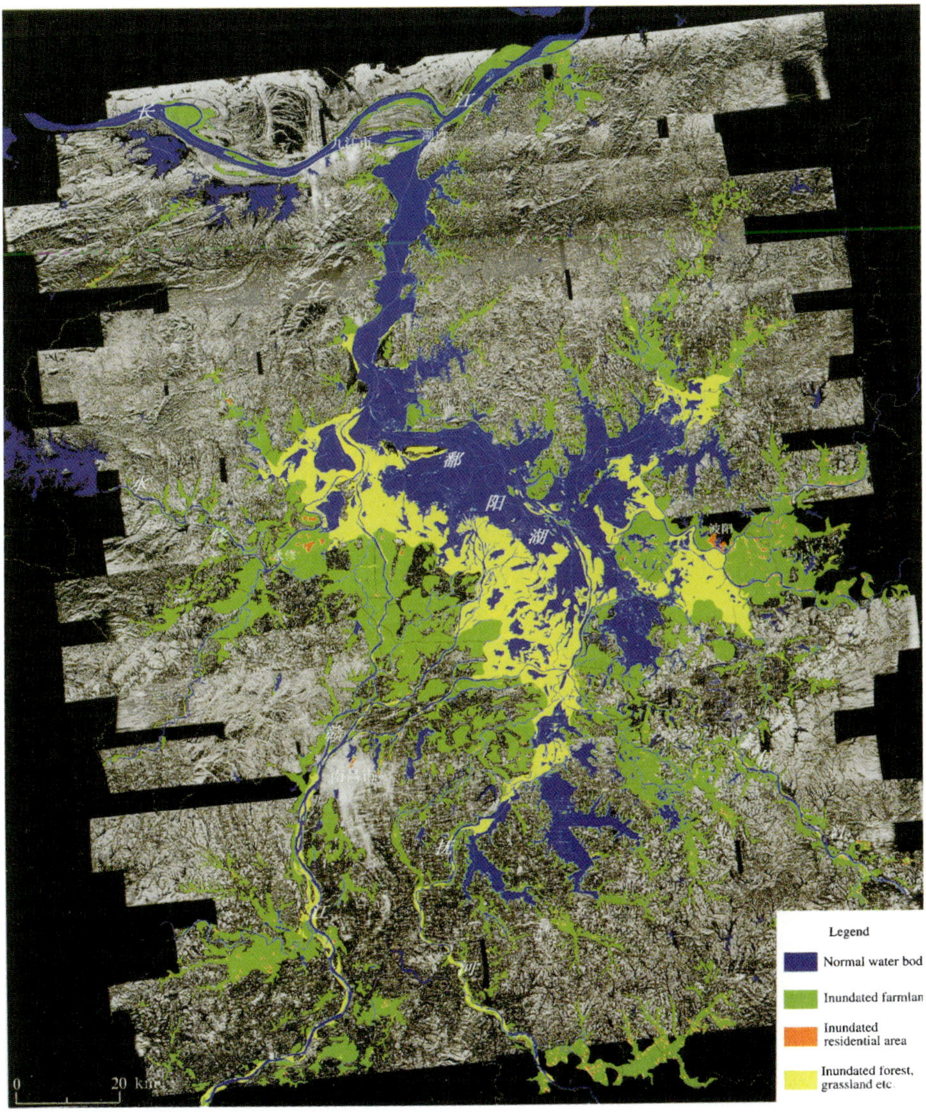

Fig. 5 L-SAR Image for Monitoring the Flood of Dongting Lake Area in 1998

2.3 Remote Sensing Satellite Ground Receiving Station Network and Image Processing System

The China Remote Sensing Satellite Ground Receiving Station, established in 1986, has the capacity of receiving Landsat, SPOT, RADARSAT, ERS-1/2, JERS-1 and CBERS data, and is able to distribute data for the Quickbird, IKONOS, IRS, and EROS satellites based on the agreement signed with the relevant agencies of other countries. Thus, it can provide a complete range of high, medium and low resolution as well as multi-spectral, panchromatic and radar image products, in essence realizing the goal for an all-weather, all time, near real-time, and multi-resolution satellite Earth observation information center. Combining the China Resources Satellite Application Center, Meteorological Satellite Center, and Ocean Satellite Application Center, China has constructed a networked operational system for acquisition of remote sensing satellite data keeping abreast of the most current international levels of system engineering and design.

China has developed its own meteorological satellite data receiving stations that are distributed in Beijing, Guangzhou, Urumqi, and other locations which can receive both polar-orbiting meteorological satellite and geo-stationary meteorological satellite data. The Haiyang (Ocean) Satellite Ground Application Systems were built and operate in Beijing and Sanya in Hainan Province. The MODIS data ground receiving stations were installed in the National Satellite Meteorological Center of the Bureau of National Meteorology, the Institute of Remote Sensing Applications of the Chinese Academy of Sciences, and the Guangdong Meteorological Satellite Ground Receiving Station.

In October 2002, the first "radiometric calibration field" for remote sensing satellites in China was formally operated, demonstrating important advancements in improving satellite measuring accuracies and quantitative remote sensing analysis. International cooperative relationships have already been established with the experimental calibration fields of both the White Sands in USA and the Toulouse in France.

Capability for processing multi-source remote sensing data, for example, data derived from meteorological satellites, Earth resources satellites, airborne remote sensing and aerial photogrammetry in China is well provided through these integrated and networked systems. A strong capacity has also been built for data storage, rapid processing, transmission, information extraction, application software design, reproduction and generation of output images. These critical processing systems ensure a wide operation of applications for the array of remote sensing data. A number of sophisticated software systems have been developed by Chinese scientists and engineers for the different applications using new remote sensing technologies.

3. THE LOCALIZED DIGITAL EARTH STRATEGY

Digital Earth is the inevitable outcome of the space era and the information society evolutionary developments. At the present time of the 21st century, the Digital Earth environment is characterized by global coverage from satellite data, extremely rich array of remote sensing data resources, highly advanced computers, multi-media and virtual reality technologies, and the in-depth maturity of global positioning systems, geographic information systems, and wideband network communication technologies. Digital Earth can now be represented through the integration and full use of 3-D descriptions and data integration of multi-scale, multi-resolution, multi-temporal voluminous Earth information resources, including remote sensing information and socio-economic spatial statistical information, in applications or stakeholder scenarios which benefit mankind.

From the viewpoint of science and technology developments, Digital Earth can be regarded as a natural and progressive extension of the functions of remote sensing applications. Therefore, the following processing steps provide perspective on the concrete technology components. The initial step is "digitalization", i.e., capturing digital Earth data from the Earth observing data and parameters which are directly received from multiple satellites and digital images through the process of rapid transmission, geometric correction, including use of large data storage and database construction to support data query, analysis, and exploitation services. The second step is "information creation", i.e., extracting effective information from vast amounts of data through statistical analysis, image and graphic information analysis, and dynamic change detection and analysis of spatial and temporal data transformation. The third step is "knowledge building", through applying various kinds of professional application models and objective understanding extracted from graphs, images, or visualization data to provide a scientific basis for engineering design and quality management or operating management information systems. The fourth step is "reproduction" , based on engineering precision and using reference boundary conditions, virtual reality modeling techniques to reconstruct natural or social-oriented historical processes, or predict future developing trends and propose multiple alterative scenarios or solutions. The fifth step is "decision-making" , i.e., to select the best one from several possible solutions by department managers and decision makers.

China has already completed national basic geographic datasets at scales of 1 : 4,000,000, 1 : 1,000,000, and 1 : 250,000, as well as regional basic geographic datasets for key areas of flood prevention along seven of its big rivers. A spatial information sharing and service platform, "China Spatial Information Network," is now being constructed. As part of the basic infrastructure for the national geographic spatial information, the high precision national digital terrain model (DEM) at a scale of 1:1,000,000 has been updated twice, and the DEMs at scales of 1 : 250,000 and 1 : 50,000 are under construction. National digital geological maps of 1 :500,000 scale have been produced, as well as the national database of landcover and landuse, forest, pasture field, lake, glacier, and historical records of earthquake. These activities are already involved in the international science database program (Codata), and have been continuously updated with the FY, HY and CBERS satellite data systems.

Seventeen provinces have already initiated "digital province" programs. The provincial programs take up one half of the total number of China's provinces. In addition, more than 60 cities, representing 1/10 of China's 668 cities, are building "digital city" . The "digital river" program, related to many provinces and some big rivers such as the Yangtze River, Yellow River, Hai River and Liao River has developed an impressive agenda to be applied to cross-catchments water transportation engineering, water resources distribution, and flood prevention and mitigation. "Digital China" can be seen to be vigorously developing with its composition of hundreds of localized and mutually supporting programs within the provinces and cities.

The Chinese Academy of Sciences and Peking University have established "Digital Earth Science Laboratory" and "Digital Earth Studio" respectively, to conduct the relevant research on the methods and theories of the "Digital Earth". A prototype system of Digital Earth has been designed which consists of storage, inquiry and index functions for remote sensing data and the creation of terabytes of relevant information providing a common, shared platform for multidisciplinary Earth science study.The prototype system allows for dynamic correlation analysis and data fusion of the Earth spheres, data mining and knowledge discovery, and simulation and prediction. The system could also serve as an "incubator" for making breakthroughs in interdisciplinary cutting-edge research, thereby establishing intersection fields and producing new ideas, and exploiting new areas in the remote sensing knowledge innovation program. Currently, the system utilizes approximately three terabytes of data, several database systems, with significant growth anticipated through incorporation of numerous connection ports preserved for connecting additional databases nationally and internationally, as well as for long distance geosciences processing. This prototype system also provides services to support the digital Olympic promotion model, digital archaeology research, virtual tourism, military simulation, and contributes to promoting"Digital Province", "Digital City" and "Digital Olympic" series.

4. DEVELOPMENT OF SPATIAL INFORMATION APPLICATIONS

China adopted a strategy as the first step towards a remote sensing applications program in China. Experience with the use of panchromatic aerial photos for visual interpretation and series map production was gained during the period of 1963 to 1975. Landsat MSS images were introduced during 1975 to 1980. During 1980 to 1985, China built remote sensing instruments and equipped aircraft with home-made sensors.

From 1985 to 1990, retrievable satellite imagery and global positioning system data was obtained, and integrated using the data fusion capabilities of geographic information systems. Chinese software engineers created remote sensing and cartographic map production software, such as GeoStar, City Star, MapGIS and Super2002, providing new entries into the international market. Since 1999, remote sensing data from Chinese satellites have provided significant support for applications and promoted the integration of spatial geographic information infrastructure and information systems. The exploitation and application of remote sensing information has been vigorously developed resulting in a number of application systems, including the ancillary decision-making geographic information system for the national economy, the resources environment and regional economic information system, the national major crop growth monitoring and yield estimation operational system, and urban GISs. Capacity for this expansive information support was targeted for the country's high-level governmental affairs and multi-level ancillary decision-makers, as well as regional sustainable development activities. The extent and depth of China's remote sensing applications has far exceeded the earliest estimations for application areas when the Earth Resources Satellite was launched. Applications are summarized in five fields as follows.

4.1 Global Change and Oceanographic Investigation

Using meteorological and oceanographic satellite data in the "Codata" scientific database exchange program, China has actively joined international global change cooperative programs, with attention focused on climatic change, greenhouse effect, ozone hole and other related fields. Significant achievements have been made on the discovery of ozone slot over the Tibet Plateau, the influence of South Oscillation and El Nino on the flooding and draughts in China, and the inversion of palaeo-monsoon of loess plateau etc.

The program entitled "Study of life substance cycle and the driving mechanism of Chinese terrestrial ecological system" is on going. This program will adopt the integration way "from top to bottom" for remote sensing inversion models and "from bottom to top" for process models to solve the scale transformation problem, reduce the uncertainty of carbon source evaluation, and evaluate the historic processes, present status, and future trends of carbon sources in terrestrial ecosystems. The research could provide a scientific basis for national ecosystem construction and negotiations on climatic change treaty.

Multi-spectral remote sensing data and its derived vegetation index were used to classify global vegetation and landuse situations, and monitor the process of land desertification, forest logging, pest disaster, urbanization, and environmental change. The method developed for crop growth monitoring using multi-temporal RADARSAT data in the Zhaoqing area of Guangdong Province has been internationally recognized. Research was also conducted on seasonal variations of vegetation index in China, simulating the vegetation alteration models of greenhouse effects, both in China and East Asia, and discussing the correlation of global forest resources and wheat growth conditions.

A "Ground and Space Monitoring System for Ocean Environment" has been established. The system includes automatic monitoring of offshore environment, high-frequency penetrating radar for monitoring of the ocean environment, remote sensing monitoring of the ocean environment, and a system integration and demonstration experiment. This work has focused efforts to provide dynamic information for offshore aquiculture and open-sea fishery resources investigations, in support of China's role as a major fisheries nation.

Remote sensing applications in life sciences are in ascendance. Vegetation indices in terrestrial ecosystems have been widely used in qualitative and quantitative assessment of vegetation coverage and its growth vitality, and used to diagnose a series of biophysical parameters. Vegetation indices can be derived from NOAA, SPOT and meteorological data over the same period for generating remote sensing monitoring forecasts of major crops growth status of both China and all other areas of the world. The AVHRR vegetation index database has been used to analyze the characteristics of the phenological seasons for east China.

4.2 Land Resources Reconnaissance

Comprehensive investigations for land resources remote sensing have been made for every province. By utilizing remote sensing imagery as a major information resource, professional visual interpretation and a series of thematic maps were prepared at scales of 1 : 500,000 or 1 : 1000,000 (1 : 100,000 to 1 : 50,000 for metropolis). Geographic information systems have been established, and linked with the construction of "digital province", "electronic governmental business" or "ecological province".

Geological mineral exploration has demonstrated impressive progress. Airborne remote sensing and satellite image applications have become conventional operations in geological mapping and mineral reconnaissance. There are new discoveries for studying phenomena such as volcano, loess, glacier, and other Cenozoic geology features. A national geological map database, with a scale of 1:500,000, has been compiled which represents the overall results of remote sensing for geological mapping.

The "Tengchong" airborne remote sensing since 1980 has initiated remote sensing for hydrothermal uranium exploration.Its techniques were then applied to uranium deposits exploration in Inner Mongolia and other areas. The successful applications in the above-mentioned areas have not only laid a firm foundation for the integrated airborne remote sensing exploration for uranium deposits, but also made contributions to nuclear investigation techniques and the peaceful use of nuclear power.

Remote sensing for heavy metal and non-ferrous metal resources has delineated many sites of large ore deposits based on the framework of ore-control structure, and relied on the multi-spectrum of lithological recognition and the tracks of mineralization dispersion halos. In the projects of remote sensing for gold exploration, it was predicted that the scientific reserves for gold deposit was 18.8 tons, and the prospecting reserves was 70 tons. This discovery has enriched two counties in north Xinjiang. Good news frequently arrives with remote sensing applications in fire extinction of coal mines in north and northwest China as well as exploring underground water and oil-gas resources.

4.3 Environment Protections and Disaster Monitoring

Valuable results have been achieved when using airborne imaging spectrometric and satellite multispectral data to analyze systematically and quantitatively the current environment and historic evolution of Taihu Lake.

The 863 program "Golden Watch for Western China" has established a remote sensing monitoring network system and dynamic database for ecological environments in west China. The program aims to oversee the ecological effect of the west China development and put forward strategies for constructing and strengthening the capacity for sustainable development at four levels of the west, regional, provincial and typical areas.

The "Remote Sensing Monitoring System and Disaster Influence Evaluation System of Sand Storms" have recognized the need for an integrated network to link remote sensing, GIS and network communication system to realize a real-time operational, qualitative, quantitative, and positioning system for remote sensing monitoring and disaster evaluation of sandstorms. At present, meteorological "Fengyun" satellite data has already been used for weather forecasting of sand storms. International cooperative programs on sand storms have been conducted with Mongolia, North Korea, South Korea, Japan and other countries.

Remote sensing monitoring and disaster evaluation system for grassland fire has been developed to promptly identify fire localities and accurately locate the coordinates of fire spots, measure burned areas, and calculate the loss of grassland. The established forest monitoring and management information system has achieved the operational capacity for rapidly informing agencies of fire status and thus avoiding fire disaster. The research on remote sensing operational monitoring for red-tide disaster was examined in the practical operations of red-tide disaster monitoring in Bohai, Yangtze River, and Pearl River estuary areas.

The integration of an established disaster macrodynamic monitoring system, airborne SAR data real time transmission system and flood disaster monitoring and evaluation system have been put to good use. Sudden flood occurrence information about the flooded extents and inundated areas can be provided within two days; and a detailed report about the information of the disaster affected population and damaged houses can be supplied within a week. The report on monitoring of drought conditions for the country can be submitted every ten days. During the devastating flood period of 1998, data were used from six satellites and three airborne remote sensing systems. The airborne L-band SAR system proved especially valuable for imaging the flooded area and providing more than 100 scenes of images related to the flooded area. The damaged areas identified by remote sensing represented half of that reported by local statistical data. Therefore, remote sensing provided more accurate scientific data for disaster monitoring and evaluation as well as for making plans about reconstruction after the flood event. The remote sensing experimental information system of the Dongting Lake area, sponsored by Canada CIDA, has been applied to 21 other countries.

4.4 Urban Planning and Engineering Construction

Since the rapid development of urbanization in China, the urbanization index has increased from 26.7% in 2000 to 37% in 2003. High resolution satellite and airborne remote sensing data provide abundance of information for urban planning and construction of "digital city". For example, the imaging flight over the High-Tech Science and Technology Garden of Zhongguancun area in Beijing provided updating data for "Digital Beijing" and"Digital Zhongguancun"; the airborne remote sensing experiment for Olympic Games sites in 2008 provided data for the bid and future planning and designing of the game sites. The 3-D imager was used for acquiring data when Macao returned to China. The airborne Earth observing experiments have been performed for Hongkong, Macao, Beihai, Shanghai and other cities, which provide a multitude of map and data, as well as for urban application research.

Remote sensing image data such as CBERS, SPOT, and IKONOS have also played an important role in urban construction applications, e.g., applications in Beijing's expansion and Honkong's district map updating. Remote sensing monitoring of urban expansion and farmland occupation has achieved significant results. In the early period of environment remote sensing of Tianjin city, automated computer map production was first used and books were published on environmental resources investigations, which received high praise by the World Bank, and provided a scientific basis for obtaining the loans from the World Bank for Tianjin's development.

Remote sensing technology has provided surveying and management information support to large national engineering construction. In the programs of Three

Gorge hydrological engineering, Qinghai-Tibet Railway, Ertan and Longtan hydro-electrical reservoirs, Three-north forest shield, the west line engineering of river water migration, and natural gas transportation from the west to the east of China, remote sensing played a special role in site selection, surveying and mapping, and evaluation of engineering impacts on the ecology and environment. In the beginning of preliminary research for Three Gorge engineering, the cultivated land in the reservoir area was estimated three times using remote sensing data. The results helped to reject the proposal of locally migrating farmers and provided data for government decision-making about migrating farmers to other provinces. The Chinese Academy of Sciences in cooperation with the Three Gorge Headquarters has established a long term resources and environment monitoring center.

Prior to the construction of Qinghai-Tibet railway, the relevant agencies and the Chinese Academy of Sciences did research on plateau tundra and route selection, and made a comparative study using remote sensing data and other sources of data on route selection. Valuable experiments have been obtained related to tunnel engineering for the survey of large tunnels in Nanling, Yanshan, and Qinling areas.

4.5 Remote Sensing Archaeology and Reconstruction of Ancient Environment

Remote sensing technology is being fully exploited for archaeological research in the humanities. The first national conference on remote sensing archaeology, held in December 2002 in Beijing, exhibited many achievements in this field. The application of satellite remote sensing data in Yangtze River delta area has revealed 28,087 sites suspected of containing historic remains of the Wuyue culture from Neolithic period to the “Spring-Autumn” period. Archaeologists investigated these sites as determined from the remotely sensed data and discovered an accuracy of 95% correct detection. The Great Wall segments of Ming and Sui dynasties in Shaanxi Province and Ningxia Automonous Region were identified from SIR-C/X-SAR images. In Inner Mongolia and North Shaanxi, many aerial remote sensing experiments were conducted with air photo images of Yuandadu, Liaozhongjing, Liaoshangjing and other areas. Several atlases on “airborne remote sensing archaeology”have been published as a result of these investigations. In 2002, over 10,000 air photos taken by Japanese during the Second World War were purchased from the USA. Currently these data are being analyzed by Chinese scientists with the objective of discriminating additional archaeological sites that were likely modified or built upon by urbanization developments over the past few decades. Spatial analysis and historic evaluation are being applied by archaeologists to link research on land use, natural environment change, and the tracking of human activities for the last 50 years.

The first regional archaeology information system for major sites in China was built for Linzhi area of Shandong Province. This system integrated satellite and airborne remote sensing images with excavation reports, investigation reports, archaeological maps and terrain maps.

Digital photos, a form of remote sensing, of Mogaoku frescos in Dunhuang area have laid a preliminary foundation to recover sculpture and frescos. With laser and holographic imaging, the plundered and damaged cultural relics of the silk-road could be vividly recaptured and displayed by using virtual reality techniques. Remote sensing provides an innovative information technology for repairing and protecting cultural relics, and recording the excavation scenes and reconstructing the environmental setting of ancient sites.

5. PROSPECTING OF EARTH INFORMATION SCIENCE

By the end of the 20th century, remote sensing, GIS, GPS, and network communication technologies, under the framework of Earth system science, had entered into a comprehensive integration phase. Newly developed multi-media, virtual reality and other visualization technologies, supported by high-speed intelligent technologies, such as parallel computation and network computation, are changing everyday to the benefit of managers, scientists, and government leaders.

Remote sensing applications in the 20th century were primarily focused on nature, inorganic environments, mineral resources, and baseline measurements, while remote sensing applications in the 21st century will focus more on human study, ecology and environment, and dynamic monitoring and assessment. Research methods will range from reconnaissance and recognition of single items to quantitative and comprehensive integration, and from basic remote sensing data analysis to the advanced multiple data source fusion.

Earth information science as it had emerged at the end of the 20th century had evolved to meet the demands of that era, but is now progressing forward with new demands. It is an intersection that incorporates fusion of space science technology, information science technology and Earth system science. Its aim, through comprehensive data sharing is to play a fundamental role in organizing and managing large amount of data from which to extract effective information for discovering knowledge, mastering regulations, and guiding government and industry produc-

Changchun and Shanghai. All of them are able to produce over ten optical-communications devices in scale. While occupying 30% of the domestic market, their laser devices made from semiconductor have also entered the international market.

3.3 Recent Major Achievements in R&D on Computer and IC in China

3.3.1 High-Performance Computer and Grid

Starting from the late 1950s, China began to devote itself to the development of computers, and for over 40 years the development has been proceeding without interruption. By now, the research and development in high-performance computers has reached the advanced world level. The Institute of Computing Technology (ICT),CAS, is the first institute in China to focus on the development of large-scaled general-purpose computers, and for this it won the title as cradle of China's computer industry. In recent years, ICT has successfully developed Dawning-1 SMP, Dawning-1000 MPP, Dawning-2000 and Dawning-3000 super servers. These products have been sold to a number of sectors, including biological information, petroleum, meteorology, taxation, education, etc. This has greatly pushed forward the development of a high- performance computer industry in China. Sales of Dawning superserver systems have accumulated to over 200 sets, and the total computing capacity has reached 20 Tflops. At the beginning of 2001, Dawning-3000 was successfully developed and mainly used in the sequencing of rice genome. This contributed greatly to Chinese scientists mapping out the working draft and finished rice genome for the first time in the world. In March of 2003, Dawning-4000L data processing intensive superserver was successfully built, with a storage capacity of more than 100TB and a peak performance of over 3Tflops. Apart from ICT, National Engineering Center for Parallel Computer and National University of Defence Technology (NUDT) are two main forces in the development of high-performance computers. Shenwei supercomputers developed by NECPC in 2000 have already been installed in the National Meteorology Center and the Shanghai National High-Performance Computing Center, and their peak performance can reach 384Gflops. In 1997, NUDT successfully developed Galaxy-III supercomputers, which adopted the scalable, distributed shared memory, and its peak performance has reached 13Gflops. Both of these two institutes are strong in scientific research and engineering, and they both have the ability to develop parallel computers whose performance exceeds 10 Tflops.

Studies in grid technology began in China at the end of the 1990s. With the support of the 863 Program, the national high performance computing environment (NHPCE) has been established in Beijing, Shanghai, Changsha, etc., and based on it, the primary grid computing backbone has been set up. Experiments such as remote login and single system image have been realized. In the last two years, more importance has been given to grid computing. The 863 Program has initiated a grid project to establish demonstrative system of a national high-performance computing grid, whose congregate capacity is around 10 Tflops. Several demonstrative application grids are expected to be established in such sectors as biological information, land resources, product design, etc. The Vega Grid Project addressed by ICT has become quite influential both at home and abroad. It includes not only the study of grid-oriented superservers, grid browsers, grid protocols, but also study on semantic and knowledge grid. Dozens of research papers on the knowledge grid have been published in important international publications such as CACM and have attracted the attention from their international counterparts.

3.3.2 Chinese-character Recognition

Input, output and recognition of tens of thousands of Chinese characters, which are different from the 26 letters of the English language, on the computer is difficult. The difficulty tends to pose as an obstacle to popularizing the use of computers in China. Optical character recognition (OCR) and recognition of handwritten Chinese characters are important Chinese-character input technologies apart from the encoding input. It can be said that China is leading the world regarding OCR and handwriting input of Chinese characters. As early as in 1985, the Institute of Automation of CAS (China Academy of Sciences) successfully developed the first set of software in China for online handwritten Chinese-character recognition. Later it established HanWang Technology Co., Ltd. Over a period of a decade, this software has been upgraded from its 1.0 version to the 10.0 version. The currently promoted 10.0 version can completely recognize handwritten script characters and more than 10,000 different Chinese characters. Microsoft, Motorola and a number of other big companies have all employed the technology of Hanwang Co.Ltd. In addition, the error rate in large-scale input of OCR software is as low as 1 per 10,000.

3.3.3 Expert System for Agriculture

Under the leadership of the Ministry of Science and Technology (MOST), the 863 Program started the demonstrative project of intelligent agriculture in 1990.

The project has made some great breakthroughs in knowledge acquisition, model construction, knowledge representation, reasoning, and the design of five high-quality development platforms of agricultural expert systems. 156 experimental expert systems of "High Yields", "Economical type," and "High Quality" have been developed. These systems involve a number of fields and crops such as cereal crops, fruit trees, vegetables, stockbreeding, fishery, aquatic products, etc., and the systems have been put into application in about 20 demonstrative areas in Beijing, Jilin, Anhui, Yunnan, and so on. According to statistics from 1998-2000, demonstrative land had reached 2,530,000 hectares with radioactive area of 5,670,000 hectares. Increases in income with reductions in cost had added up to several billion in RMB. Economic benefits and social benefits are enormous, especially in Yunnan Province where technology is comparatively less advanced. So, on the question of how to disseminate high technology to technology-poor areas and regions, China's practice can serve as a reference for the third-world countries.

3.3.4 Research on CPU and Other Core Technologies

As is well known to all,core technologies, such as CPU chips and operating systems in China's information field, have long been controlled by foreign companies. Since the early 2000s, we began to master CPU and other core technologies. In two years' time we have achieved some positive results. In August 2002, ICT successfully developed Godson-I CPU Chip, the first general-purpose CPU with independent intellectual property. This chip has 64 bit floating-point components. Its performance has reached the level of foreign general-purpose CPUs, which were widely used in the latter half of the 1990s in the world. In the process of its design, 10 patents were granted to this technology. The hardware support for the protection against hostile attacks has an especially great prospect in areas like e-government. Godson-II CPU Chip, with a peak performance of 2 Gflops is expected to come out at the beginning of 2004. We hope and we believe that with a few years'efforts, China will reach the world's advanced level regarding the core technology of CPU design. Apart from Godson CPU, ARCA series and Zhongzhi series of 32-bit embedded SoC have also been developed by Arca Technology Corporation and Beijing University. Both of these products are expected to enjoy some market share of network computers (NC).

In the mid 1990s, China developed and patented COSIX, a UNIX operating system (OS). However, marketing promotion was not quite successful at first. In recent years, the rise of Linux operative system provided precious opportunity for us to develop OS. MOST has supported some enterprises in developing Linux-based desktop office systems. Some enterprises and research institutes have already developed embedded-Linux and other embedded OS for different applications. Development of high security server OS compatible with Linux applications is also under way. In a few years, the situation of lacking our own OS will be changed.

4. ACHIEVEMENTS IN BASIC RESEARCH ON INFORMATION SCIENCE

Basic research in China are mainly financed by the National Natural Science Foundation of China (NSFC) and Key Basic Research Plan of MOST (973 Program). Beginning from 1998, 17 basic researches on information science have been included in the 973 Program, such as high-performance algorithm, image understanding, mechanization of mathematics, mass information organization and processing in network environments, large-scale scientific computation, very high density and high-speed optical information storage, quantum communications and quantum computing, agent-based software middleware, virtual reality, etc. In short, considerable and influential achievements have been made during the past few years in China in the basic researches on information science.

4.1 Research on Software Theory

The first group of scholars engaged in the research of software in China is mostly mathematicians. They are all well accomplished in the area of software basic theory, and have made some scientific research findings that are influential globally.

4.1.1 Executable Temporal Logic Language and XYZ System

Software easy to program is usually inefficient in execution. It is exactly because of low efficiency that the functional language that used to be popular eventually withdrew from the market. In the 1990s, Institute of Software of CAS, initiated and realized the executable temporal logic language and the XYZ system, which effectively solved the dilemma of being easy to program and being low in execution efficiency. In recent years, XYZ system has been advanced in application of real-time control, multimedia, etc., as well as visualization tools in software architecture combination of artificial intelligence, computer graphics and movie technology helps to advance artificial intelligence in creative art.

Beijing Institute of Electronic System Engineering started mechanism studies on knowledge discovery and has proposed "control flow / data flow diagram comparison" method and "cloud model" that could uniformly represent and deal with random and fuzzy uncertainty. Furthermore, they have also successfully applied this innovative model to data mining, knowledge acquisition and other new fields. In addition, ICT first proposed establishment of "national knowledge infrastructure", and has built the multi-discipline professional knowledge base, which includes over 3 million pieces of knowledge. And this has been put into application in natural language processing, agricultural information system, and so on.

4.3 RESEARCH ON DESIGNING EFFICIENT ALGORITHMS

Algorithm and computational complexity is the core of computer science. Many domestic scholars have devoted much to the seeking of efficient algorithms, including approximation algorithms, to cope with NP hard problem, which is a kind of difficult problem which computing time increases by exponential function with scale of the problem. In such fields as integrated circuit design, electric power schedule, transportation, securities and information search, the 973 project of studying high performance algorithm has obtained a number of very efficient algorithms that are highly applicable.

Manned by 400 professional over the past six years, ICT has made great breakthroughs in key technology of digital visual broadcasting. The research team proposed the "Fast and Robust static Sprite generation algorithm", which has been accepted by MPEG-4, an international standard. In July 2002, "code rate control algorithm" designed by this team was again accepted by international JVT (Joint Video Team) standard. As to recognition of Chinese sign language, this team studied the large vocabulary list (more than 5,177 Chinese sign phrases), and correct rate of recognition has been reached at 92%-95%. On synthesis of sign language, display of 5,596 basic Chinese sign phrases has been implemented with virtual synthesis technology. This is the first time in the world that we realized a real-time communication system between normal people and deaf-mute.

The Institute of Software has successfully developed parallel computation library, such as PQR. These software subroutines have been used in 20 universities and research institutes in the United States, Europe, Japan and so on. Meanwhile, they have also developed parallel solver for large-scale oil well numerical simulation called PRIS. On domestic high-performance computers, they have developed parallel software for refined numerical simulation of oil well with a million nodes. The software has enabled them to achieve significant economic results.

4.4 Research on Quantum Computation and Quantum Communications

Quantum information technology can perform new functions which traditional information technology cannot reach. Therefore it stands at the forefront of modern information science. University of Science and Technology of China (USTC) has made great progress in key issues of quantum information technology. Researchers of this university put forward first the probability quantum cloning theory, which produces perfect copies of non-orthogonal states with some probability. The most successful cloning efficiency has been achieved. By using a linear optical method, they successfully realized the universal quantum-cloning machine in experiment. In addition, during the process of the study on the collective decoherence, they found special quantum states, which can always conserve their quantum coherence even under the interaction with the environment. Based on this important discovery, they further present quantum error-avoiding code which is one of three different kinds of quantum coding principles. This principle of coding has been confirmed in three experiments by American scientists and provided a new method to overcome decoherence, the major obstacle, in practical application of quantum-information technology. They also proposed a new practical quantum-information processor. Moreover, scholars from Tsinghua University brought forward automata theory based on quantum logic. This theory proved that some basic quality of automata rely on the distribution rate of the logic. As a result, it unveiled some essential difference between quantum logic automata theory and classical automata theory.

4.5 Research on Information Security Theory

Chinese researchers independently studied and realized our information cryptographic algorithm and system. At present, our study in cryptology has attained

Progress of Materials Science and Technology in China

Shi Changxu Guan Dehui

Shi Changxu, *born in 1920, member of Chinese Academy of Sciences and Chinese Academy of Engineering, fellow of the Third World Academy of Sciences. He is honorary member of Materials Research Society and Institute of Metals of Japan, TMS fellow in USA, and has many honorary positions in China. He was the Director of Institute of Metal Research, CAS and now is the honorary director. He was Director of Division of Technological Science, CAS(1984-1996), Vice-President of Chinese Academy of Engineering(1994-1998), Vice President of National Natural Science Foundation of China(1986-1991) ,etc. He is a materials scientist and one of the pioneers in China in the research and development of superalloys. He has been very active in participation of national R&D program planning and many times joined in national materials science and technology development and planning committees.*

Abstract: The following contents are discussed in this paper: the infrastructure and sources of funds of materials research, the status and prospects of conventional materials, highlights of some materials science and technology in China, including IT materials, medical biomaterials, advanced inorganic nonmetallic materials, metals and metal science and finally materials design. However, this is far from complete due to limited space and the interdisciplinary nature of the subject.

The trends of development of science and technology in 21st century are as follows: the progress of information technology will be continuously growing, the future of life science and technology is boundless; the technology of new energy resources will mature; and space and marine technologies will be further developed. Along with progress of material civilization and improvement of living standard of mankind, manufacturing and architectural industries will be more prosperous, environmental and ecological situations will be universally emphasized, and so on. All the above are closely related to materials and materials science and technology, because materials represent the fundamental base of human civilization and new materials are forerunners of the development of high technologies. Therefore, materials are put in a high priority for most developed countries. Together with information and life science, materials are considered one of the three top areas to be developed in the 21st century.

China continues to put great emphasis on the development of materials, thereby the development of national economy and the needs of the people have been ensured and the self-reliance policy of national defence has been implemented.

1.THE INFRASTRUCTURE AND SOURCES OF FUNDS OF MATERIALS RESEARCH IN CHINA

Materials science research is mainly carried out in universities,research institutes of the Chinese Academy of Sciences, and some industrial ministries. Some applied materials researches are usually going on in enterprises related to their improvement of the quality of products or development of new process, similar to that in some other countries. However, in order to reinforce the input of manpower and funds for basic research, a system of national laboratories, called "state key laboratory" was established in 1983 in different fields at some universities and research institutes. Currently there are about 160, among which about 1/4 are related to materials science and technology. Considering the singularity of disciplines of the state key laboratories, it was decided that a laboratory was established by combining several key laboratories at Institute of Metal Research, CAS, in 2001,called "Shenyang National Materials Science Laboratory".

Funds for fundamental materials research in China came from different sources. Besides ministries, local governments and enterprises, have several sources on the national level.

As to basic research, the nation set a policy of "the majors should be centralized, and the small ones should be unrestrained". For the basic research items, funding is provided by the National Natural Science Foundation of China, and for key topics support is provided by Major State Basic Research Development Program (973 Program). There are six research areas of this program, materials science is one of them. New materials are one of the major fields of National High Technology Research and Development Program (863 program). The major projects of "863 program" in materials field are advanced structural materials, advanced functional materials, opto-electronic materials, and integrated devices. The purpose of this program is to provide a technological basis, and some applied basic researches are involved, to enhance the upgrading of enterprises. In order to facilitate the transformation of research results into production, some programs have been established, such as "Torch Program", "High Tech Industrialization Program," with materials being the major content in all the programs, usually, however, as matured achievements, and not with basic research involved.

2. THE STATUS AND PROSPECT OF BASIC MATERIALS IN CHINA

Basic materials usually refer to conventional materials, such as iron and steel, nonferrous metals, organic materials, constructional materials, etc. These materials, related to large quantities and extensive applications, are the bases of national economy and pillar enterprises. Hence, the status of basic materials can represent the national power of a country.

The increase of production of basic materials in China was rapid in the past 20 years[1]. Taking iron and steel as examples, the quantity of crude steel production exceeded 100 million tons in 1996 and ranked No. 1 in the world, with more than 180 million tons of crude steel being produced last year, as shown in Fig. 1. The increase of steel production does not rely only on building new plants but on regulation of superstructure and improvement of technologies:

◆ Refined raw materials, injection of coal powder and enlargement of blast furnace;

◆ Change open hearth to L.D process for steel making;

◆ Refine process outside the furnace to remove sulfur and gases;

◆ Continuous casting and continuous rolling;

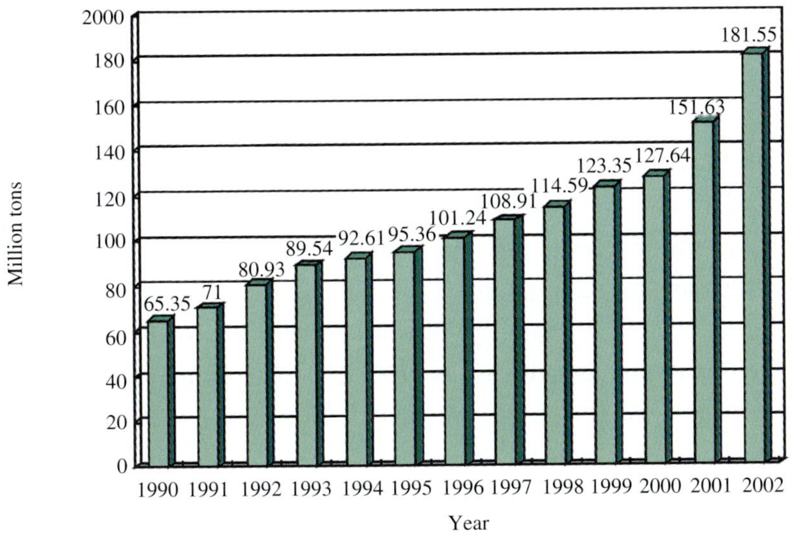

Fig. 1 The increase of steel production since 1990
(from Prof.R.Y.Yin Beijing Central Iron &Steel Research Institute)

The reserve of nonferrous metals is relatively abundant in China and the production of 10 common nonferrous metals is only next to that of the USA, but the amount per eapita is only 1/10 to that of USA. The production of Mg is very attractive among the 10 common metals, that is, 210,000 tons were produced last year in China, and 450,000 tons in the whole world. This is due to large reserve and the low cost of production.

As to the polymeric materials, the quantity of synthetic fibers produced in China has ranked No. 1 in the world since 1998, but the production of plastics is lagging far behind, about one half has to be imported from abroad and the volume ratio of plastics to steel is only about 0.5 compared to that of over 2.0 in the USA.

The amount of production of constructional materials is even more attractive due to vast market in China. For instance, the quantities of glass and domestic ceramics produced amounted to 1/4 of that of the world in recent years, and cement produced in 2002 was $7.4x\ 10^8$tons , it was nearly 1/2 of the whole world.

Above all, through many years of R&D in materials science and engineering, China has formed a comparatively integrated system of basic materials, and has become a great country for basic materials production. However, the competitive power is still lagging behind due to less innovation ability. For this reason, there were several subjects related to steel, aluminum, plastics and cement in "973" program, such as "basic research on new generation of steel" (1998), "Basic research on promotion of quality of aluminum" (1999), "basic research on the scientific problems of cement" (2001), etc. Some of these projects have acquired noticeable results. Taking steel as an example, the quantities of low carbon steel, low alloy steel and alloyed structural steel are over 70% among the total amount of steel produced in China. Through refinement of grain size, homogenization and purification of the steel produced, the yield strength can increase from 200 to 400 MPa, 400 to 800 MPa and the ultimate strength from 800 to 1500 MPa respectively. The above targets have been achieved through three years efforts. The low carbon steel has passed the production test in a large steel plant with a coil size of 50 tons each and extrusion test in an automobile factory[2].

The key point for increase of strength of low carbon steel is to induce ferrite transformation through deformation at appropriate ranges of temperatures and formation of fine grains by dynamic recrystallization. It has been proven that as long as the grain size is below 6μm, the properties of the rolled steel can be ensured, as shown in Table 1.

Table 1 The ferrite grain size of low carbon steel vs mechanical properties
(Steel composition: 0.18C, 0.21Si, 0.6Mn, 0.016P, 0.02S, 0.008Al, 0.017O, 0.005N)

Grain size (μm)	4.0	5.6	6.7	8.0	9.4	11.0	16.4
σ_b (MPa)	605	558	505	518	525	520	500
σ_s (MPa)	483	438	358	368	380	390	353
φ (%)	48	60	58	58	60	55	54

For low alloy steel, the ferrite grain size must be below 1μm, and this requires heavy deformation (> 85%) and steel of high purity. Experiments showed that if the contents of S, P, O, and N of the steel decreased from 255 to 78 PPm, the rate of nucleation of ferrite could be doubled, and consequently, fine grain size would be resulted. The yield strength can reach 1000 MPa if the carbon content is decreased by 0.003%.

The most important achievement in aluminum research is the breakthrough of de-silica from diasporic-bauxite by flotation. The reserve of aluminum ore in China is ranked the third in the world. However, 98% of the reserve is diaspore which requires almost double energy consumption to extract due to high content of silicate. Through large amounts of basic researches, Chinese scientists designed appropriate reagents based on quantum chemistry approach, then a flotation process was developed to reduce the silicate content to such a degree that the Al/Si ratio is from 4 to over 10, to yield a grade of treated ore is then economically suitable for normal Bayer process[3].

3. HIGHLIGHTS OF MATERIALS SCIENCE AND TECHNOLOGY IN CHINA

Chinese scientists and engineers have been involved in all fields of materials research and development. It is impossible to describe all the achievements of materials science in China in this paper, therefore, only a few highlights of materials science progress are presented. Today's important very active areas are not included, such as organic functional materials (molecular materials), novel energy materials, and ecomaterials, etc. However, nano-structured materials are scattered in the text because it is inevitable to describe the modern advanced materials of today, but no special paragraph is arranged.

3.1 Development of IT Materials Still Has a Bright Future

Information on functional materials is naturally very important in the information epoch. China has lagged behind in this aspect as a whole, for instance, the diameter of silicon single crystals produced in China is currently at the level of 6 to 8 inches, although production lines of crystals with diameter of 12 inches have been established, the quality remains to be improved[4].

Along with the development of high frequency communication, compound semiconductor materials are more and more important. The quality of compound semiconductors, such as GaAs and InP, is comparable to that of international standards, however the pullers of crystal growth and wafer processing equipments for mass production are backward, and cannot meet the requirement. A plant is under construction for production of 2-3 tons of semi-insulating GaAs materials with a diameter larger than 4 inches and the above problems are expected to be solved. The research on space materials science, growth of GaAs single crystals under reduced gravitation condition was carried out in 1987 using Chinese recoverable satellites[5].It was demonstrated that perfection and homogeneity of the crystal were manifestly improved and the performance of devices and ICs made from the crystal were greatly enhanced.

As for low dimensional semiconductor materials such as GaAs, InP and GaN based superlattice and quantum well, quantum wires and quantum dots etc., a lot of work is ongoing. For example, the shape and density controllable In(Ga,Al)As quantum dots (QD) and quantum wires (QWR) of free defects growing on both GaAs and InP substrates using SK growth mode and MBE technique have been obtained, and the vertical anti-correlated InAs/InAlAs/InP QWR super-lattice was discovered at Institute of Semiconductors, CAS. In addition, 960-1020 nm QD lasers with the light output powers as high as 3.62W and red light emitting QD laser continuous-wave (CW) operated at room temperature have been successfully fabricated. SOI technology is under development, and some noticeable achievements have been obtained.

Memory storage is critical for information materials. Scanning tunneling microscopy is used not only for surface analysis, but also for nano-processing or manipulation of atoms. Clusters of diameter with nm size have been made in Nanoscale Physics and Devices Laboratory, CAS, in the 90s[6].

A new concept called tip-chemistry was proposed by scientists of Peking University. Using single walled carbon nanotube as STM tip, they have written memory holes as small as 4nm on charge transfer complex by inducing localized thermochemical decomposition of the complex. A storage density of 10^{12-13} bits/cm^2 can be achieved, which is about 10^3-10^4 times as high as that of the present optical storage[7], as shown in Fig. 2. These techniques are far from practical applications.

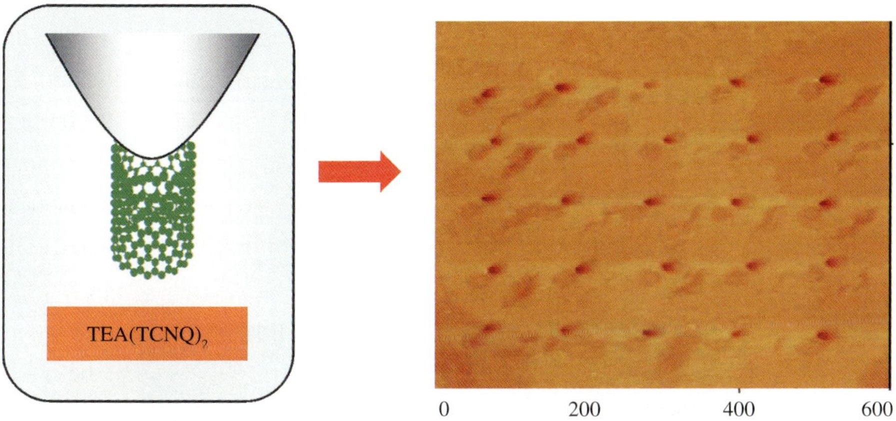

Fig. 2 Thermochemical holes burning on TEA(TCNQ)$_2$(from Prof.Z.F.Liu Peking Univ.)

As to display materials, Chinese scientists are now engaged in the development of organic light emission display material(OLED), and it is now on the verge of industrialization. The thickness of the device is only a few hundred nanometers and can be rolled. The most important feature is that OLED has wide angle of vision compared with that of existing display materials, and is now on the verge of industrialization.

Fig. 3 Organic light emission display(from Prof.S.T.Li,HK City Univ.)

A new type of cold cathode lighting element has been successfully developed by utilizing field emission material invented by scientists of Zhongshan University. This type of lighting element could be used in large area display boards. This product has been put into production in small batches and this novel cold cathode lighting element reaches the world advanced level[8].

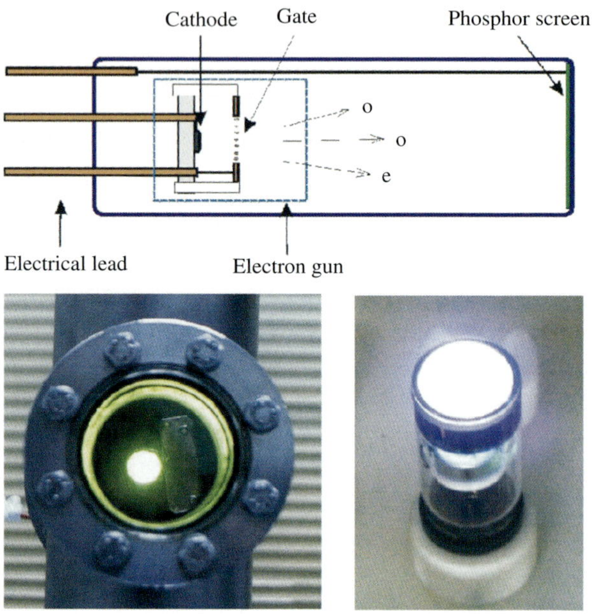

Fig. 4 Cold cathode lighting element(from Prof.N.S.Xu,ZhongshanUniv.)

The most noticeable information of the functional materials in China is probably growth of artificial inorganic crystals, such as quartz, lithium niobate($LiNbO_3$), lithium tantalite ($LiTaO_3$), etc., especially the nonlinear optical (NLO) crystals. As early as the late 1970s scientists at CAS discovered BBO ($BaBeO_4$), then a molecular design system was proposed for searching for new NLO crystals[9], and developed LBO (LiB_3O_5) with even superior properties. These crystals can be used for ultra-violet or deep violet region, and they were ranked high in the world market in the 1990s. Then CBO (CsB_3O_5), KBBF ($KBe_2BO_3F_2$), KABO ($K_2Al_2B_2O_7$) were developed successively. These crystals led the wavelength from ultraviolet to deep and vacuum ultraviolet region (200nm-157nm), and thus NLO crystals developed in China are still ranked high. Scholars of Nanjing University developed dielectric superlattice from semiconductor to dielectrics[10]. Dielectric superlattice showed plenty of new phenomena that are promising in applications.

3.2 R&D on Biomedical Materials Will Be Emphasized

Biomedical materials are one of the most important areas of advanced materials, and it has been widely emphasized in China. This field is even more active since the nanotechnology has stepped in. The following two examples are results of research on artificial bone.

One is that the discovery of non-living porous calcium phosphate ceramics can induce bone formation[11]. This is realized mainly by mesenchymal cells and intrinsic bone growth factors. These cells may come from the capillary growing into the pores, and the marrow stromal cells in blood. The bone growth factors may be concentrated by the materials from the body fluid and secreted by the stimulation of the materials to non-osseous cells. Clinical tests have proved that the materials with osteoinductivity could promote the new bone formation and are excellent scaffold materials for bone tissue enginnering and have the potential to induce the generation or reconstruction of damaged bone. Fig. 5 is histological sections of calcium phosphate ceramics 1 and 8 months after implantation in the muscle of dogs.

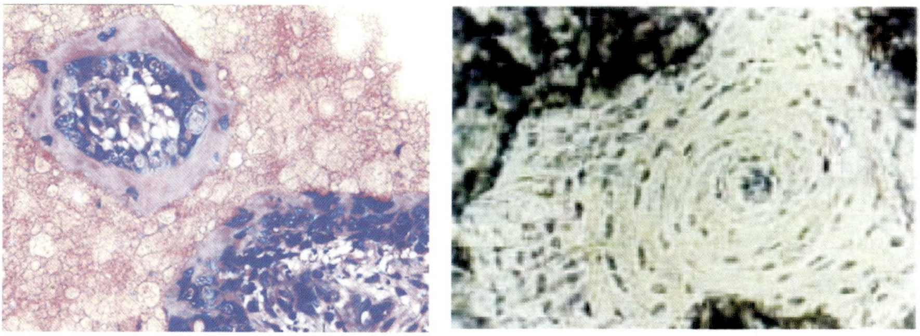

Fig. 5 Histological sections of the calcium phosphate ceramics implanted after 1 and 8 months in the dogs.(from Prof. X. D. Zhang, Sichuan Univ.)

The other is bone materials prepared by self-assembly. Actually all the biomaterials are formed by self-assembly. Based on the principle of self-assembly of natural bone and microstructure analysis, take type-I collagen as matrix,which is mineralized in calcium phosphate solution, thus nanocrystallized apatite contains carbonate radical, then nano-HA/collagen(nHAC) is self-assemblized with a structure of nano-periodic layers of 6 nm. It has been proven that the material has excellent bio-compatibility, degradability and ability for healing. It only takes 10 weeks to cure if this material is embedded into an injured ulna. The embedded material is gradually degraded while the new bone is growing[12], as shown in Fig.6.

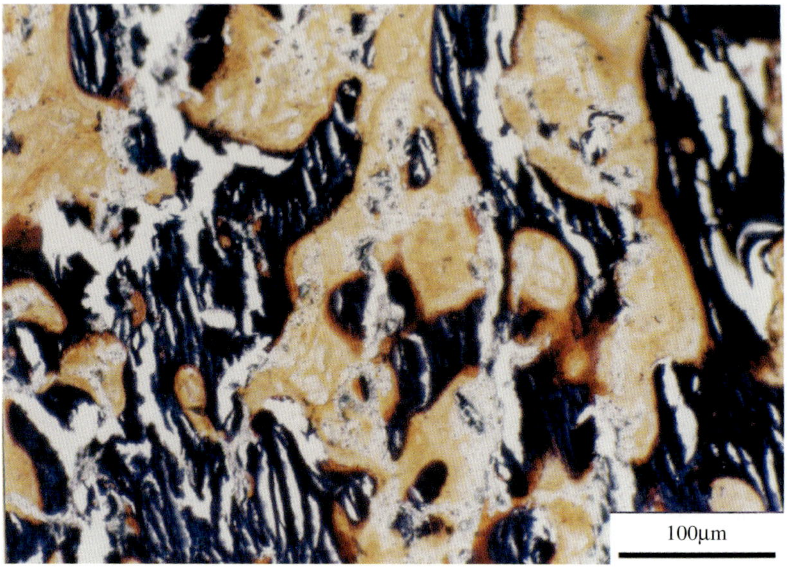

Fig. 6 New bone is growing embedded with nHAC artificial material,40% has replaced by growing bone.(dark color)(from Prof E.Z.Cui.Tsinghua Univ.)

3.3 Recent progress of R&D in advanced inorganic nonmetallic materials

China has paid much attention to advanced ceramics research since the late 1950s, as exemplified with the founding of Shanghai Institute of Ceramics, CAS. This institute has made important contributions on the development of ceramic materials for national security and waterless cooling of internal combustion engines for automobiles. At present there are many research institutes and universities engaged in R & D of advanced ceramics in China, who have abtained many good results.

Chinese scientists have made lots of efforts on improving the toughness of ceramics. For instance, a breakthrough has been made on injection molding of nonplastic study with the invention of a process of colloid injection molding[13].Another development is a ceramics of Ti_3MC_3 (M=Al, Si, Ge) with multilayered structure which can be plastically deformed[14]. As to the functional ceramics, Chinese scientists developed a system of PMN which can be used for multilayered capacitors with a low sintering temperature. These devices have been sold widely on the world market. Another system of BZN dielectric ceramics which can be applied to high frequency with low energy consumption has been developed[15], having promising for use in mobile telephones, digital televisions, etc. It has been proposed that there exists a microdomain and macrodomain transformation in the relaxation type ferroelectric ceramics and this was experimentally proven. This may be one of the bases for development of relaxation type ferroelectric ceramics. Growth of relaxation type ferroelectric single crystals is usually prepared by a flux method, Chinese scientists developed a process to grow PMNT single crystals of large size and high quality by the Bridgman method[16] .

Research on carbon nanotubles has been very active in China since its discovery in 1991. Chinese scientists prepared single-walled or multi-walled carbon nanotubes with different methods, such as evaporation by arc discharge or lasers, and decomposition of organic gas phase by catalysis etc[17]. Some process has been developed with the control of diameter and orientation of nanotubes[18].

Research on applications of carbon nanotubes has been carried out in China, such as hydrogen storage, with the highest record reported to be as high as 20% (wt) in multi-walled nanotubes doped with lithium [17,19]. If carbon nanotubes are used for cathodes in lithium batteries, long life and fast charging are expected. Carbon nanotubes are excellent for use as field emission material due to the high strength, high toughness, good thermal stability and electrical conductivity, and some research work is now undergoing in China. Carbon nanotubes have been used as pattern molds to prepare one-dimensional GaN nanorodes, which have perfect crystal structure and good light emission property[20]. Recently, carbon nanotubes of 30 cm long were prepared by spinning from carbon nanotubes on silicon substrate[21], this phenomenon may be an indication that carbon nanotubes can be used as structural materials.

3.4 The Progress of Metal Science and Technology is Noticeable

Metallic materials have been very important since the founding of New China, and some achievements of metal science and technology are worth mentioning.

In 1983 scientists abroard discovered the quasi-crystalline phase of 5-fold symmetry for the Al-Mn system.China discovered a phase of the same structure in the NiTi alloy system at the same time. Fig. 7 is a photo of the image of NiTi quasi-crystalline phase under high resolution electron microscope. Subsequently, 8-fold, 10-fold and 12 fold symmetry quasi-crystals have been discovered. Consequently,

China has become one of the top research centers for quasi-crystalline materials in the world[22].

Fig. 7 NiTi quasi-crystalline phase under Hrm(from Prof.H.Q.Ye,IMR,CAS)

China has focused on research of amorphous metals since its discovery by Duwez in 1960, and started to develop in the early 80s a prototype production line, equipped by on-line automatic rolling system, with a capacity of 100 tons/ year at Beijing Central Iron and Steel Research Institute (BCISRI), and a 1000 tons/year production line was established in the 90s with ribbon width of 220mm, as shown in Fig. 8. At present, a yearly production capacity of 3000 tons of amorphous and nanocrystallized soft magnetic ribbon has been reached. The products have wide use in electric and electronic industries.

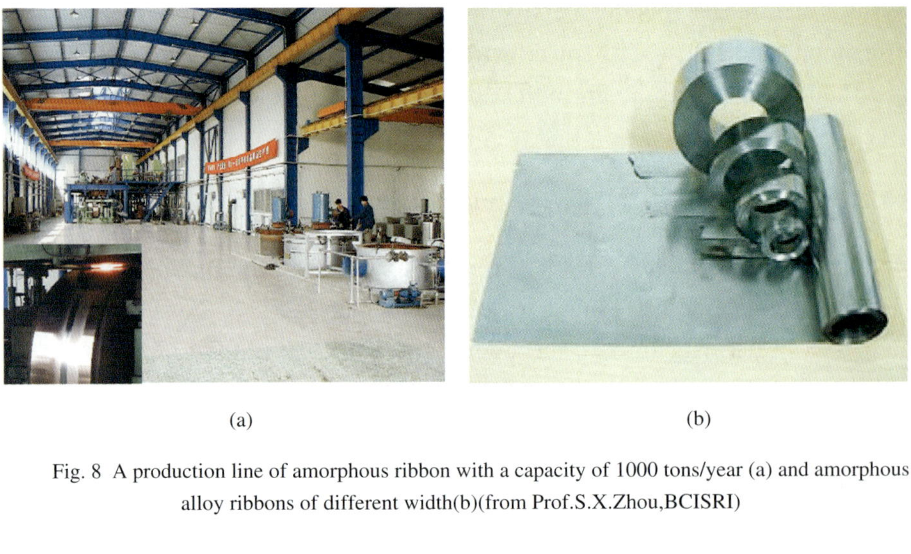

(a) (b)

Fig. 8 A production line of amorphous ribbon with a capacity of 1000 tons/year (a) and amorphous alloy ribbons of different width(b)(from Prof.S.X.Zhou,BCISRI)

Nanocrystalline materials resulting from crystallization of amorphous solids are dense and porosity-free. An abnormal Hall-Petch relationship was discovered through the study of the fully-dense nanocrystalline materials[23]. Bulk nanocrystalline copper samples with high purity and full density were prepared by means of conventional electro-deposition technique with an average grain size of 30 nm. The nanocrystalline sample can be rolled with a degree of deformation of 5100% at room temperature without strain hardening[24]. Furthermore, an abnormal tensile property indicates that with an increase of strain rate, the fracture strain is increasing accordingly. H. Gleiter considered this the first demonstration to show the intrinsic behavior of plastic deformation of nanostructured materials without porosity.

Surface mechanical attrition treatment to form a nanostructured surface layer on metallic materials was developed which exhibits significant potential for practical applications. This novel technique will not only enhance the overall properties of the materials such as wear resistance and corrosion resistance, but remarkably lower the surface nitriding temperatures of iron and steels[25]. Scientists from Shenyang National Laboratory for Materials Science(SNMSL) have made some contributions as shown in Fig. 9.

In terms of structural materials development in recent years, intermetallic compounds have received active attention in China, as seen by ten international

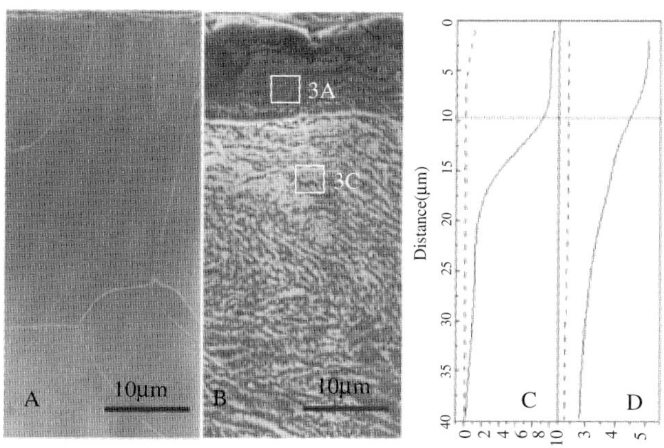

Fig.9 Cross-sectional observations of (A) an original coarse-grained Fe sample and (B) a SMAT Fe sample after nitriding at 300℃ for 9 hours.(C) distribution of nitrogen(D) the microhardness along the depth from the top surface layer in the original Fe sample (dashed lines) and in the treated one (solid lines),respectively.(from Prof.K.Lu,SNMSL)

Combination of levitation smelting and electromagnetic confinement forming parts of inclusion and pollution free can be expected.

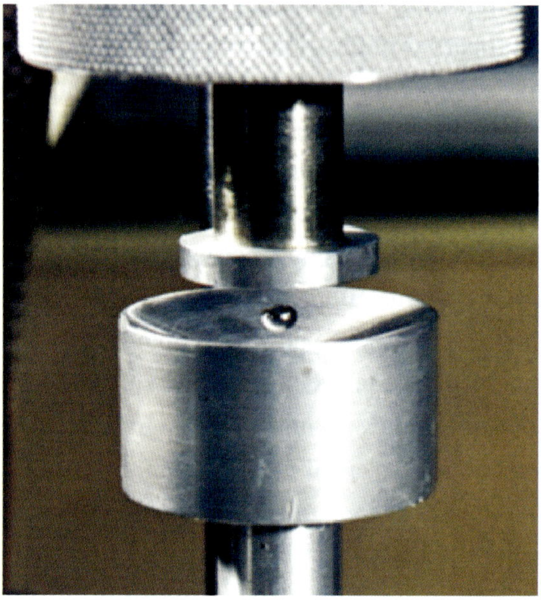

Fig. 12 An experimental setup of acoustic levitatin(from Prof.B.B.Wei, Northwestern Polytechnic Univ.)

3.5 Materials design and process simulation

Materials design has a bright future. Materials design refers to prediction of relationship of materials composition, structure and properties through theoretical calculations, including the processes of materials manufacture, henceforth new materials can be developed according to the specified properties.

It is well known that materials science and engineering are composed of composition/structure, synthesis/process, properties and performance 4 elements, thus forming a tetrahedron(Fig. 13(a)). However, considering composition and structure are not equivalent because a single composition may lead to different structures through different processing, and consequently different properties, it is reasonable that materials science and engineering are expressed by 5 basic elements, that is composition, structure, synthesis/process, properties and performance. Then, a hexahedron is formed by connecting the 5 elements as shown in Figure 13b[30].This expression vests with a proper position of theory and materials design(including process design), that is at the center of the hexahedron.

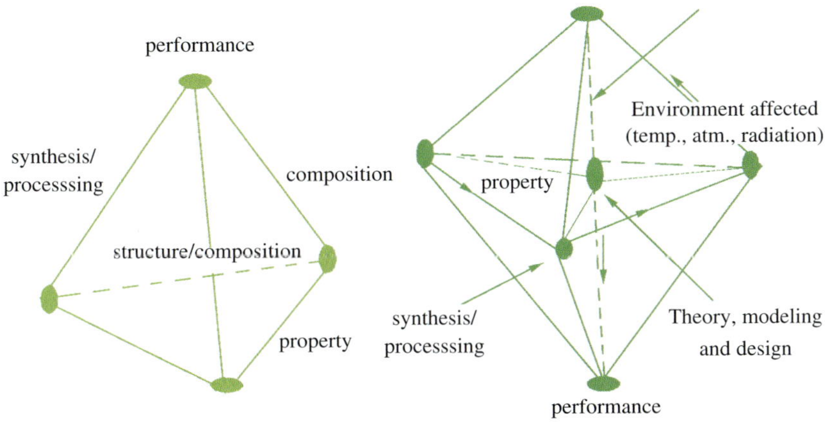

(a) (b)

Fig.13 The expression of materials science and engineering
(a)4-elements model; (b)5-elements model

Furthermore, performance of a material is explicitly expressed as properties under the environmental influence as temperature, atmosphere and state of mechanics.

Materials design started from model recognition in China. Based on semi-empirical statistical regulations, guiding materials and/or process are designed, and good results have been achieved[31]. In the meantime, materials design has also been placed in national programs, as the subject "materials micro-structure design and prediction of properties" in 863 Program in 1987, "physical foundation and applications of materials calculation" in climbing program in 1997 and "basic problems of materials design and prediction of properties" in 973 Program in 2000, etc. Now a research team of multi-specialized fields at different cross-disciplinary levels has been formed with personnel coming from the whole country. In the team some deal with the relation of elections, atoms or molecules from the first principle, some treat the real materials in a nano-, micro- or meso manner, some treat the problems occurring during production or application of materials as continuous medium. Large quantity of work is going on, and some noticeable results have been obtained. For instance, based on "anion coordination polyhedron model "which was proposed by a Chinese scientist [32] ,a series of nonlinear optical crystals have been developed. It is well known that there is a controversy about melting of metals, and it has been unified through computer simulation[33]. However, as to the structural materials, when designed by computer simulation,some difficulties may be encountered due to thermodynamic, nonequilibrium and structural sensitivity of the structural materials. But along with the progress of science and technology, deep understanding of materials, and development of colossal and intelligent computers, an era of scientific development of materials will eventually be coming up.

4. CONCLUDING REMARKS

The contents of this chapter do not give detailed scientific description only the results are presented in most cases. because of the space. Furthermore, it is sure that some important achievements have not been included for the same reason and because of inter disciplinary nature of materials science.

Acknowledgments: There are many materials scientists who have provided their research results and/or reviewed the paper.

REFERENCE

1 周廉等. 我国基础材料工业与技术的成就与展望. 两院咨询报告. 中国材料发展现状及迈入新世纪对策. 山东科技出版社，2002:93-188

2 Yuqing Weng, Changxu Shi, Research on new generation steel in China, Facets(IUMRS), No.2, 2002: 1-7

3 王淀佐，李海普，李光辉. 中国铝土矿的加工技术与科学问题. 钟掘.《铝——21 世纪基础研究与技术发展研究论文集》.2002.11: 1-14

4 王占国. 半导体材料研究新进展. 王大中，杨叔子.《技术科学发展与展望》. 山东教育出版社，2002: 317-332; Z. G. Wang. R&D of electric and optoelectronic materials in China. An invited talk at EMRS Conference in 2001

5 L. Y. Lin. Preparation and properties of GaAs single crystal grown from melt under microgravity conditions. Materials Science Forum, Vol. 50: 183-196

6 Q. J. Gu, N. Liu, W. B. Zhao et al. Regular artificial nanometer-scale structures fabricated with scanning tunneling microscope. Applied Physics Letters , 1995, 66: 1747-1749

7 雷晓钧，陈海峰，刘忠范. 用 STM 针尖诱导热致气化模式的纳米级信息存储. 中国科学(B 辑).31, 2001: 67

8 N.S. Xu, S.Z.Deng, J.Chen. Nanomaterials for Field Electron Emission: Preparation, Characterization and

Application. Ultramicroscopy , 2003, 95:19-28

9 C.T. Chen, N. Ye, J. Lin et al. Computer-assisted search for nonlineral optical crystals, Advanced Materials, 1999, 11(13):1079-1089

10 Naiben Ming, Superlattice and microstructures of dielectric materials, Advanced materials, 1999, 11(13): 1079-1089

11 Xingdong Zhang, Huipin Yuan, K. de Groot. Calcium Phosphate Biomaterials with Intrinsic Osteoinductivity, Notebook: Workshop 1#, Biomaterials With Intrinsic Osteoinductivity, The 6th World Biomaterials Congress, May 15-20, Hawaii, USA

12 C. Du, F.Z. Cui, X.D. Zhu, et al. Three-dimensional nano-HAp/ collagen matrix loading with osteogenic cells in organ culture. J. Biomed Mater Res, 1999, 44: 407-415

13 Zixiao Pan, Wei Pan, Rong Li. Preparation of SiC ceramics by spark plasma sintering. Key Eng. Mater, Vol. 224-226, 2002: 713-716

14 Yanchun Zhou, Zhimei Sun. Crystallographic relations between Ti_3SiC_2 and TiC. Materials Res.Innovations, 2000, 3(5): 286-291

15 Xiaowen Zhang, Fei Fang. Study of the structure and dielectric relaxation behavior of Pb(Mg1/3)O_3-$PbTiO_3$ ferroelectric ceramics. J. Mater. Res., 1999, 14(12): 4581-4586

16 Haosu Luo, Guisheng Xu, Haiqing Xu. Compositional homogeneity and electrical properties of lead magnesium niobate titanate single crystals grown by a modified Bridgman technique. Jpn. J. Appl. Phys., 2000, 39(96): 5581-5583

17 成会明 . 纳米碳管（制备、结构、物性及应用）. 化工出版社 , 2002: 1-47

18 Z.W. Pan, S.S.Xie, B.H.Chang et al. Direct growth of aligned open carbon nanotubes by chemical vapor deposition. Chem. Phys. Lett., 1999, 299:97-102

19 C. Liu, Y.Y.Fan, M. Liu et al. Hydrogen storage in single walleed carbon nanotubes at room temperature. Science, 1999, 286: 1127-1129

20 W. Han, S.Fan, Q.Li et al. Synthesis of GaN nanorods through carbon nanotube-confined reaction. Science,1997, 277 :1278-1279

21 Kaili Jiang, Qunqing Li, Shoushan Fan, Spinning continuous carbon nanotube yarns, Nature,2002, 419:801

22 叶恒强 . 五次对称性及镍钛准晶相的发现与研究 . 中国科学院综合计划局编 .《攀登者的报告 . 学苑出版社，1999:32-34

23 K. Lu. Noncrystalline metals crystallized from amorphous solids. Materials Science & Engineering (Elsevier), 1996, R16: 161-221

24 L. Lu, M.L.Sui, K.Lu. Superplastic extensibility of nanocrystalline copper at room temperature. Science, 2000, 287: 1463-1466

25 W. P. Tong, N. R. Tao, Z. B. Wang, J. Lu, and K. Lu. Nitriding iron at low temperature. Science ,2003, 299: 686-688. (in Reports)

26 张永刚 . 韩雅芳 . 陈国良等主编 . 金属间化合物结构材料 . 国防工业出版社 ,2001: 604-631

27 Y.X.Zhu, S.N.Zheng, L.H.Lou et al. Superalloys with low segregation, a paper on <The Innovation in Real Materials Conference>, held in Washington D.C., July 1998

28 傅恒志、沈军、郝启堂等 . 高熔点大比重合金真空电磁约束成形

29 W. J. Xie, C. D. Cao, Y. L. Lu et al. Levitation of Iridium and liquid mercury by ultrasound. Phys. Rev. Let., 2002, 89(10):104304(1-4), Reviewed by Choi, Science 802(2Aug. 2002)

30 Changxu Shi. Hightlights of materials science and technology at the turning of the 21 century. Program in Natural Science, 1999, 9(1): 2-14

31 陈念贻.数据信息采掘与材料设计的半经验方法.石力开.材料设计.天津大学出版社, 2000: 238-266

32 C.T.Chen. A localized quanta thorectical treatment, based on an anion coordination polyhedron model. Sci. China, 1979, 22:756-757

33 Z.H.Jin, P. Gumbsch, K.Lu et al. The limit of superheating, Physical Rev. Letters , 2001, 87, Reviewed by R.W.Cahn. Melting from within. Nature , 2001, 413: 582-583

The Development of Publishing Technology in China

Wang Xuan

Wang Xuan, *a computer expert, was born in February, 1937. He graduated with degrees in mathematics from Peking University. Director of the Institute of Computer Science and Technology, professor and tutor for Doctorate Students, academician of Chinese Academy of Sciences, academician of Chinese Academy of Engineering, academician of the Third World Academy of Sciences, member of the Standing Committee of the NPC, Vice-Chairman of the Central Committee of the Jiusan Society, Vice-Chairman of the NPC Education, Science, Culture and Health Committee, inventor of a Chinese language laser phototypesetting system for the Chinese Language, with one European and eight Chinese patents; presided over the research and production of electronic publishing systems which initiated a technical revolution in the Chinese publishing and printing industry also widely used abroad. He was awarded twice the First Prize of the National Science and Technology Promotion Award, the Science Prize of the UNESCO, the Golden Plate in Geneva International Invention Exhibition, the Golden Prize of Chinese Patents, the Achievement Award from the Chinese Institute of Engineers of the Americas and the State Prominent Science and Technology Award 2001.*

Abstract: Since the 1970s the printing and publishing industry in China has taken some giant steps forward. The lead type composition with over a hundred-year history was abandoned thoroughly and replaced by modern electronic publishing systems. This new pre-press technology brought China into a leadership position in the industry. With the arrival of the new age of the Internet, China's newspaper and publishing industry became a major user of modern publishing technology. With fierce market competition and continuous advancement, the publishing industry in China will face more development opportunities and challenges.

1. THE HISTORY OF THE DEVELOPMENT OF ELECTRONIC PUBLISHING IN CHINA

Great advancements have been experienced by the Chinese printing and publishing industry in the last 20 years. During these years, the lead type composition was abandoned directly by adopting the laser typesetting system. The air lifting of paper moulds of newspaper pages was replaced directly with the satellite transmission of newspaper page information. The color separated printing plates can be output directly without galley proofing. The Internet was introduced the capacity for high-speed digital communication, which made a major change in the way the printing, publishing, and newspaper industries produced their products. Among all these achievements, four innovative technologies emerged.

1.1 The First Technology Innovation: Leaving Lead and Fire and Leaping Directly from Lead Type Composition towards Laser Photo-type Composition (1987-1993)

Twenty years ago, almost all printing in China was done using lead type to compose pages. Based on an investigation of the technical features of the second generation optical mechanical phototypesetter and the third generation CRT phototypesetters, and considering the huge size of the Chinese character set and the state of technology in China at that time, we decided to bypass the second and third generation phototypesetters which were most popular at that time. Instead, we developed directly the non-commercialized fourth generation laser phototypesetting system.

As a result of this, the Chinese printing industry did not experience the popular use of the second-generation optical-mechanical phototypesetters and third generation CRT devices. The industry leaped directly toward the most advanced fourth generation laser composition.

When the project of laser composition was initiated, we found the huge size of the Chinese character set was a big problem. The huge size of the character set made the storage amount of the font information very large, so we were forced earlier than our western colleges to adopt the outline description of fonts to reduce the amount of information. There was another big obstacle when implementing the project. The problem was describing the font shape efficiently. Chinese characters are much more complex than western characters. The size of those vertical and horizontal strokes must be accurate when the font is resized. This forced us to be the first to adopt HINT technology to describe the key features of the strokes in a character.

Also, we adopted the full-page output strategy from the beginning. In other words the complete book or newspaper page is output at once. This page is then used to make the printing plate. There is no need for stripping and galley pasting. The pages are composed on the computer screen and are output as a whole.

1.2 Bypassing the Facsimile Newspaper Transmission and Adopting the Remote Page Description Language Transmission

During the 1980s, paper moulds of newspaper plates were transported by airplanes. Few cities were using newspaper facsimile machines to transmit newspaper pages. Because facsimile transmission was of low resolution and prone to data loss, the quality of the newsprint was not very good. Newspapers in China started using PDL (Page Description Language), which is now an industry standard. This makes it easy to use the unified page description language in the transmission of newspaper pages. The PDL-based newspaper page transmission was adopted in 1990. When using this kind of page transmission, newspaper pages were described in PDL format. When the recipient receives the page description in this form, they can render the page image without any loss of detail. By using page description language in the transmission, the amount of the information transmitted is reduced tremendously, and the amount is about 1/50th of the facsimile transmission. This makes the transmission speed very high. In 1992, "People' s Daily" started to transmit their newspaper pages to 22 cities in China in this way via satellite. It only took two minutes to transmit a single page. This technology was soon popularized among the domestic and overseas Chinese newspaper industry. After 1992, most of the larger Chinese newspapers were using this method to transmit their newspaper pages. Then many provincial and city newspapers were also using this method to transmit the pages, and their circulation amount increased as a result

1.3 The Third Technology Innovation: Good-bye Traditional Electronic Color Separation Scanners (starting early 1992)

Before 1992, electronic color separation scanners were used for the composition of almost all the color publications in China. It took 2 to 3 hours to complete the separation and composition for a color plate. In 1987, we brought forward a plan to develop an open color image setting system in order to replace the electronic color separation scanner. In 1992, "Macao Daily" adopted our color newspaper publishing system, which marked the beginning of the technology innovation of abandoning the electronic color separation scanner. By using this system, text and color images were processed together, and it took only 20 minutes to output a color plate (now the time has been decreased to 2 minutes). Currently there are hundreds of newspapers publishing color pages, and tens of newspapers are publishing color pages every day. Chinese systems hold a 90% market share in the Chinese color newspaper market. In 1994, we also successfully developed high-end desktop color-publishing systems. The output quality is comparable with traditional electronic color separation systems. This allowed color photography in books, magazines, and newspaper to be commonplace.

1.4 The Fourth Technology Innovation: Throwing away the Paper and Pen by Implementing the Full Control of the Workflow of News Processing

In 1994, the "Shenzhen Evening News" adopted the Founder Editorial Workflow Management System. This enabled their editors to edit their newspaper themselves and thus marked the beginning of the abandoning of paper and pen. There are now more than 200 newspapers working in this way. Besides the editorial systems, advertisement production, data management systems, distribution systems and electronic newspaper systems were also developed. These systems were closely connected with the newspaper pagination systems. This constituted an integrated information management solution for newspapers and other publications.

2. INFORMATION TECHNOLOGY IN THE NEWSPAPER AND PUBLISHING INDUSTRY IN THE INTERNET AGE

2.1 Integrated News Collecting and Editing

In the past 10 years, China's news publishing industry experienced 4 generations:

1st Generation: Data Storage System based on Novell file sharing; client workstation platform was DOS.

2nd Generation: Files saved based on NTFS file system, using Windows NT and client/server technology, with multi-thread technology adopted for server software, so as to change rapidly with client's requirements.

3rd Generation: File saving based on large relational databases, using Windows as the system platform to realize multiple periodical management and production.

4th Generation: Based on Internet/Intranet platforms and using Browser/Server architecture. Integrated workflow control in news collecting, editing, processing, manuscript transmission in the publishing industry became commonplace. The system can control all cycles from manuscript edition, checking, dispatching and composition, to picture scanning, production, dispatching, and block making of large proofs, imposition, clean proof checking, and large proof signing. It takes pages as the core of the whole production process, integrating imposition, collection and edition control, advertisement control, file control and archive indexing closely into one platform. It connects seamlessly with imposition software enabling control in real time of large proofs, monitors composition results and records automatically modifications during the composition process. Its full compatibility with indexing systems enables indexed results to be directly cut and pasted into the manuscript. The output can be printed on large proofing systems so that a client can approve it.

2.2 Media Network

As a new media method, networking has spread into each corner of people's life with unprecedented vitality. According to statistics at the end of 2002, there were 59.1 million people getting on the Internet in China, only second to the number in the USA. As the main information distributing method, the website has quickly become standard. The media website and e-commerce website construction is divided into three levels:

2.2.1 Website Application Level: Includes web news production and release, on-line comprehensive service and control integration, on-line classified advertisement, web page personalization and information indexing engine.

2.2.2 On-line e-commerce Level: e.g. on-line book and newspaper subscription, on-line information subscription, on-line personalized service and electronic payment, on-line advertisement and transaction control, etc.

2.2.3 Service Level for Media Website Construction: e.g. website design consultation and web page design, etc.

2.3 Integrated Solutions Include the Following Systems

2.3.1 Website Content Automatic Collection System

The system can get up-to-date information in real time safely, automatically and precisely, and save them into local database, enquire and index according to designated classifications, and transmit automatically the data into other applications and systems. Through this system, information databases based on enterprise can be created rapidly and effectively, and government authorities can implement secure on-line information collection solutions using page filtering.

2.3.2 Classified Information Release System

The system is specialized in handling classified information focusing on definition of system scheme for web release. It uses template technology to realize active release providing information service active release platform based on database for professional information services websites. This system is aimed for on-demand database-based information. It can extend information classification and index and control information according to customisation. It provides automatic classification, storage and content administration by various means of information collection. It indexes and releases active information based on database through the full-text indexing and customized field characteristics functions.

2.3.3 Website Content Service System

This is a software system especially used by large news websites and those providing ICP content service. It facilitates management to websites providing news information and content service in efficient content collection, production and release. It improves the automatic process of website content production, reduces technical difficulties and saves labor costs. It improves work efficiency, reinforces management control, so as to provide an excellent software system platform for the integral marking and operation of the news website. It realizes full-range digital information free flow from traditional media to network information by specific technologies. It also enlarged and completed news collection capacity and methods of the website that depend more on automatic processing by the system while processing a great deal of content and information.

2.4 Media Digital Asset Management

In fierce competition, China's newspaper and publishing industry is being converted from an industrial model to an operational model, focusing more on the construction of information technology supporting platforms, accelerating e-commerce based on information technology so as to turn itself to be a high-quality asset.

In this new information industry, media content such as manuscripts, pages, pictures, page files, background documents, audio-visual documents, are valued digital assets. How to better produce and control these high-valued digital content and provide related services? This is an imminent question for the development of the information industry or media. With the continuous increase in information collection, digital information is growing fast, and constantly. Thus, a question comes out: How to stock and control those contents? Only by effective management can people ensure a large amount of digital content to become active information. Also, technology must be able to freely access and process information, extend their life cycle, and have the content create new value.

Digital asset management includes 4 cycles: multi-channel information collection and creation, flexible information processing workflow control, content storage control, and cross-media publishing. It forms a unified platform covering collection, processing, storage and release, and accelerates the information production process with complete digital asset management. This improves production efficiency, reduces production costs of cross-media publishing, and allows for efficient utilization and value-addition of digital assets.

3. CTP AND DIGITAL WORKFLOW

3.1 Computer To Plate

Computer To Plate (CTP) systems output text and images directly to plates by use of digital workflow. After automatic processing, printing can be made directly on offset machines after automatic processing of plates, thus avoiding major processes like film output and manual PS printing. It improves production efficiency and reduces costs. Traditional methods use ultraviolet rays exposed to film already sensitised by laser phototypesetter,. This process causes certain degradation in image quality. However, CTP images digital information directly onto plates, which reduces intermediate processes so as to maintain quality. This is no doubt a great development of the printing technology. Peking University was ever highly attracted by CTP's prospect while selecting the laser phototypesetting direction in 1975. Around 1990, CTP was first put into practical use, and was rapidly developed in the 1990's. Eventually, CTP will eliminate laser film imagesetters.

The light source and corresponding mechanism of CTP is closely linked with plate requirements. Compared with phototypesetters, that is the biggest difference. Therefore, the development trend of CTP plates should be first discussed. Through the utilization and development in the 1990's, three plate types emerged: argentic salt plates, photosensitive high molecular plates, and thermal-sensitive plates.

Argentic salt platesuse mainly the diffusion and diversion process of argentic salt. There are physical developing core layer and a photosensitive halogen silver emulsion layer on the processed aluminum plates. Developing is done after exposure, and argentic salt will be diffused and diverted to form images. Then, the plates are ready for printing after being water washed and processed by a plate-curing agent. The advantages of argentic salt plates are high photosensitivity (photosensitive power less than 10μJ/cm^2), high contrast, high resolution and good print running ratio (above 100,000 copies). With the above strength and mutual technology, argentic salt plates are the most widely used CTP plates taking the biggest share in the market at the present time.

Photosensitive high molecular plates consist of three layers: processed aluminum base, photosensitive high molecular layer and polyvinyl alcohol layer. The photosensitivity is much lower than that of argentic salt plates (requiring 50-300μJ/cm^2 , 100 times of the power for argentic salt plates), with high print running ratio (above 200, 000 copies).

The CTP equipped for the above plates use generally visible laser source, which are mostly argon laser (blue, wave length 488cm), double-frequency YAG laser (green, wave length 532cm), and occasionally nitroneon laser.

Thermal-sensitive plates have appeared in recent years. These consist of an unprocessed aluminum base, ink absorber and infrared absorbed image layer. The image layer is scanned by high-power infrared semi-conducted laser and melted to expose a lower ink-receptive layer. The biggest strength of thermal plates is that no development is necessary and no wasted liquid is to be processed as far as environment protection is concerned. It can be operated in the sunlight. However, the photosensitivity is rather low, requiring power 100-300mJ/cm^2, around 1,000 times of that for visible photosensitive high molecular plates. This makes it difficult to compare the plate's output CTP speed with the previously mentioned two systems.

All thermal-sensitive CTP plates use infrared semi-conductor laser devices. Up to now, thermal plate CTPs are rarely used by newspaper publishing. Their main users remain commercial printers due to the fact that the output speed of this kind of CTP cannot be compared to those of the above-mentioned two visible lights CTPs. While for newspaper houses, handling capacity and timeliness are extremely important. The future models of infrared thermal plates may eliminate steps, which is highly desired. Therefore, all big companies are turning to infrared thermal platemaking equipment, and add infrared thermal plates as an option to visible lights equipment so as to make double use of one machine.

Recently, CTP technology based on traditional plates appeared, which can get high-quality images on traditional plates, and are already used in Germany and Japan.

There are more than 5,000 CTPs installed around the world, but only 100 in China, showing its slow development in this respect. Presently, CTP plates are more expensive than PS plates, which is a main factor to affect the rapid promotion in China. With more volume, the price will be reduced to be close to that of common plates.

3.2 Digital Workflow Control

With the rapid development of digitalisation and network technology, traditional block making and printing are facing new challenges and opportunities. New industrial norms, new software technology, new hardware, new solutions, new service ideas, new commercial modes and new management thoughts are being accepted and applied more and more by the enterprises in this industry. We can clearly see that digital fast printing and printing-on-demand is profitable when using advanced CTP technology including imposition and digital proofing. Network remote proofing/output and printing e-commerce are about to arrive. Advanced workflow systems will help the printing and block making enterprises improve their core competitiveness in the fierce competition, bring them full-range improvement in quality, efficiency, management and profit.

ElecRoc was researched and developed by Beijing Founder Electronics Co. Ltd., primarily used in production environments such as publishing/printing enterprises, commercial printing houses and plate-making/printing centers, and magazine publishers. It adopts the most advanced database and Internet technologies with open format at the industry's standards to the satisfaction of new requirements by clients in the new Internet age.

Features of ElecRoc:

3.2.1 Uses JDF Workflow

Beijing Founder Electronics Co. Ltd is a member of CIP4 — a global organization in the printing industry. JDF is the most up-to-date, open, scalable (expandable) standard for publishing and printing workflow. JDF adopts XML for coding, covering all continuous controlling information during the whole workflow, such as content making, pre-press, in-press, after-press and release etc., with excellent scalability. ElecRoc adopts JDF as an electronic job ticket, realizing highly efficient and smooth, automatic, and manageable workflow. It connects industrial workflow with a management system, providing a solid basis for publishing enterprises to develop on-line business and implement e-commerce.

3.2.2 Uses PDF Workflow

It adopts PDF as the internal standard file for workflow. PDF has excellent features satisfying high-end printers. One feature is the ability to embed pictures and words, stable, open, highly compressed, able to be previewed and fit for network transmission. Also, its reliability and editability enables ElecRoc to create digital real-time ganging and imposition, as well as last minute content modification, a feature asked for by users..

3.2.3 Internet based, supports remote submission, remote proofing and remote control

The open structure based on the Internet and browser page can be operated on any platform and Internet terminals. Clients need only to be equipped with standard PC/MACs that are used for getting on the Internet for any job to be submitted, proofed and controlled smoothly and remotely. Therefore a remote output and on-line business is created.

3.2.4 Job tracking and checking, business management system based on large database

Jobs can be queried, tracked, checked according to customized job tickets. XML database reinforces business tracking and data statistics capabilities of network, and outputs relevant data to MIS or ERP systems used by printing enterprises. Then, industrial and managerial workflows can be linked automatically to help enterprises realize full digital management.

4. NETWORK PUBLISHING (E-BOOK)

e-book has unanimous acceptance by the publishing industry as it can publish and sell digital publication via the internet at low cost, without logistics, inventory shipping charges, being duly updated and good for environment protection. Founder created the Apabi e-book series of solutions from April, 2001 to the end of 2002. Now there are already hundreds of publishers, book suppliers, libraries and websites using Apabi to promote e-book business.

The five letters of APABI represent: "A" for Author, "P" for Publisher, "A"for Artery,"B"for Buyer, and "I" for Internet. The true e-book methodology incorporates all areas of traditional publishing into its new technology.

Apabi consists of the following core software:

4.1 Production and Publishing Software (Apabi Maker & Apabi Writer)

Apabi Maker is production software for ebooks, enabling all formats to be converted to e-book format. It keeps all information like the characters, fonts, pages and colors of the original files, including pictures, mathematic formulas, chemical formulas, forms, chess and cards as well as music scores. This format compresses words and images efficiently, producing small files.

Apabi Writer is a content editing tool for e-books. It allows addition of superlinks and elemental data information (including authors, publishers, prices, content brief introduction and classification, etc.), enabling reading e-books more conveniently and faster. It can add multi-media links to e-books and produce those with sounds, cartoons and multi-media, making e-books more colorful with richer content.

4.2 Security Release Software (Apabi Rights Server) & Transaction Processing Software (Apabi Retail Server)

These are system programs for digital intellectual property control and protection, avoiding illegal copying and transmission, as well as securely distributing and selling e-book. It's the core of digital rights maintenance (DRM) provided to publishers (distributors) and on-line bookstores.

4.3 Reading Software (Apabi Reader)

It is used to read various file formats like CEB, PDF, HTML, TXT and OE etc., to enlarge or reduce fonts, change background colors, rotate images and maximize / minimize pages. It has lining, lightening, circling, bookmarking, annotating and commenting functions. It can turn up and down, jump directly to designated pages, turn to the first or last page, and has its specific function of semi-page and full-page turning, which is protected by a patent. This software can also be used to classify, sort, seek, and administer e-books. Also, it has full text seeking, content extraction, full-text reading aloud; It supports cartoon broadcasting and audio-visual files like sounds, and translates by reading directly from the screen.

Apabi solutions can make publishers, newspaper and magazine publishing houses enter into network publishing rapidly at low cost. It allows the user to quickly publish e-books, and establish websites with great amounts of books, and e-libraries.

China's news publishing industry has adapted very quickly to new technologies. With the increasing market competition and continuous emerging of newer technologies, China's news publishing industry will face greater opportunities and challenges.

Energy Technology for Sustainable Development in China

Yan Luguang

Institate of Electrical Engineering ,Chinese Academy of Sciences

Yan Luguang, *born in 1935, Dongyang City, Zhejiang Province. Graduated from Moscow Power Institute in 1959. Member of Chinese Academy of Sciences, Foreign Member of Ukraine Academy of Sciences, Member of the Third World Academy of Sciences. Present positions are: Research professor and Chairman of Scientific Committee of Institute of Electrical Engineering, Chinese Academy of Sciences , President of Ningbo University, Chairman of Energy Research Council of Chinese Academy of Sciences, President of Chinese Solar Energy Society; Vice-President of China Electrotechnical Society and China Energy Research Society. For a long time he has been working on the development of special electrical equipments and on the research and development of new technology in electrical engineering in China. His main working areas include: High Pulse Power; Fusion Electrical Engineering; Superconducting Electrical Engineering; Magnetohydrodynamic Power Generation and Ship Propulsion; Renewable Energy; High-speed Magnetic Levitation Train, etc. Many important achievements have been made in these corresponding areas. In 1990s he won six first-class or Second-class science and technology advancement awards issued by provinces or ministries of Chinese Government.*

Abstract: Sustainable energy development is a main task human faces in the 21st century. Because of rapid economic development and low energy consumption levels per person in China, the energy scale will grow significantly during the first half of 21st century. Therefore it is very important to establish a sustainable system. Considering that coal is the main resource, oil resources are in shortage, potential of water energy resources is limited and efficiency of renewable energy is low in China, attention is now being paid to making greater effort to develop multifold energy technology. Emphasis is to develop the technologies of high efficiency and clean utilization of coal, new energy development, and oil and energy saving. In this paper the progress of these main areas during recent years in China is briefly presented.

1. INTRODUCTION

In the 21st century human being expects to enter a new era of continuous, harmonious and steady development of economy, society and environment. The sustainable development of energy and electrical power has an important significance. China is a big and rapidly developing country; its sustainable development will have important influence on the world. Table 1 shows general situation of energy and electrical power development in China and forecast for the first half of 21st century. The data about 20th century are based on the related statistical materials [1,2]. The forecast is chosen from different estimated values which seem more close to the actual situation。

From these data, it can be seen that the main features of energy development in China are: ① energy requirements will increase rapidly since China is a big developing country whose economy is developing rapidly, from 1949 to 2000, the total production of commercial energy increases more than forty times. Comparing the total production in 2050 with that in 2000, it will increase continuously 3-4 times. Although energy consumption level per person is not high compared with advanced countries, but its absolute value will increase from about 9% of total world energy in 2000 to about 15%in 2050. It is very important to determine guiding principles for sustainable development according to the requirement, resource and technology. ② Coal is the main resource in the country's energy structure. Coal accounts for above 60%of total energy sources currently. It was estimated that coal will amount to above 40% in 2050. However, coal accocnts for only 27% of primary energy sources in the world. Utilization of a large quantity of coal causes serious environmental pollution and ecosystem destruction problems. The utilization efficiency has to be increased greatly.

Therefore it is critical to devote major effort to developing the technologies of high efficiency and clean utilization of coal. To reduce the percentage of coal utilization gradually, China needs to develop new energy,i.e. nuclear power and renewable energy. ③ Oil production and its resources are limited. The production of crude oil amounts to about 20%of total primary energy products in China. After 1993 it became an oil net import country. The imported oil exceeded 70 million tons in 2000. The imported oil probably will amount to 40% in 2020. Therefore, the energy security becomes a priority problem of consideration. Many ways are needed for saving oil, including strengthening oil prospecting in many ways, developing the oil industry and taking an active part in international development. ④ The utilization efficiency of energy is low. Energy intensity (energy consumption per GDP) is high in our country, it is 4.7 times higher than the world average value and close to the level of UK, USA and Canada at the beginning of 1970s. The conversion efficiency of energy is only 29%. Comparing with world advanced level it is ten percent lower To study energy saving technology in every area is a focal point of energy development. ⑤ Population in rural areas amounts to above 65%(860 million), therefore solving the energy problems in rural area is an outstanding task.

Table 1 General Situation and Forecast of Energy and Electric Power Development in China*

			General Situation of Development					Forecast of Development		
	Year		1949	1980	1990	1995	2000	2010	2020	2050
	Total Population (billion)		0.542	0.987	1.14	1.21	1.275	1.4	1.5	1.5~1.6
Energy	Total Production	Mtce	23.7	637.4	1039	1290	1070	~1900	~2500	~4000
	Coal	Mt	32	620	1080	1361	1000	~1500	~2100	~2500
		%	96.3	69.4	74.2	75.3	63.8	53.8	54	42
	Crude Oil	Mt	0.12	106	138.3	150	163	280	360	740
		%	0.7	23.8	19.0	16.6	21.8	20	18.7	25
	Natural Gas	billion m^3	0.007	14.27	15.3	17.95	27.2	100	160	250
		%	—	3.0	2.0	1.9	3.5	6.7	7.8	8.0
Electrical Power	Total Power Generation	Twh	4.3	300.6	621.3	1007	1369	2600	3800	~7000-8000
	Total installed capacity	GW	1.85	65.87	137.8	217.2	319.3	570	~800	~1500-1800
	Fire-electric Installed Cap.	GW	1.69	45.57	101.9	162.8	237.3	440	~600	~1000-1300
	Hydroelectric installed Cap.	GW	0.16	20.3	36.0	52.2	79.4	115	160	~250
	Nuclear Power Installed Cap.	GW	0	0	0	2.1	2.1	10	30	~200
	Renewable Electric Installed Cap.	GW	0	0	0	0.1	0.5	3	20	~100

* The data are from *Energy Policy Research, 2002, No 1; Energy Data 2002; China Energy Annual Review, 1997*[1,2]

Considering the energy requirement and its features, energy technology development is significant for sustainable development in China. The focal points are : ① high efficiency and clean utilization of coal, ② development of new energy, especially development of nuclear power and renewable energy, ③ oil and energy saving. In this paper the related progress during recent years in China is presented briefly.

2.HIGH EFFICIENT AND CLEAN UTILIZATION OF COAL[3,4]

Coal is the main energy resource in China's energy structure. This situation will not be changed until the middle of the 21st century. The widespread attention to the environmental pollution problems caused by production and utilization of coal is increasing day by day. The reason of causing pollution is not coal itself, but the utilization method. Since the 1980s the main countries producing coal started actively the "clean coal program." Since the 1990s China formulated and approved a related program to develop clean coal technology, that is a technology for processing, burning, conversion and pollution control of coal to obtain high efficiency and clean utilization. It received great attention and related arrangements were made. People gradually recognized that coal is a reliable, low cost energy source which can be used cleanly.

Recently coal-fired power generation is the main consumer of coal. The coal used for power generation amounts to 50%of total production in China. The percentage will increase continuously in the future. In the USA the coal used for power generation is 85%of its total production. Therefore to develop advanced, high efficiency and low pollution technology for coal-fired power generation becomes the central point of the whole clean coal program. In this respect, recently many new technologies have been developed including circulating fluidized-bed combustor (CFBC), pressurized fluidized-bed combustor combined-cycle (PFBC-CC), integrated gasification combined-cycle (IGCC) , magneto-hydrodynamic-steam combined cycle (MHD-CC). The studies of all these four technologies have been carried out in China and certain progress has been made.

In CFBC, different kinds of fuel, especially low quality coal, can be burned with high efficiency. In the burning process the SO_2 exhaustion can be reduced by adding desulfurizer. Low temperature combustion in fluidized-bed can control the production of NOX. In the world, CFBC power station with 250MWe electrical output has been put into commercial operation in France, and the contract for 300-400MWe boiler has been signed. After many years development, China has created a quite strong R&D foundation. The design and manufacture of boilers with the capacity lower than 410t/h has been accomplished. The production occupied most of boiler markets where boiler capacity is lower than 75t/h. The engineering demonstration projects have been started. In these projects one boiler is 150MWe in which our own technology will be used and the other boiler is 300MWe in which introduced technology will be used .

The advantage of PFBC-CC is similar to CFBC. After going through dust trap, the high temperature exhaust gas produced in PFBC-CC enters into the gas turbine to produce mechanical power, then make gas-steam combine cycle. Its power generation ability increases 20% comparing with pure steam turbine with same steam parameters, its efficiency increases 3%-4%. In the world, eight electrical power stations have been built, the one with the highest capacity is 360MWe, the others are 80-100 MWe. In China, the research and development for PFBC technology have already about 20 years' history. During "9th Five-Years Plan" period a PFBC-CC intermediate experimental power station with 15MWe output was built at Jiawang Station of Jiangsu Province where domestic technology and equipment were used. A demonstration power station with 100MWe output was planned to build in which most of equipments will be made in China except gas turbine equipment and some introduad advanced technology. To further increase efficiency up to 45%-48%, a method using part gasification and pre-combustion has been studied in the world to increase the inlet temperature of gas turbine up to 1100-1300℃, it formed the second generation technology of PFBC-CC with over-critical vapor parameters, it is still in the testing stage now. In China the primary study for some key technology and basic studies for this process are also underway.

In IGCC, coal is gasified to be fuel gas, then drives gas turbine to generate electrical power, its exhaust gas goes through a heat recovery boiler and produces steam to drive steam turbines to generate electrical power. Its advantage is high efficiency and low pollution. IGCC has passed conception verification and technical demonstration operation, and entered into commercial demonstration stage with 250-300MWe capacity. Since the whole system is quite complicated with high investment, its further development will combine co-production of electrical energy, heat energy, city gas and chemical products to form multi-product industry in which the base is coal gasification. High efficiency, clean, comprehensive utilization coal engineering started quite late in China. Key technology studies just started in "9th Five-Years Plan" period. During "10th Five-Years Plan" period The State Electrical Power Company is going to introduce technology by public bidding to build a demonstration power station with 300-400 MWe in Yantai Power Station of Shandong Province.

MHD power generation is a high efficiency, low pollution and new direct heat to electricity conversion method. Since principle experiments demonstrated successfully at the end of 1950s, this technology obtains quite wide attention. Since the beginning of 1970s the research, development and experiment were in progress actively in the former Soviet Union, USA and Japan with expenses of several hundred million US dollars. Its object was to develop the MHD-Steam Combine Cycle power station with efficiency up to 50%-60%. 11.7MW electrical output and 250 hours operation duration have been achieved. Since 1980s the former Soviet Union and USA began the test power station program using natural gas and coal as fuel. After obtaining some intermediate successes, almost all countries in the world stopped their national development program in the mid 1990s because there are still great technical difficulties in MHD and other coal-fired combine cycle obtained good progress. China began to study MHD power generation in 1962. The Institute of Electrical Engineering of the Chinese Academy of Sciences, the South-East University and the Shanghai Institute of Power Equipment have made continuous research and development work for more than 30 years. 12 experiment facilities were built for experimental study altogether. In 1986 the coal fired MHD power generation technology was decided as a subject to be included into the National High Technology Research and Development Program ("863" Program). The work has concentrated on the R&D for intermediate test power station. Some good results were achieved: 130 KW electrical power output, two hours continuous operation were obtained from a coal fired experiment and a saddle-shape superconducting magnet system with 0.44 m bore diameter and 4 Tesla central field has been constructed and tested. Because international development has stopped, the whole MHD program stopped at the end of 2000. Some people are still working on MHD thruster and MHD high pulse power supply.

In general, the technology of high efficiency and low pollution coal-fired power generation in China follows all the international progress. Quite a good foundation was created, but the support for R&D is too small and the work is only at the starting stage. It still has to be further strengthened.

Except the clean coal technology for power generation, much works has been done to reduce pollution such as coal washing, selection and fabrication, exhaust gas purification. Coal washing and selection are used to reduce the content of ash and sulphur, then to reduce the exhaust of ash, SO_2 and so on. In developed countries the coal washing and selection are used already up to 50%-90%of the total coal. The coal washing ability of factories built in our country is about 500 million tons. But because of small scale of factories, poor reliability of facilities and policy reasons the cost of washed coal is still high, the percentage of washed coal is still low, being about 20%-30%. Using shaped coal and water coal slurry is an important measure to increase burning efficiency and reduce pollution. The coal shaping technique for civil usage in our country is on quite high level. The sale volume in cities and towns is about 40 million tons per year. Nine experimental factories producing water coal slurry were built. Their total production abilities are 1.76 million tons per year. Five power stations burning coal slurry are in commercial operation. In the area of exhaust gas purification at power stations, certain efforts were made and the situation is improving in recent years, but still of a starting stage. In general, new built big coal-fired boiler has electrical dusters with 3-4 electrical field, its efficiency is lower than 99%,the facilities in most of middle and small boilers fall behind. Denitration control does not exist in the facilities lower than 300MW. Many studies and developments of desulfuration technology were made, but compared with advanced country we still have a long way to go. The coal liquefaction problem will be mentioned later in section "Oil Saving."

3.NUCLEAR POWER[5]

Nuclear energy includes fission energy and fusion energy. The fission energy is proven to be able to replace fossil fuel in large scale and is a commercial competitive clean energy source. After a rapid development in 1960s-1970s it has become a mature industry. At the end of 2000 a total of 438 generator sets are in operation with total installed capacity 350 GW, producing 17% of total electricity in the world. Facing challenges from safety, economy, nuclear waste treatment and nuclear non-proliferation, since 1990s the development of nuclear power entered into a stagnant period. According to the forecast, up to 2020 the total capacity will still be 330-360GW. In China the program for development of commercial nuclear plants started in 1980s. The brief information about nuclear plants built and under construction in China is listed in Table 2. After all construction finished in 2005 there will be 11 sets in operation with total installed capacity of 8.7GW. The next construction program is under active preparation. It is estimated that thc total installed capacity in 2010 will exceed 10GW and in 2020 will exceed 30 GW.

Table 2 Brief Situation of Nuclear Power Plants Built and Under Construction in China

Name of Plant	Installed Capacity and Type	Time of Construction and Going online
Qinshan,Zhejiang	300mw, pressurized water reactor,made in China	1985-03 starts construction, 1991-12-15 in operation
Daya Bay, Guangdong	2 × 900MW,French pressurized water reactor	1986 starts construction, number 1 and 2 facility in1994-02-01and 2005-06 in operation respectively
Qinshan phase 2	2 × 600MW,pressurized water reactor , made in China	1996-06 starts construction,number 1 and 2 facility in 2002-06 and 2003-06 in operation respectively
Qinshan phase 3	2 × 700MW,Canada,CANDU-6 heavy water reactor	1997-05 starts construction,number 1 and 2 facility in 2003-02 and 2003-11 in operation respectively
Lingao, Guangdong	2 × 984MW,similar to Daya Bay,	1997-05-15 starts construction, number 1 and 2 facility in 2002-07 and 2003-03 in operation respectively
Tianwan phase 1 Jiangsu	2 × 1000MW,Russian VVER-1000 pressurized water reactor	1999-05 starts construction,number 1 and 2 facility in 2004 and 2005 in operation respectively
Total : 7 plants , 11 facilities	Total power : 8668MW	

It seems that real and significant development of nuclear power probably will happen between 2020 and 2050. If nuclear power will consist of more than 10% national installed capacity, the total installed nuclear power will reach 200GW in 2050. In addition, it was estimated that the uranium resource in our country can only guarantee to provide U^{235} for 30 GW pressurized water reactor for 30 years' operation. Therefore research and development of new advanced nuclear reactors become very necessary and urgent. The encouraging progress has been achieved in fast breeder reactors and high temperature gas-cooled reactors. The research on a fusion-fission hybrid reactor is also carried out.

In fast breeder reactors fast neutrons initiate fission, causing breeding of nuclear fuel. It can increase the uranium resource utilization rate from 1%for pure pressurized water reactor to 60%-70%. In China the basic research of this technology started at the middle of 1960s. It was put into the National High Technology Research and Development Program ("863" Program) in 1986. The project of Chinese Experimental fast breeder reactor with 65MW heat power and 20 MW electrical power was approved formally in 1995. The construction started in 2001. The first critical condition will be achieved at the end of 2005.

High temperature gas-cooled reactors are advanced reactors with good safety, high efficiency of power generation and simple system. It was put into the "863" Program in 1986. A project was approved to build a high temperature gas-cooled experimental reactor with 10 MW in 1992. The construction started in 1995. The critical condition was achieved in December of 2000. It is the only constructed module pebble bed reactor with intrinsic safe characteristics in the world.

The fusion energy is an inexhaustible and sustainable energy. Humans have made continuous effort for controlled release of nuclear fusion energy for more than half of century, and encouraging progress has been achieved. It was known that magnetic confinement and inertial confinement are two main approaches. Ignition was achieved in experiments. The design and development of experimental test reactor have started. In China the research of controlled nuclear fusion started in 1950s. The work concentrated on research and experiment of magnetic confinement Tokomak facility since 1970s. Table 3 lists the main parameters of the Tokomak facilities in China. Now we are preparing to join the international cooperation program of international fusion experimental reactor (ITER). About inertial confinement, the "Shenguang" laser facility was built and efficient research was carried out.

Table 3 Main parameters of the Tokomaks in China

Name of facility	CT-6	HT-6	HT-6M	HT-7	HL-1	HL-2A	HT-7U
Location *	IP	IPP	IPP	IPP	SWIP	SWIP	IPP
R_o m	0.45	0.45	0.65	1.22	1.02	1.64	1.70
a_o m	0.10	0.10	0.20	0.30	0.26	0.40	0.4 × 0.8
B_o T	1.5	1.0	1.5	2.5	3.5	2.0	3.5
I_p KA	50	20	100	300	400	350	1000
Year of operation	1974	1978	1982	1995	1984	2003	2004

*IP-Institute of Physics, Chinese Academy of Sciences, Beijing, IPP- Institute of Plasma Physics, Chinese Academy of Sciences, Hefei,Anhui province, SWIP-South West Institute of Physics, Nuclear Industry Company,Chengdu,Sichuan province

4. RENEWABLE ENERGY

Renewable energy, including solar, wind, biomass, water, geothermal and ocean energies, is an abundant, widely existing, clean primary energy source. With the gradual exhaustion of fossil fuels, its role in the sustainable energy development of the whole of mankind will be higher and higher, its percentage in the energy supply system will grow gradually. The renewable energy will be one of the main supports for the sustainable, coordinated and stable development of the mankind. Fig.1 shows the well-recognized historical development of total usable energy on earth. It is clear that we are entering into the phase of transition from fossil and fissile energy to the sustainable renewable energy, this process will prolong for 2-3 centuries. So, as for the whole world, the renewable energy is also very important for China for future sustainable energy supply.

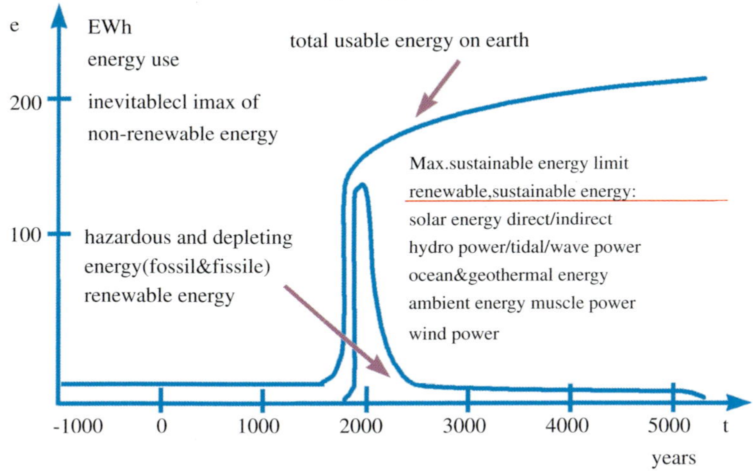

Fig.1 Historical Development of the Earth Usable Energy

As a non-commodity energy source, the thermal utilization of the biomass, solar and geothermal energy and the mechanical utilization of the wind and water energy have several thousand years' history in China. Up to now, the non-hydro, non-commercial renewable energy source, mainly the biomass energy, supply about 260 Mtce of energy for rural areas, which is about 20%of the national total energy consumption. Except that hydropower is already an important support for the electric power industry, the relatively large-scale commercial utilization of other renewable energiy and formation of their new industry are still at the starting stage. As commodity energy sources, the practical needs for development and application of renewable energy in China are coming from three main considerations: ① supplying energy and electricity for rural and remote areas without electricity and energy resource, ② fulfilling the rapid development requirement of the electric power during the 21st century, and ③ changing the energy structure to guarantee sustainable energy development in next 2-3 centuries. Correspondingly, the development and application of renewable energy in China can be divided into three main stages.

Since the 1970s the Chinese government has already well recognized the importance of the active development and application of the renewable energy for rural and remote areas. The work has being included in the national five-year plan. It was determined that the guide principles for active development for rural areas are: ① planning in accordance with actual needs, ② using several complementary energy resources to cover different needs, and ③ integrated development and efficient utilization of various forms of energy resources. Through more than 20 years' continuous effort, significant progress has been achieved in the field of research and development, application and industrialization, and quite good foundation was created. In the near future the main work will continue in this direction. With the technological progress of the renewable energy power generation, its commercialization development and successful utilization, since the beginning of 1990s, people gradually recognized that they could play an important role for sustainable development of electrical power industry for 21st century and change the energy resource structure in our country in the future. The related work has already started. In 1994, the State Science and Technology Commission, State Planning Commission and State Economy and Trade Commission have jointly formulated "The Development Program of New and Renewable Energy in China for years 1996-2010." The further development of technology, market, and industry was further strengthened under active government support. Table 4 summarizes the present general status of some important renewable energy development and application in China. In general, at the present the importance of renewable energy for rural and remote areas is quite well recognized. Applications in these areas are still the main focus in the near future. The importance for electric power industry is gradually recognized. The process has already started with the rapid development of wind farms. The discussion on the future sustainable energy development gets more and more attention. Quite good foundation for renewable energy development, manufacturing and application has been laid. The market is expanding faster than expected, especially for solar heaters, small PV power supply and wind generator. As an example of rapid development, Fig. 2-4 shows respectively the increase of solar heaters, PV power and wind farms during 1990s in China. According to the forecast, in 2005 the total installed capacity of wind farm will reach 1.2 GW, total installed power of PV will reach 80 MW, total installed area of solar heaters will reach 64 Mm^2.

Table 4 The Development Status of Renewable Energy in China

Energy	Item	Present Situation
Biomass	Biogas digesters	About 6.9×10^6 sets, $2.09 \times 10^9 m^3$/year
	Firewood forest	About 1×10^7 hectares
Tidal	Power station	8 stations, 11MW
Geothermal	Power station	5 stations, 27.78MW (except Taiwan)
	Direct use	1.698×10^4 TJ/year
Wind	Mini-Generators	2.1×10^5 sets, ~25MW
	Wind farms	28 sets, ~400MW
Solar	PV cell	~22.8 MW
	Hot water heaters	$\sim 3.2 \times 10^7 m^2$
	Solar house(passive)	$7.5 \times 10^6 m^2$
	Green house	4×10^5 hectares
	Cookers	3×10^5 sets

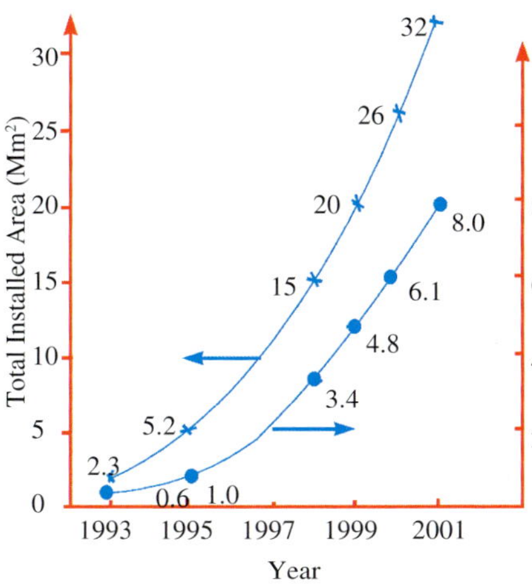

Fig.2 Development of solar heater in China

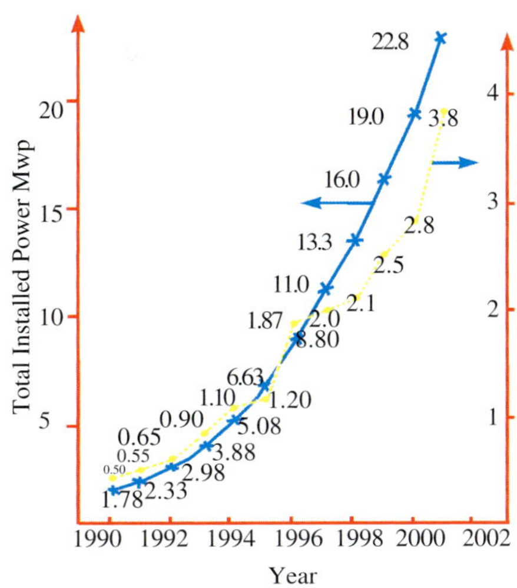

Fig.3 Development of PV power in China

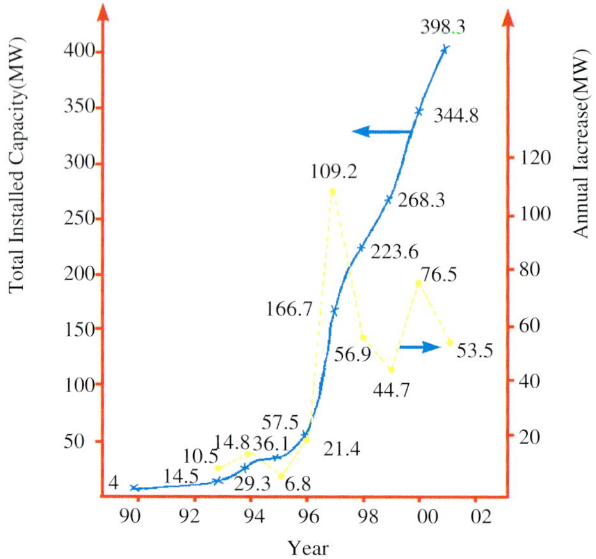

Fig.4 Development of wind electrical field in China

The main problem of the renewable energy development and application in China is that although the development and formation of the corresponding industry have obtained quite good achievements, in general they are still at the starting stage. Reducing costs and nurturing the market are major tasks. It is needed to enhance the technologies and the manufacturing capability and also to develop the market infrastructure during the transition to a market economy. The main difficulties we are facing are: ① many small, relatively isolated research and design institutes and companies are doing research, development and manufacturing work, while institutional fragmentation and uncoordinated efforts hinder the technological progress, quality enhancement, service improvement and accumulation of commercial experience, ② due to the high cost, small market demand and lack of the economic competition capability, the application scale is small and it is incapable to achieve economic scale, and ③ although the Chinese government has taken some favorite policy to promote the development of renewable energy industry, it is still insufficient, especially the funding for research and development is inadequate and the credit and venture capital are too small, and therefore, to go forward smoothly on existing basis, it is needed to make great and continuous effort. The main effort are as follows: ① the R& D and demonstration work should be further strengthened to improve the performance and quality of the corresponding equipment and systems, to increase its life, especially to reduce the cost, ② the corresponding industry, such as large wind generator factory, PV-cell factory and etc., should be developed significantly, so that the future development can rely on our own industry, ③ since the capital investment and cost of electricity for power generation are still high in near future, while they supply electrical power to customers or to electrical system a favorite policy from government is needed to ensure some profit, and ④ the international cooperation should be strengthened.

5. OIL AND ENERGY SAVING

In 1949 the output of crude oil was only 0.12Mt in our country. Production of oil received rapid development in 1960s and 1970s. It reached 104Mt in 1978, but since the middle of the 1980s the growth slowed down. The output went of from 130Mt in 1986 to 165Mt in 2001. According to the extractable resource, it was estimated that output of crude oil will be 170Mt in 2010 and 180 Mt in 2020 in our country. The practical consumption of oil increases rapidly in China. Our country became an oil net import country in 1993. The net import already reached 70 Mt in 2000. Imported oil will reach 40%in 2020. This makes oil supply be an important problem closely connected with national energy security and import dependence. Research and development of technology for coal liquefaction and oil saving have important meaning, except for strengthening prospecting of oil and natural gas, and development of domestic powerful oil industry.

Under the national support, the coal indirect liquefaction and direct liquefaction technology had quite good progress. Commercial development is accelerated. The coal indirect liquefaction is a technology that coal passes through gasification to get synthetic gas and then gas synthesizes to be oil. In China the key technology research was started in 1980s, the middle scale testing with hundreds tons and the industrial demonstration with thousands tons were completed. The "development of key technology of synthesis from coal to liquid fuel with several ten thousands of tons and its industrial software" was completed in 1990s. This technology has developed and reached industrialization stage. Now the effort is directed to built a synthesis oil industrial demonstration factory with several ten thousands tons capacity. The coal direct liquefaction is a technology that converts coal to liquid fuel by adding hydrogen under high pressure. In China the study of technology for coal direct liquefaction was started at the end of 1970s. Now all flow charts from coal to qualified product have been passed. With international cooperation three pre-studies of feasibility for industrial demonstration projects for coal direct liquefaction were completed including: “Xianfeng” project in Yunnan province with Germany, “Shenhua” project in Shanxi province with Japan and “Yilan” project in Heilongjiang province with USA. The engineering construction for Shenhua direct liquefaction of coal project has been approved.

Transportation is the main oil consumer. Automobile, airplane and ship all use oil as the energy source. With the rapid development of transportation the percentage of oil consumption for it increases day by day. Therefore oil saving in transportation should be a focal point. The main measures of oil saving in transportation include : ① to develop city public transportation to slow down the rapid increase of private car, ② to increase operational speed of rolling stock on railroad to maintain its important position in passenger transport between cities, and ③ to actively develop new technology for electrification of transport means, such as electric cars and Maglev. Hopefully these transport means will play an important role in future transportation system. Recently certain common recognition has been reached and quite good progress has been made.

The energy consumption of automobile amounts to about one fourth of total energy consumption in the world. The electric automobile has many advantages, that is high efficiency, low environmental pollution, low noise, ability to use different energy resource and no restriction of oil resource. Its study and development is getting wide attention. In our country during recent years some works are in progress. Several prototypes of electric automobile including net electricity driven, mixed fuel driven and fuel cell driven have been made. A good foundation for further development is established .

High-speed Maglev is the only ground transport means which has highest operational speed up to 500 km/hour[7]. It is suitable for high-speed, long-distance large-volume passenger transport between big cities. Its practical use will keep the important position of ground passenger transport, and well coordinated with development of civil aviation to save aviation oil. By continuous effort in recent years, certain common recognition that our country needs high-speed Maglev has been reached. With China-Germany cooperation and technology transfer a Maglev operation line from Pudong airport to downtown of Shanghai has been built within two years. The design speed 430 km/h was reached. It laid the foundation for practical use. Now people are making active effort for its long line use. It is possible that China may bring Maglev to practical use and accomplish industrialization first in the world.

Energy saving, increasing efficiency of energy utilization actively, and reducing energy strength are important parts of whole energy work. Measures of energy saving and R&D of technology are conducting in everywhere according to their own real situation. High temperature and low temperature heat recovery technology has certain universal significance, to which special attention was given and certain achievements have been obtained.

6. CONCLUDING REMARK

Sustainable development is a current significant task for the whole mankind, therefore it is needed to study and development of new energy technology in many aspects. According to real requirements and energy features of our country, the study, development, application and industrialization of high efficient and clean utilization of coal, nuclear power, renewable energy, oil and energy savings have received high priority during recent years. Certain achievements were obtained, quite good foundation was established, but in general, it is still at a starting stage. The differences between our country and advanced countries in the world are still large. The funds for science and technology development are not enough, it is expected to be further enhanced. The main feature of energy technology development is long-term, so, it is necessary to have a long-term national program, with continuous and steady support. While in our country the related work has started already, which is necessary to insist and enhance. Pushing forward the technology development, good international cooperation has also important significance.

REFERENCES

1 Wang Qing, Energy Data 2002, Energy Policy Research, 2002, No.1

2 The Department of Resource Saving and Comprehensive Utilization of State Economy and Trade Commission of the People' s Republic of China, China Energy Annual Review, 1997

3 Cai Ningsheng, Study and Industrialization of Clean Coal Technology, Report of High Technology Development of Chinese Academy of Sciences, Science Press, Feb. 2002, p119-128

4 Yan Luguang, Ju Zixiang, Sha Ciwen. International Progress and Chinese Strategy of MHD Power Generation, Advanced Technology of Electrical Engineering and Energy, No.1 1994 p9-15 & No.2 1994 p11-17

5 Duan Yibing, Current Situation and Prospect of Nuclear Power,Report of High Technology Development of Chinese Academy of Sciences, Science Press, Feb. 2002, 238-247

6 Yan Luguang, Li Anding, Meng Xiangan, New Energy and Renewable Energy in China, Solar Energy, No.6 2002 p3-7

7 Yan Luguang, Thoughts on Development Strategies of China's High-Speed Maglev in China, Science and Technology Review, No.11, 2002 p3-6